Autism Research between Psychology and Neuroscience

Michele Di Salvo

Autism Research between Psychology and Neuroscience

From Leo Kanner to New Frontiers

palgrave
macmillan

Michele Di Salvo
Naples, Italy

ISBN 978-3-031-68337-4 ISBN 978-3-031-68338-1 (eBook)
https://doi.org/10.1007/978-3-031-68338-1

Cover credit: creation, persistence and deletion of memories—about a cherry tree in winter (white, black and red) #8 by Sergio De Rosa

This Palgrave Macmillan imprint is published by the registered company Springer Nature Switzerland AG
The registered company address is: Gewerbestrasse 11, 6330 Cham, Switzerland

Dedication

This book, written by a man, owes a great deal and would never have been published without women, thanks to their vision, dedication, stubbornness and determination. But, above all, love and passion.

First of all, my thanks go to Loredana Al Ghazi and Linda Gilmore, for their academic support and backing. And to Clelia Petrecca, who has followed as commissioning editor of this project from the beginning, believing in it with enormous determination and continuously discussing and building solutions. Without her, this work would never have been born.

A special thanks is to my wife Geppina, who patiently read the Italian drafts, but above all put up with me and supported me during the writing and revision period.

I also owe thanks to my mother for making me love science as a child.

Most of all I am indebted to the autistic children I met, to their eyes, their searching gaze, their talking hands. And to their parents who lovingly ask themselves what all parents ask themselves: *I would like to know what my daughter thinks.*

My dedication goes to my daughter Shanti, who in recent months has chosen to study history at university: not an easy path in a discipline that we need more and more. With the wish that she will always be rigorous and creative, two gifts that are needed today maybe more than ever.

Contents

CHAPTER 1

Introduction: Autism Research Between Psychology and Neuroscience: From Leo Kanner to New Frontiers

Different from Whom?

My best friend was called Bruno. We met in first grade at about 5 years old and remained friends until about 27, when he decided his journey had come to an end.

Bruno was a wonderful person who taught me to have faith in life.

A concept I learnt from him through assimilation, and which I found again at the age of 45 in the words of Alexander Lowen (1972).

There is a profound difference between 'faith in life' and 'trust in life'. Faith is something external, which we acquire from our experience in the world, from the feedback we receive to our actions. Faith is an internal energy, which lies within us and gives us the life force to do things and lean towards the future, to build, to be social, to search and to discover. He taught me faith in life. And I believe that for a child who later becomes a man there is no greater gift. Because faith in life, something that is inside you and not dependent on the outside world, is that strength that makes you face challenges and overcome difficulties.

Bruno had severe lateral hemiparesis caused by anoxia at birth. A very severe anoxia.

Anoxia, in the clinical definition, occurs when the infant suffers a condition of decreased oxygen supply to the brain, resulting in irreversible and severe damage. Specifically, Bruno had a twin sister, Simona, who was born first. A beautiful girl and I am sure Bruno would have been an

M. Di Salvo, *Autism Research between Psychology and Neuroscience*, https://doi.org/10.1007/978-3-031-68338-1_1

equally beautiful man. The doctors simply 'forgot about him': the sister was born and the birth was over. After seven hours they realised from the mother's pain that something was wrong and they noticed Bruno.

His faith in life was already powerful and mature. He wanted to be born. He did not want to give in to medical error. The severe, very serious damage he suffered was fortunately confined to the motor part and clearly fundamentally musculoskeletal development. His strength, vitality, lucidity and ability to think, reason, speak and communicate were intact and always were, accompanied by a subtle irony that showed, behind the motor difficulties, a rare vitality.

He never gave up. He never stopped studying. He could have applied for and received a disability pension right away, but he did not. He had to work, and he did.

He also won a public competition and for a while worked in the City of Naples. He did not want to be a burden on either the family or society.

Once he said to me over coffee, "why do you have to work and I don't?".

Bruno also had a great love, a girl confined to a wheelchair. As if life with Bruno was not cruel enough, Bruno's almost girlfriend (she never officially declared herself, but they were always together and you could see they were happy and loved each other's presence) died, run over by a hit-and-run car.

As Temple Grandin (1995) notes: "A marriage works best when two people with autism marry, or when a person with autism marries a person with a disability or an eccentric (…). They attract each other because their intellects operate on similar wavelengths".

[*Personally, I find this the best possible definition of attraction between two people and it is puzzling that it was made by an autistic person, who is supposed to have low levels of empathy and poor theory of mind, but we will come back to that*].

I believe that this loss irreparably undermined Bruno's great faith in life. And depression mixed with that sense of not weighing on others did the rest. I am certain that his great soul is now free and can give vent to that great desire to run that he was never able to unleash.

My Encounter with Science (and Family Matters)

We are not the product of a design, just as we are not the immediate product of the expression of our genes. After years of research, we have discovered (re-discovered and re-evaluated actually) something we have called epigenetics. Psychology has discovered and re-evaluated on a scientific basis what homo sapiens already knew within himself: the influence of culture, environment, ecology, society and social dynamics in helping to determine who we are and who we will be.

I grew up in a house full of books. My grandfather read two newspapers, and at least an hour of his day was dedicated to books. My mother was a science teacher who devoted herself, before teaching, to university research in two seemingly distant areas: biology and geology.

Even as a child, I was fascinated when we went to the mountains by the immediate transitions between animal life and orogenesis: a whole.

Two of my cousins are doctors, one a paediatric cardiologist (it's more than that, really) and one a paediatrician for children. Another cousin of mine is a research biologist. An uncle of mine was among the fathers of the Genome Project in Italy, the titanic worldwide project of mapping human DNA. And as you can imagine, I was often in his lab (*I built my first barometer with laboratory parts and 'the vacuum machine' at the CNR in Naples*).

I remember one time they 'discovered an enzyme' and my uncle even went on the news. Of course, I was about ten years old and I couldn't understand that they didn't discover an enzyme, but something more complex, and that it had to do with enzymes that repair DNA and stitch up severed ends, and that their discovery was part of the CRISPR sequences.

But that is another story.

What it has to do with my story is that enthusiasm for discovery, if it enters you at the age of ten, the spirit of research aimed first and foremost at understanding the phenomenon, is a powerful spark for a child: there, you realise that there is a place where the whys are discovered.

The other side of my family is made up of jurists and I have dedicated myself to this branch of study (*I later discovered by chance that Alexander Lowen also had a degree in law before taking up psychology; perhaps this is also why I share his 'broader approach'*).

I was fascinated by the rigour, the need for justice, the assertion of principle and the defence of law. And above all, the legal concept of 'fact' (which has enormous relevance in the criminal field).

"Facts, as such, are not and cannot be true or false. They may or may not have happened; they may have happened in one way or another. True or false are the factual statements, i.e. the statements we make about how past events may have unfolded. One can, therefore, preach the truth or falsity of a factual statement, but certainly not of a fact. A fact is or is not. An account of past facts may be, in whole or in part, true or, in whole or in part, false. Truth and falsehood have to do with the words one uses to describe, recount, reconstruct facts. And it is in the antithesis truth/falsity that one can find a common element between the very different areas of word usage" (Carofiglio, 2015).

I was the only fool in my classical high school who chose history and physics as examination subjects for his diploma (*the comment 'what a bizarre choice' by a teacher has become my interchange*). During my first year at university, I decided to prepare myself and obtained a second degree in pedagogical sciences.

My whole professional life has been in the groove of these two halves.

I have been involved in communication in the broadest sense for more than 25 years.

I am a member of the Society for Neuroscience and the European Federation of Neuroscience Societies, and my writings, which I have had the privilege of publishing in academic journals on three continents, concern neuroscience and criminal law, the concept and forms of human intelligence and neuroplasticity.

I have written articles and books on relationships, the psychology of relationships, language, educational and communicative dynamics, consciousness and free will, and the law articles I have published all focus on the protection of the 'weak position' and the legal hermeneutics of people's rights.

My approach to autism fits into this context and into the overall framework of these subjects of study and research, with a cross-disciplinary and international group in CrossMedia Labs, a multidisciplinary meeting and discussion group.

It is within the context of these views and this approach that the considerations you will find in this book about the talents of autism and neurodiversity and issues of neuroethics and bioethics in general fall.

As we will see in this story, concepts such as theory of mind, empathy circuitry and systemisation skills, are all essential to understanding the communication systems of human beings, and the encounter with the world of autism is an opportunity for all of us to discover a great deal about these relationship systems.

In the neuroscientific field, my approach is neurocircuit and not locational, so the picture of autism characteristics from my point of view is an expression of a phenomenon that creates a broader, less localised and more integrated disorder.

In this sense, I believe that this book, in itself, can make a contribution in terms of analysis and approach, perhaps also because it steps outside the normal academic specialisation of reference and uses—I hope with adequate rigour and scientificity—an approach that can be of value.

As Groddeck (1920) wrote, "it is worth considering, for once, universally known things through the eyes of an outsider, and even if the result were only to provoke a more lively clash of opinions, I would already be satisfied".

As Uta Frith (2003) wrote, "Autism is too fascinating to be treated by scientists alone; after all, it is one of the most powerful reflections one can make on the human condition".

But in this sense a valuable lesson from one of the fathers of quantum physics should be read and above all remembered. Richard Feynman said in a lecture in 1963, addressing an academic audience of lecturers and students:

"In these times of high specialisation there are very few who have such a deep understanding of two different fields of knowledge that they do not make fools of themselves in at least one of them. [...] I have said that observation is the judge of the truth of an idea, but where does the idea come from? Indeed, the rapid progress of science requires human beings to constantly invent new ideas to verify. [...].

In the Middle Ages, people thought that mere observation of nature suggested laws, but it doesn't work that way: it takes a lot more imagination. So the next question to talk about is: 'Where do ideas come from?'. Actually, it makes no difference, as long as they come. To check whether an idea is correct or not we have a method that has nothing to do with the origin of the idea itself: we simply verify it by observation. [...] There is no authority to decide which idea is good and which is not: we no longer need revealed truths. We can consult the luminary on duty and ask him to tell us his point of view, and then do the appropriate experiments and find out whether what he says is true or not. If it is not true, too bad for him: that is how 'authorities' lose some of their 'authority'. [...] Relations between scientists—especially at the beginning, when a discipline is young—are not always rosy, and in the past, in fact, there have been rivalries and polemics; as between normal people, after all. One example is the heated disputes that marked the birth of modern physics. Such a mass of observations has accumulated in our discipline that it is almost impossible to find an original idea, different from everything that has been thought up to now, and at the same time in agreement with all the observations made. So if someone, somewhere, somehow gives us a new idea, welcome! [...] We are certainly not going to argue why and wherefore he came up with it. Who cares how much that guy has studied, or why he wants to be heard. In this sense it makes no difference where the ideas come from. Their true origin is unknown, we call it 'imagination', 'creativity' [...] Strangely enough, many do not believe that there is a place for imagination in science. It is imagination of a special kind, different from that of the artist. The difficult thing is to try to imagine something that nobody has ever thought of, that agrees in every detail with what is already known, but is different; and is also well-defined, and not a vague statement. It is not at all easy. [...] Doubt pushes us to look in new directions and seek new ideas. The progress of science is not only measured by the quantity of new experiments, but also, much more importantly, by the abundance of new hypotheses to be tested. If one could not, or would not, look in new directions, if one had no doubts, or did not recognize the value of ignorance, one would not be able to come up with new ideas. There would be nothing worth verifying, because we would already know what is true and what is false. [...] In science, doubt is clearly a value. ... doubt should not inspire fear, but should be welcomed as a valuable opportunity".

Science and Knowledge in a Complex World

The article in which Einstein described the photoelectric effect is less than 8500 words long, contains just five notes, four of which are for terminological specification and one for bibliographical reference, and was immediately published.

It was 1905, he was 26 years old and a 'third-class technical expert' at the Patent Office in Bern, the Annalen der Physik was the most prestigious physics journal of the time, and the theoretical section of the journal was headed by Max Planck, the best-known physicist of the time, the man who in 1900 had discovered the 'elementary quantum of action' and inaugurated the greatest revolution in physics since Galileo and Newton: the quantum revolution.

A few different cases are less well known but more interesting.

In 1934, Enrico Fermi published an article in the German journal *Zeitschrift für Physik* entitled 'Towards a theory of beta radiation' in which he described the weak interaction. The year before, the article had reached the editors of Nature. Once revised, the article was rejected because, recalls Frank Close, 'it contained speculations too far removed from reality to be of interest to readers'.[1]

Rosalyn Yalow won her Nobel Prize in Physiology and Medicine in 1977. Years later she proudly displayed the *rejection* she received from *The Journal of Clinical Investigation*. The *referees* had been sceptical that humans could produce antibodies small enough to bind to insulin.

Prior to the famous paper that is now recognised as the decisive insight describing the standard model in the presence of the boson, Higgs had already written a second short paper describing what was later called 'the Higgs model'. That contribution was sent to *Physics Letters* but was rejected on the grounds that it "did not justify rapid publication" (Intini, 2017).

On 9 August 2017, I was particularly struck by the title of an article 'Einstein was right. The star and the black hole confirm the law of Relativity'.

The article began "it is still the stars that reaffirm how solid the theory of general relativity is, even more than a century after its formulation" (la Repubblica, 2017). As soon as I read the headline I thought 'of course he was right', but then I thought 'imagine if he had to wait a century for someone to finally prove him right'.

In Chapter 3 I will discuss bias, error and noise in scientific research. In this introductory context, it is worth mentioning the case of Robyn Dawes (1974), a researcher from the Eugene Group of Excellence who devoted himself to the study of judgement in the 1960s and 1970s. In 1974 he reached a breakthrough in the simplification of predictive tasks with an astonishing and almost heretical idea: instead of using multiple regression to determine the exact weight of each predictor, he proposed giving them all the same weight. Dawes called this formula an improper linear model. Surprisingly, he found that these models were as accurate as 'proper' regression models and much more so than clinical judgements.

At first Dawes and his assistant Bernard Corrigan had difficulty getting their article published in scientific journals: the editors did not believe them.

We live in a complex time. Our life is in the dimension of complexity (see Morin, 1990; 1999).

"Abandoning once and for all a locationalist idea of the brain, we are finally moving towards a systemic and interrelated vision of cooperating and integrated brain circuits and systems. In this revolution in our view of the mind (which goes far beyond the brain), we cannot but be astonished at its extraordinary complexity. We are like astronomers seeing beyond the solar system for the first time. This new orientation is a debt we all owe to autistic people, and an extra motivation to study it, for all that these extraordinary people can enable us to understand about our minds and our world, by gently and humbly entering their minds and their worlds" (Di Salvo, 2024).

In this globalised complexity, we should orient ourselves by considering a model of shared intelligence and knowledge, and the road must be travelled together.

As Steven Sloman and Philip Fernbach (2017) have argued, "Sharing skills and knowledge is more sophisticated than it may seem. Humans do not merely make individual contributions to a project, like machines operating on an assembly line. Rather, we are able to work together, aware of others and what they are trying to do. We pay attention together and share goals. In the language of cognitive science, we share intentionality. This is a form of collaboration that is not observed in other animals. We actually like to share our mental space with others. In a knowledge community, what matters, more than possessing knowledge, is having access to it. A scientist studying rocks cannot store in his memory everything that is known about geology and related disciplines, but the scientist can keep track of reference texts, websites and other experts who can provide the information needed. A more familiar example is medicine: there has been such an explosion of medical research that general practitioners can no longer know everything they need to know about all the various diseases and concerns that patients present. Fortunately, they now have access to electronic databases, which allow them to search for what they need, when they need it".

In this common path, divergent opinions are an indispensable asset.

In the words of Beau Lotto (2016), "the brain does not always want toys… it needs to learn that it can pick itself up after being knocked down, and in doing so become more resilient... but it is not just [*people*] who have to take risks, our culture has to do so too. The past of our brain also includes the ecology of our culture. After all, culture itself is simply another product of the brain, a collective manifestation of thought and behavior, so it too grows and adapts as it faces challenges... it is not only the human brain that changes: the collective cultural brain does the same thing. Both are constantly redefining the concept of normal by creating a new one with each passing second".

Science in the age of the web needs its claims to be increasingly supported by research and its results. In turn, research needs funding to move forward. And in this circle, the risk is that a short-circuit is created between research centres and funding, apex academic positions and mutual referrals.

The web has infected us in this too. A bit like likes on social media, a study counts as a function of the number of citations and 'where' these citations occur. And resumes, also as a function of academic career, count as a function of 'how many' articles you have written. In this reputational short-circuit, one runs the risk of 'analysing data' without having a theory, in the eternal debate between having a theory first and looking for empirical evidence, or collecting data and then imagining a theory to explain it. A bit like the search for primacy between the chicken and the egg.

Without a theory, we cannot explain the data, but neither do we know what to look for and how to order and understand it. But the risk is to only look for and consider data that validate (or not) a theory.

As Lowen (1972) put it: "The same process that wreaked havoc in our culture and corroded every faith of the past, has given modern man a new vision of himself. And this is the idea of his own omnipotence. (...) Modern man no longer recognizes a supreme authority. He believes that nature is governed by physical laws and that, by being able to decipher those laws, he can control it. This is a bold vision and science seems to provide man with the means to realize it. (...) Can it be said, then, that man has gained a new faith, faith in science or in the ability of the rational and inquiring mind to unravel all mysteries, to overcome all obstacles? The fact is that many truly believe in science and in that possibility. But to believe is not to have faith; a belief or conviction is subject to verification, faith does not need to be supported by verification. A belief is the product of a conscious mind, sincere faith is a matter of the heart. I feel that the balance has shifted strongly against the forces of natural life: human understanding, pleasure, the body, nature and the unconscious. We are destined to see the amount of information increase without any safeguards to protect our sensitivity. Performing research that consists solely of gathering information and processed statistical data has become the supreme aim of our higher education programs. (…) The insidious effect of this tendency to focus all attention on data is a progressive loss of faith in the natural capacity of human beings to understand themselves, other human beings and the world in which they live. (…) We can and must trust our senses if we are to find meaning in the confusion of our existence".

The consequent risk is to forget the person as a function of research, focusing on pathology as if it were something else, separate and isolated from people.

The mind of a scientist is peculiar[2] and generates typical reasoning in which the enthusiasm and euphoria of discovery, the perceived 'within reach' goal, often overshadows other essential factors. It must always be remembered that not everything that is possible, or sometimes even conceivable, is also permissible or legitimate to do. History's excesses, even recent ones, have been curtailed in this respect with bioethics committees and protocols. But it is becoming increasingly evident, especially with the technological advances taking place with unprecedented acceleration, that they are not enough.

One of the great lessons of the neurodiversity movement is precisely the centrality of the person as an individual and not as a subject, the need to go back to talking and confronting neurodiverse people: there is no better source of information than their narratives and those of their families. Even when these narratives disprove the laboratory data. Indeed, perhaps precisely when they disprove the data.

A Question of Language

Choosing the word is not just an exercise in style, but a precise social and political choice.

"I like people who carefully choose the words not to say", said Alda Merini,[3] one of the greatest Italian poets of the twentieth century, diagnosed with schizophrenia.

Through the words we choose and those we choose not to use, through syntax, we narrate the reality around us, but we do more. We educate those who read us and those who listen to us, and we help build their perception of the world.

This is why, in an increasingly fast-paced world in which content and information flood us seemingly unfiltered, the writer has an ethical and professional duty of care and attention to language, word choice and selection and the syntax with which he or she constructs the story.

It often happens that one reads important books and articles, on autism and on any neurodiversity, in which very worthy, well-prepared and very human persons express themselves 'without special care' with expressions such as 'autistic subjects', 'these subjects' or '...compared to normal subjects'.

But even the phrase 'affected by autism' can be dangerous: because being 'affected' communicates in itself the disease, and in the uncaring and untrained imagination also an idea of 'virality', almost a transmissibility by contact.

They may seem like minor or negligible details, but they are not, because diversity arises from the perception of difference, and this perception arises in language, in the syntax with which we construct, narrate and describe the world and our reality.

I well understand that clinical language has these characteristics. Medicine very often—in research and not in the encounter with the frail person[4]—needs a 'neutral tone' detachment. But we have to consider that describing neurodiversity with expressions similar to those we use for cancer does not help research but creates a world that is not only unhelpful, but objectively negative for people living in a neurodiverse condition.

"A language does not reflect reality, but rather creates a reality" stated Paul Watzlawick (1980).

A message we must not forget.

Why autism instead of ASD Acronyms are important both for reasons of synthesis and for uniformity of language and reference in international scientific research. In this book, I have tried to talk about people, human people without labels, and I have chosen to eliminate them as much as possible.

As I will illustrate in Chapter 15 even from a definitional point of view, I propose something different, broader, in a dynamic view of autism, and eliminating a static label can help us get started on this broader reasoning.

As Loredana Al Ghazi (2020) wrote:

"The first large-scale study to examine the terms that members of the UK community use to describe autism (commissioned by the UK National Autistic Society) showed that there is no single accepted term used when talking about people with autism. We use terms such as 'autistic person', 'person with autism', 'on the autism spectrum' or simply 'people on the spectrum'".

The term 'on the autistic spectrum' was preferred by a significant number of autistic adults (45%), parents (38%), family/friends (48%) and especially professionals (60%). The term 'autistic' was endorsed by a large percentage of autistic adults, family/friends and parents, but by considerably fewer professionals; 'person with autism' was endorsed by almost half of the professionals, but by fewer autistic adults and parents (Kenny et al. 2016 p. 446).

The Scientific Method and the Need for Grand Theories

We need to go back to theories, the big theories.

We must all decide to abandon the search for the 'one cause' that will lead us to the 'one pill' that will solve and change the state of things.

This is the last remnant of the delusion of omnipotence of twentieth-century man, who has abandoned all religion, the animist one of the god and magic in all things, the poly and monotheist one, of the anthropomorphic god, powerful but 'at hand', to the 'god himself', capable of modifying every aspect of nature: the man–God Almighty.

According to Schrödinger (1946) "*Knowledge, or at least the feeling that this is the case, is undoubtedly as old as the stimulus that drives us towards knowledge. [...] The enormous successes of science, which advanced by leaps and bounds, led some of the best minds astray, to the point of making them believe that science was about to clarify everything worth knowing, that outside of it there would soon be nothing interesting, especially that it would soon solve the problem of spirit, giving an absolutely objective explanation of the true nature of thought. [...] The ideal reference of knowledge to lo was replaced by a physical reference. With its technical by-products, often bordering on the miraculous, knowledge of nature proved useful to the physical ego. Thus a material 'egoism' took the place of the ideal one, perhaps making the exclusion of the latter less sensitive*".

Returning to the humanity of man means looking into the eyes of the other who is different from us, as equals, at the same height, in the same position, without any delusions of omnipotence, giving us the opportunity of the discovery that perhaps our greatest superpower is our humanity.

What weapons does science have, then, today, in order not to fail in its mission to provide reliable knowledge? Obviously I have no solutions. But I do believe that a path that is at least acceptable for now is that of constant verification of research results through independent experimental repetition on the one hand, and double-blind evaluation of contributions on the other.

Evaluate theories and research results for what they are not based on 'who says so'.

There is no certainty of evaluation. Those who rejected Yalow's and Higgs' articles were certainly just as knowledgeable as those who accepted Einstein's article. They were expert astronomers who denied Galileo's discoveries, and equally Newton, in his unquestionable genius, knew nothing about atomic physics (which does not make him any less of an amazing or brilliant physicist—he was apparently autistic, but we will come back to that).

Even the scientific world has its biases.

In addition to all those concerning human beings (and scientists are human beings like any other), there are the 'biases of my faction' (Pinker, 2021) and some objectively typical of the field, such as the conflict between statistical significance and Baynesian verisimilitude.

While it is difficult for them to be defeated, it is also true that knowing them and taking them into account can often help.

What This Book is About

If we were to define, in the final analysis, what Homo Sapiens has always done and what distinguishes him from every other living species, with which he shares everything else, is a twofold activity: research and storytelling. To do this, he developed spoken language and writing.

This book and what it contains is that research and the telling of its story.

What do we talk about when we talk about autistic people? We have moved from childhood schizophrenia to psychic pathology, from psychic and neurological retardation to autistic spectrum syndrome. We question what autism is and often forget that we are talking about people.

So it should be made clear, before we begin, that this is not a book about autism, although it is a book about autism, but first and foremost it is a book about people.

Whenever I talk about a symptom, a state and a complex condition, I would like you to close your eyes and see the person. I assure you that it will be a different reading, and your life may also change.

This book is not about disease, defect, impairment, retardation. It is about people who are in a condition, in a certain state. It is, therefore, a book that does not talk about pathology, but about neurodiversity. It deals with it, it tries to describe it, but from two specific points of view. It is a book that speaks of love, the driving force of every being, and it speaks of faith in life, without which there is no life to be lived.

This book is first and foremost intended to be a general recapitulation of approximately one hundred years of research on autism.

Why? Because as Jung (1929) said, "Here, where nature is greater than man, no one can escape it".

According to Sturgeon's famous law (Dennett 2014), ninety per cent of everything is rubbish (and in this particular case it is not): the challenge I have set myself is to save that (more than) ten per cent of every front and proposal for research and explanation. But it cannot contain everything, because not everything is correct, as Gerald Edelman claimed.

The study of autism is riddled with errors, but many of these have opened up new avenues and highlighted aspects that had not received enough attention.

Perhaps this requires a systematisation of at least the basic assumptions and a look at the directions of study of greatest interest and best overall perspective.

This is my intention, without presuming to be exhaustive on every single aspect, but through a description that is, I hope, intellectually honest and rigorous in scientific terms.

At the end, I will propose a common element that underlies many of the theories that have been proposed. But I would not like to anticipate anything, and above all, I would like to make it clear that this is not a 'theory of everything'.

As David Disalvo (2011) wrote, 'I *am not the kind of person who shouts "I have the answer.", and this is not that kind of book. (…) We must always keep in mind that science is not about answers. Science is made of questions. If we want to use the tools of science, then of science we must also accept the rules, the most important of which is that we do not fool ourselves by convincing ourselves that we have hit the nail on the head once and for all. Does this mean that we cannot discover truths and, on a practical level, exploit them in our lives? Of course not: if it did, books like this would never have been written. It simply means that we must be careful not to reason as one would in a court of law, concluding that the case is settled and that we can move on. On the contrary, I think we should hope to resolve some elements of the 'case' but also open up new ones—which, going forward, will engage us as much and more than the previous ones"*.

We are used to playing the 'smartest guy in the room' game, a typical position in which people such as von Neumann, Fermi and Majorana have often found themselves, but it is a game they never played in the first place.

An Eastern wisdom saying attributed to Confucius states that 'if you are the smartest person in a room, then you are in the wrong room'.

Perhaps it is time to try another point of view. When we are all in the room the smartest is the room. And try to see what comes out (of the room).

Playing in this new way, I met people who were better than me, more knowledgeable than me, who had read 'that book' more than me, who corrected a line, a paragraph, a chapter, seeing things I had not seen and noticing gaps that I did not know.

[*Of course, I have not always been so lucky. I have also encountered what I call 'soul killers', people I would never want a young person to meet at university as a professor, people who judge and reject ideas on the basis of a few lines, or often just one word, who evaluate without having read, who think they know what you are going to write, and comment without offering any improving contribution or constructive criticism. But as I have stated my intention is to save the good. What do I save of these, for me fortunately rare cases? Well the fact that they too are a test for the strength and soundness of ideas and arguments. According to some anecdotes, Walt Disney knocked on the door of 300 banks before he found one willing to finance his first playground. Think how much those 299 who said no lost, but also how convinced and determined and motivated Walt must have been. But if by chance a 'soul killer' should prevail, console yourself: it means that your idea was fragile and you needed to consolidate it. Which is a good thing regardless (For example, having met them has allowed me this note, which I hope will be useful and give someone strength and courage.)*].

This story and this research would not have been possible without the valuable advice of everyone in the room (and outside). If the text has any value, the credit is theirs. Listing them all, as is often done, runs the risk of being a long list 'of references', forgetting that smile, that joke, that comment, that emoticon accompanying an email or message, that make 'the person' beyond the name. For the certainly present shortcomings or aspects that have not been fully clarified, the responsibility is mine. Now we have to see what happens 'outside the room'.

I hope that this contribution will not be exhaustive, that it will raise questions, perplexities and doubts that will lead to new analyses and critical comments and that it will stimulate the refutation of the theses presented and the extension of the arguments and objects of analysis. If so, this contribution will have achieved its author's intended purpose.

If he has contributed even a drop to seeing the human person 'inside' autism, and has been able to transform us into humans behind our apparent normality, he will have made us all closer and more equal, like drops in the sea.

To paraphrase Mother Teresa, without that drop the sea would still be the sea but with one less drop.

An Initial Map

A story is also a journey. Even more challenging if it is a journey of research.

For orientation, it is good to have a map, something more than an index.

It is not a rigid map; if you wish, you can skip or reverse stages or go back, each following their own route.

In Chapter 2 we will address the history of autism, its historical presence and its "discovery" in the modern era, with the first definitions. In this chapter, we will go through the various defining phases up to the DSM-IV.

In Chapter 3 we will introduce the first psychological theories and approaches, up to the theory of correlation with vaccines.

We will, therefore, take a break from the historical discourse and in Chapter 4, the characteristic and symptomatic picture of 'autism spectrum disorder' will be described from a medical-clinical and psychological perspective.

Chapter 5 is devoted to a description of neurogenesis and brain formation, which will serve us for a better understanding of the discourse in the following chapters.

Chapter 6 is devoted to the description of cranio-cerebellar features.

Chapter 7 focuses on the definition of the DSM-5 Ter criteria (the most recent version in force since 2023) currently used for diagnoses.

In Chapter 8, the story will begin again with the 'second definition' of autism, the one introduced by Lorna Wing and which has formed the basis of diagnostic definitions since the 1980s.

In Chapter 9 we will recount the extraordinary adventure of the neuro-scientific research behind the psychological theories related to the 'second autism'.

Chapter 10 is dedicated to the account and synthesis of research in the field of genetics.

In Chapter 11, we will return to etiological and co-etiological factors, focusing on environment and epigenetics.

Chapter 12 is provocatively titled 'Autism Global Emergency' and will address the epidemiological data and evidence also through a statistical and methodological critique. It will be an extensive chapter in which we will compare the American and British paradigms and learn about the parenting and self-advocacy movement as well as recent definitions of neurodiversity.

In Chapter 13 I will describe the framework of currently recognised and adopted therapeutic and support interventions.

In Chapter 14 we will discuss the relationship with sound, hearing and new neuroscientifically based approaches and interventions.

Chapter 15 brings our journey to a close, with a 'journey within a journey' in seven stages, in and out of the laboratory, and I will close our research and our story by proposing a theory that somehow brings together the evidence gathered, and that attempts to describe a unified picture. In this context, I will also propose a new definition of autism.

Without anticipating anything, I hope that this map will be useful to you and that this journey will stimulate more questions than certainties, that it will show wonderful sights, and that I can humbly be an acceptable guide.

Bibliographical Notes

I have included in the bibliographical references of each chapter only those directly cited.

It is common experience, however, that when writing many are the stimuli, ideas and suggestions that come from different and transversal readings, which somehow 'are present' in the book, even without direct quotation.

In Chapters 5, 9 and 10 I am unquestionably indebted, even when not mentioned in the bibliography, to authors such as Daniel Dennett, Giulio Tononi, Steven Pinker, Antonio and Hanna Damasio, Mary Hellen Immordino-Yang, Christof Koch, Stanislas Dehaene, Anil Seth, David Perlmutter, Richard Davidson and especially Gerald Edelman (who wrote the most complex book I have read so far, an opinion I share with Oliver Sacks).

In Chapter 12 on epidemiological perception, I am indebted to readings of the works of Des Fitzgerald, Jessica Leveto, Sarah Hendrickx, Barry Prizant, Sami Timimi and Robert Chapman, but also to readings that apparently have nothing to do with this topic such as Paul Glimcher, Alexandr Lurjia and Charles Darwin, even though they are not directly cited in the references in that chapter.

For Chapter 14 but a little bit everywhere, my debt of greatest personal affection goes to Oliver Sacks, and this is far beyond the single quote that appears a little bit everywhere.

The idea, where present, of a historical narrative came to me from Gabriella Greison and Yuval Noah Harari: it is not new and will never die, because we are made to tell stories.

How can you say, after all, that that essay or novel did not influence what you will write one day even if you do not remember it? For example, I will quote a poem by Eliot, which I know because my high school Latin and Greek teacher, Italo Sarcone, from whom I continue to learn a lot to this day, quoted it one day.

Bibliographical references are indicated in direct form, e.g. (Kanner, 1943) or discursive with expressions such as "*…Leo Kanner, who in 1943 published…*" (which clearly indicates 'Kanner, 1943' in the bibliography). I hope this choice makes reading more fluid without losing rigor.

To lighten the reading I have included notes at the end of the chapter indicated with numbers that elaborate on the point with separate bibliographical references. I hope this choice proves useful.

Furthermore, I am happy if I was able to include new references in the bibliographies, in the sense that they are not always found in specialist autism publications: this is also part of that opening of the mind and broadening of horizons necessary for a journey and research.

Notes

1. Things would have turned out differently according to David Schwartz (2017).

   "...following the rise of Hitler, the Via Panisperna group had in fact decided to boycott the German magazines, although they remained the most prestigious in the field. (...) Some historians question the veracity of the matter: they point out that at the time 'Nature' only accepted short notes on such topics, and was not at all suitable for the detailed presentation of a new quantum field theory. A more suitable British journal would have been the 'Proceedings of the Royal Society of London', which had published all of Dirac's fundamental articles on quantum electrodynamics; it would have been logical to send the article on Fermi's beta decay there. These historians speculate that Fermi wanted his German colleagues to read the article first, in particular Born, Heisenberg and of course Pauli himself. With a little lie - the failed attempt to publish in 'Nature' - he would have silenced the young colleagues who were against publishing in German journals, and at the same time achieved his main aim".

   Schwartz D (2017) The Last Man Who Knew Everything—Basic Books.
2. It is interesting here to note what Robert Wolke, Professor of Chemistry, writes about the mind of a scientist:

   "Scientists objectively describe what nature provides; they therefore think of colour as a fundamental characteristic of light: for a scientist, light of different colours is radiation of different wavelengths. Artists, on the other hand, create their own interpretations of nature, and thus tend to think of colour subjectively, as something that can be processed with paints and dyes, rather than accepting light in its natural state.

   Scientists dealing with light claim that they can produce any perceptible colour by combining red, green and blue light of varying intensities; on the other hand, artists dealing with pigments claim that they can paint an object with any colour by combining red, yellow and blue pigments in varying amounts. And they are both right, because there is a fundamental difference between the colour of light and the colour of an object.

The light-based primary colour system is called additive because different combinations of wavelengths are added together to produce different colours of light. The pigment-based primary colour system is called subtractive because different combinations of wavelengths are absorbed, i.e. subtracted from light, to produce different colours of paints and dyes".

Wolke R. L. (2000) What Einstein told his barber. More Scientific Answers to Everyday Questions.

3. Quoted by Mastroianni B. (2020)—Litigando si impara—Florence—Cesati Editore—p.13 consulted in Italian translated by the author.
4. See in this regard Borgna E. (2017)—Le parole che ci salvano"—Turin, Einaudi.

References

Al Ghazi L. (2020) From schizoid psychopathy to a natural human variation. Deconstructing Autism in Perspectives on Autistic Spectrum Disorder, Loredana Al Ghazi and Tamara Zappaterra, ETS Editions.

Carofiglio G. (2015) Con parole semplici- Laterza Bari - p.136 consulted in Italian, translation by the author

Dawes, R. M., Corrigan, B. (1974) Linear Models in Decision Making, in Psychological Bulletin, 81, no. 2, pp. 95-106.

Dennett, D. (2014) Intuition Pumps and Other Tools for Thinking - W W Norton & Co Inc.

Di Salvo, M. (2024) Psychotherapy and cognitive and neuronal plasticity. Social brain and post-traumatic stress disorder. Psychoanalytic Research Journal of Relations in Psychoanalysis

Disalvo, D. (2011) What Makes Your Brain Happy and Why You Should Do the Opposite. Prometheus Books

Feynman, R. (1963, 1998) The Meaning of it All. a collection of three lectures in April 1963 at the University of Washington as part of the John Danz Lectures Series.

Frith, U. (2003) Autism: Explaining the Enigma. John Wiley and Sons Ltd, consulted in Italian, translation by the author

Grandin, T. (1995) - Thinking in Pictures: and other reports from my life with autism - consulted in Italian, translation by the author

Groddeck, G. (1920) Psychoanalytische Schriften zur Literatur und Kunst

Intini, E (2017) 8 scientific papers rejected before Nobel - Focus - https://www.focus.it/scienza/scienze/8-articoli-scientifici-rigettati-prima-del-nobel
Jung, C. G. (1929) Wirklichkeit der Seele
Kenny, L., Hattersley, C., Molins, B., Buckley, C., Povey, C. ,Pellicano, E. (2016). Which terms should be used to describe autism? Perspectives from the UK autism community. Autism, 20 (4), 442-462.
Lotto, B. (2016) Deviate: The Science of Seeing Differently. Weidenfeld & Nicolson London UK - consulted in Italian translation by the author
Lowen, A. (1972) Depression and the body. The biological basis of faith and reality. Penguin books - consulted in Italian translation by the author
Morin, E. (1990) Introduction à la pensée complexe
Morin, E. (1999) L'Intelligence de la complexité. Épistémologie et pratique
Repubblica - Sciene - 2017 -https://www.repubblica.it/scienze/2017/08/09/news/einstein_aveva_ragione_di_nuovo_la_stella_e_il_buco_nero_confermano_la_legge_della_relativita_-172699126/
Schrödinger, E. (1946) Der Geist der Naturwissenschaft, Eranos-Jahrbuch, vol. 14, 491
Sloman, S., Fernbach, P., (2017) The illusion of knowledge. Why we never think for ourselves - ed it. Raffaello Cortina Editore 2018 - consulted in Italian translation by the author
Stanovich, K. The bias that divedes us - cited in Pinker, S. - Rationality - 2021
Watzlawick, P., The Language of Change. Elements of therapeutic communication, Italian translation Feltrinelli, 1980, p. 24 - consulted in Italian translation by the author

CHAPTER 2

The First Autism: From Kanner's Definition to the DSM-IV

Autism Before Autism

What we now call autism existed before the invention of its diagnostic label by modern psychiatry.[1]

Some mediaeval tales tell of 'strange' children found in forests or hidden in ditches. For example, William of Newburgh, a twelfth-century English historian and clergyman, in his Historia Rerum Anglicarum tells of two 'green children', so-called because of the colour of their skin, who appeared in the village of Woolpit in Suffolk, England, spoke an unknown language, ate only raw broad beans and did not respect social rules.

In the Fioretti of St. Francis, the character of Brother Juniper appears, incapable of understanding the point of view of others and assessing the consequences of his own behaviour. In Russia, between the sixteenth and nineteenth centuries, the figure of the 'holy or blessed madman' (jurodivyj) recurs. Its main characteristics include mutism, echolalia or incoherent phrasing; insensitivity to pain, hunger and cold; bizarre, repetitive and stereotyped behaviour; convulsive attacks; inability to understand points of view other than one's own and indifference to social conventions.[2]

Since the early seventeenth century, stories of encounters with 'wild children' have been circulating in Europe and North America, an expression denoting three different types of childhood experiences: living in solitude in the wilderness, being raised by animals, and growing up

M. Di Salvo, *Autism Research between Psychology and Neuroscience*,
https://doi.org/10.1007/978-3-031-68338-1_2

confined in enclosed or underground environments for long periods of time. The common denominator of such stories is the experience of almost absolute social isolation and, often, the inability to speak. The most famous cases are those of Peter (in Germany in 1724), Victor (whose story became the film l'Enfant Sauvage by Francois Truffaut), Kaspar Houser (1828) and the little girls Kamala and Amala (found together in India in 1920).

In her book Mitzi Waltz (2023) extensively covers social and medical history and I refer to his work for a rigorous and timely history.

Characters with eccentric and bizarre behavioural patterns reminiscent of autistics also appear in literature. For example, Sherlock Holmes, Nero Wolfe and Hercule Poirot represent the prototype of the genial detective, who focuses his attention on small details, cultivates selective and circumscribed interests, sometimes with almost maniacal dedication and attention to a personal order, follows a strict daily routine and acts in a detached manner, without allowing himself to be influenced by emotions.

It should also be borne in mind that autism, until its 'discovery', was subsumed under the category of 'mental retardation'.

In 1887, the British physician John Langdon Down (1887), after whom the well-known chromosomal syndrome is named, called idiot savants certain 'feeble-minded' children who presented extraordinary abilities in calculation, drawing and music.

In the twentieth century, the British neurologist and psychiatrist Alfred Tredgold (1908) reported cases of children with repetitive behaviour and attraction to objects or their components. It can be assumed that many protagonists of these stories were autistic and that the legends or tales about them evoke the real experiences of those who encountered them. After all—even in literature—when narrating and describing aspects of a human person's behaviour, in order to be credible—as the characters we have mentioned are—such a description must have a reference to a 'common' behavioural experience, in the sense of known, even if extraordinary (Frith, 2010).

The 'Discovery' of Autism

Prior to the twentieth century, there was no clinical concept of autism; among the forerunners of research on the subject in the nineteenth century was Ludwig Binswanger, for whom "autism consists of detachment from reality, together with a more or less marked prevalence of the inner life" (Mistura, 2006).

The term autism derives from the Greek αυτός [aw'tos], meaning "same" and was initially introduced by the Swiss psychiatrist Eugen Bleuler in 1911 (Stotz-Ingenlath, 2000) to denote a behavioural symptom of schizophrenia[3] on the basis of earlier work by Emil Kraepelin.[4]

The term autism in the modern sense was first used by Hans Asperger in 1938.[5]

It later came to denote a specific pathological syndrome in 1943[6] by Leo Kanner who spoke of 'early childhood autism' (Fendrik, 2005).

Prior to Leo Kanner, Melanie Klein described a case in the 1930s that she called child psychosis and that today would be diagnosed as autistic. It is worth remembering that in that historical period the diagnosis of mental illness was ineluctably accompanied by even forced institutionalisation, with the violence that we have only recently come to understand of this condition. In 1920 (Mostert, 2002), two German professors, Karl Binding and Alfred Hoche, caused a sensation with the publication of an essay entitled Die Freigabe der Vernichtung lebensunwerten Lebens (The Permission to Destroy Lives Unworthy of Life) in which they argued that it was necessary to kill 'incurable idiots, empty human shells who eat their fill', sacrificing their lives like ballast to enable the nation to rise even higher. As a conclusion to their thesis, the two professors reported the lengthy calculations made to estimate the cost of caring for these people, which, in conclusion, represented 'an enormous capital of food, clothing, and heating, which is taken out of the national product for wholly unproductive purposes'. As Patel (2009) comments, "their calculations were unimpeachable, their conclusions repugnant." If we think these are Nazi theses, the dates say something else: it is Nazism that fed on theses put down on paper by respected academics thirteen years later! Karl Binding was a jurist (first professor and then rector of the University of Leipzig) while Alfred Hoche was a psychiatrist (professor in Freiburg and director of the city's psychiatric clinic). On this reflection I refer to the beautiful book by Kandel In search of memory: the emergence of a new science of mind which addresses the topic in chapter 29 entitled "Rediscovering Vienna Via Stockholm".

Following Kanner's and Asperger's work, interest in autism increased, but at first confusion between autism and schizophrenia continued to be great: even the publication of the DSM-I (1952) did not provide clarity. Indeed, autism did not find a place in the manual as an independent disorder but rather as a representative symptom of a primary manifestation of schizophrenia. A further important aspect was that the research guidelines, inspired in part by psychoanalysis, had taken into account the possibility that this condition might depend on some parental psychopathology. For a long time, parents were blamed for their child's condition, and a major factor in this was the work of psychoanalyst Bruno Bettelheim (1967), in which parents of autistic children were held responsible for their children's condition. Bettelheim argued that the parents of autistic children had a genuine psychopathology that manifested itself through a behavioural and communicative modus operandi that the mother enacted towards her infant child and that induced the child to a reaction of withdrawal from the relationship. The mother, faced with this attitude of withdrawal on the part of the child, generated a further pathological response, thus defining a vicious circle that produced what Bettelheim called 'chronic autistic disorder', treatable, in his opinion, by separating the child from the parent.

In this same historical period, however, it was thought that other possible causes of autism also needed to be researched, especially by virtue of an observable and objective fact: a considerable number of autistic persons could show a concomitant condition of epilepsy. This led numerous scholars to consider the possibility that a neurobiological mechanism could underlie autism.

In 1964, Bernard Rimland published his book in the wake of the then-dominant current of Skinner's behaviourist psychology.

Rimland claimed that autism was not a psychological manifestation caused by insensitive parents, a widely held belief popularised by Bettelheim. Instead, Rimland suggested, autism was the result of biochemical defects 'triggered by environmental aggression'. He recognised that there could also be a genetic component that predisposes children to the disorder. He argued that autism could "be treated, or at least ameliorated, with biomedical and behavioural therapies" (Venables, 2006).

Childhood Autism challenged the medical establishment's perception of autism: Rimland was outspoken about what he believed to be the main causes of autism: environmental pollutants, antibiotics and vaccinations (Maugh, Thomas, 2006).

At times, this put him at odds with the established medical community. In a letter to the editor of the Washington Post in 1997, Rimland wrote: "*The reason the public—and Congress—supports alternative medicine is that conventional medicine, despite its arrogance, is too ineffective, too harmful and too expensive. Unconventional medicine is a rational alternative to a much greater evil: conventional medicine*".

Rimland has advocated chelation therapy, a typical treatment for lead and heavy metal poisoning, for some children with autism. Neither the American Academy of Pediatrics nor the FDA supports its use for the treatment of autism.[7]

Rimland supported research focusing on "natural, non-toxic ways" to deal with the symptoms of autism. He believed that vitamins (particularly B6 and magnesium) and minerals could help change body chemistry and bring about behavioural changes.[8]

Rimland also advocated the use of secretin, a natural gut hormone, stating that it was "perhaps the most important discovery in the history of autism". He claimed that children treated with the hormone 'showed sudden and dramatic improvement'. However, researchers from North Carolina and the University of Chicago, in separate studies, showed that children treated with secretin showed "no improvement" compared to those treated with a placebo. This treatment has not been recommended by the National Institute of Child Health and Human Development.[9]

Another step towards more contemporary ideas on autism came in the early 1970s with the work of Kolvin (1972) and Rutter (1972), who emphasised that there were clear differences between autism and schizophrenia in terms of onset, clinical features and family history. During these years, despite the fact that the idea that autism was a condition assimilated to schizophrenia and that it was caused by parental psychopathology was still very strong, the seeds were beginning to be sown for a conceptual change of great importance.

With the end of the 1970s, autism began to be studied independently of schizophrenia. Within this framework, Rutter (1978) proposed a definition of autism that was intended to refine the one presented by Kanner years earlier: the symptoms, according to Rutter, had to be present from an early age and refer to problems of a social nature, communication difficulties (in both cases not due to intellectual retardation), the presence of unusual behaviour and rigidity. Rutter's proposal was very relevant, strongly influencing the decision to include the category of 'childhood autism' in the DSM-III (1980), defining it as a generalised developmental disorder and at the same time creating a dedicated section where autism could find a place. The new category of generalised developmental disorders, constructed ad hoc in the DSM-III, thus included childhood autism, the onset of which had to occur in the first 30 months of life, and the generalised developmental disorder of early childhood.

From their first description of autism, both Leo Kanner (1943) and Hans Asperger (1944) had guessed that it was a syndrome due to an organic condition. Unlike Asperger, Kanner later hypothesised that autism was caused by psychodynamic causes.

Kanner claimed that his patients were different from individuals with schizophrenia because the latter disorder rarely manifested itself in early childhood and, above all, was characterised by a gradual change in behaviour. The children examined, on the contrary, exhibited extreme isolation (extreme autistic aloneness) from birth, which excluded everything that came from outside. Almost all of the mothers reported, for example, the absence in their children of the anticipatory attitude of reaching out to be picked up and, once picked up, clinging to the person holding them. Instead, the children were able to establish and maintain a good relationship with the objects available in the environment. Their world consisted of elements that, once experienced in a certain sequence or manner, had to remain absolutely unchanged (anxiously obsessive desire for the maintenance of sameness). Moreover, most of them possessed normal or above-average intelligence. Parents reported, often with pride, on their children's excellent memory, which enabled them to repeat lists of rhymes, animals, plants, names of presidents and even lullabies in French. Kanner (1943) concluded by asserting that, just as other children had been born with physical or intellectual handicaps, those he observed had come into the world with an innate inability to establish emotional contact with people.

In 1943, Leo Kanner had first described the autistic syndrome in a medical journal as a neurological (organic) disorder. In the following months, dozens of families from all over the United States of America came to him for consultation with a child matching his description of autism.

Kanner observed that these were middle and upper middle-class families, with an educated and often 'career' mother, and he considered that these were the characteristics and, therefore, the causes of all autism cases, underestimating the fact that only people from the higher socio-economic classes could turn to him, because they had heard about his article and because they had the means to pay the associated, substantial medical expenses.

Later Kanner realised that autism was equally widespread in the poorer classes and in 1969, during the first meeting of the National Society for Autistic Children (now the Autism Society of America), he acknowledged the limits of his explanatory hypothesis, reducing the stigma that had been created regarding the excessive responsibility of parents for the onset of the disorder. Kanner bequeathed the editorship of the autism journal he founded (the Journal of Autism) to Eric Schopler, who was one of the first to realise the considerable explanatory limitations of his original hypothesis.

Kanner's initial hypothesis that the child with autism was neurologically healthy and that the cause of autism could only be found in a hypothetical 'inadequate relationship' with the mother, was based on this error.

For nearly two decades, this hypothesis, now considered incorrect, dominated the international clinical scene, often directing children and families exclusively towards treatments of dubious therapeutic value. The psychiatrists Bettelheim and Tustin (1972) were among the main exponents of this approach derived from Kanner's reflections, which they disseminated internationally and now considered obsolete.

Anna Freud and Dann (1951), with an investigation of children who survived the Nazi concentration camps at the end of the Second World War, had shown that even those extreme conditions of deprivation of affection could not induce autistic pathology.

Observation of the epidemiological data, which often reveals more than one case among members of the same family and a high prevalence of autism in males (*3 or 4 times higher than in females, or even 20 times higher in Asperger's syndrome*), also provides evidence that autism is generated by causes other than inadequate maternal love.

When epidemiological research and scientific observation clearly revealed that a neurological deficit lies at the root of the autistic syndrome, many parents began desperately seeking pharmacological and dietary remedies.

The desire for a cure leads many parents to mistake for positive results of drugs and diets those positive changes in the state of health that could also be achieved by a placebo. Many 'sellers of illusions' or pseudo-scientific treatments have played on this terrain, who, taking advantage of the anguish of families, propose 'new and miraculous cures', but in reality without verifiable effects.

Asperger's, like Kanner, was also convinced that autism was a different disorder from schizophrenia, congenital and characterised by a deficit of empathy or what Simon Baron-Cohen later called 'theory of mind', i.e. the ability to infer the mental states of others.

Asperger's (1944) definition was broader than his colleague's because it included cases in which the symptoms faded almost into normality. Moreover, he had a more optimistic view: he wrote that only a minority of these children, those with severe intellectual retardation, would have a sad fate; the others possessed compensatory abilities that could counterbalance inadequacies and even lead them to success, thus to greater social integration.

Criticism of Classical Psychoanalysis

Classical psychoanalysis has been accused of blaming parental figures, in particular, the female figure referred to as the refrigerator mother, attributing the cause of the syndrome to a disturbance in the primary relationships with those who assume the role of caregiver (caring work) (Powers, 1994).

Bettelheim even went so far as to propose as a 'rehabilitation therapy' the detachment from the family unit, the so-called 'parentectomy' (Chess, 2002; Arpinat et al., 2009).

> *"Until not so many years ago, there were those who, guided by the psychogenic theory that blamed parents for autism, recommended the removal of children from their families. With the refutation of this theory, and the unjust accusations against parents banished, 'therapeutic removals' have also disappeared, and parents are now seen by doctors and psychologists as a resource of great value not only in the diagnostic phase, but also in the rehabilitation phase"* (Surian, 2005).

This old explanatory and therapeutic model has become the target of criticism and ostracism, first in America and then in Europe, partly due to the increasing prevalence of biological theories in the aetiopathogenesis of mental disorders, as opposed to the psychogenic theories that had previously dominated the field.

Current Aspects of the Relationship Between Psychological Sciences and Autism

At present, the issue of the relationship between psychology and autism is complex, but the classic dichotomous oppositions do not represent the state of the art of current reflection on the relationship between psychological sciences and autism spectrum disorders.

First of all, one should not confuse the broader clinical psychology with psychoanalysis (which is a particular theoretical direction of psychotherapy, which is itself a part of clinical psychology); moreover, the interpretative assumptions of psychoanalysis that are contested are more than half a century old.

On the contrary, research and intervention in clinical developmental psychology have produced a significant amount of verified scientific data on the various aspects of the genesis, clinical assessment, functional characteristics, and possible lines of rehabilitative intervention and support for autistic persons and their families.

Secondly, psychoanalysis itself, in parallel with its clinical and theoretical development, has abandoned many of its original hypotheses on the subject of fifty years ago, revising the old hypotheses on the role of parents in the genesis of autism spectrum disorders.

Indeed, a great deal of research, from John Bowlby[10] onwards, has shown how the family environment greatly influences children's development and characteristics, and how family dynamics and parent–child relationships can be subject to dysfunction, becoming a source of discomfort and severe distress.

The Internet has helped many autistic people overcome the obstacle of not perceiving non-verbal signals and emotional exchange, which they find so difficult to manage; it has given them a way to form networked communities and work remotely (Biever, 2007).

The sociological and cultural aspects of autism have developed: some in the community seek treatment and others believe that autistic neurodiversity is simply another way of being (Harmon, 2004; Silverman, 2008).

Neither Kanner nor Asperger's had discovered autism, they had simply observed, described and codified it and included it in the scientific literature. They had done so at a time when psychoanalysis was still the dominant model of psychological thinking.

The psychodynamic approach to autism, which attributed a relational cause to it, was gradually abandoned in favour of a neurobiological approach. It is due to a British psychiatrist, Michael Rutter, that a rapid step was taken towards the clinical diagnosis of autism. In several articles published in the 1970s, he proposed four major characteristics to identify it: onset before the age of two and a half years, social developmental deficits, inadequate communication and unusual behaviour. By means of the simple classification scheme, Rutter established a starting point for a universally usable diagnosis. The latter came in 1980, when the American Psychiatric Association (APA) granted autism an official, validated status and included it within the third edition of the Diagnostic and Statistic Manual of Mental Disorders (DSM-III) under the category 'pervasive developmental disorder', independent of schizophrenia.

The DSM is the most widely used nosographic system in the world for research purposes as well as a taxonomy for everyday clinical use. The first version dates back to 1952 (DSM-I). To understand the evolution of the manual up to the third edition (*which marks the turning point away from psychodynamic psychiatry*), one has to go back to 1973, the year in which the journal Science published an article by psychologist David Rosenhan (1973) discussing the validity criteria for psychiatric diagnosis. The paper recounts the author's direct experimentation and subsequent vicissitudes, together with those of a heterogeneous group of collaborators: having presented themselves at the admissions service of several psychiatric hospitals and complaining of 'hearing voices', they had all been admitted and detained for a long time despite the fact that, once they had arrived at the ward, they had confessed to pretending. They were eventually discharged with a diagnosis of 'Schizophrenia in remission'. Also in 1973, the gay rights movement forced the APA to admit that homosexuality was not a mental disorder and to remove it from the DSM.

Kendell (1979) published an important scientific article with some of his colleagues on the unreliability of psychiatric diagnoses. Finally, from 1965 onwards, within Anglo-Saxon psychiatry and psychoanalysis, the so-called 'anti-psychiatric movement' began, which radically criticised the concept of mental illness, considering it as an individual's response to unfavourable social and life circumstances rather than as the result of disease, dysfunction and disorders.

These events challenged the credibility of psychiatry as a science and as a legitimate field of medicine. Consequently, in 1973 the APA appointed a distinguished psychiatrist and scientist, Robert Spitzer, to lead a task force that would draw up an 'atheoretical' manual of mental disorders (*i.e. not inspired by a specific aetiological or etiopathogenetic theory*) but capable of describing symptoms and signs to be attributed to the field of mental disorders. The method chosen for its construction implied an agreement between the specialists' opinions, derived from direct observation (behavioural manifestations) of discrete symptoms (ordered to form precise diagnostic categories) and punctual symptoms (recurring in the various patients ascribable to the corresponding chosen clinical category): the aim was to create a reliable instrument for research and dialogue between professionals of different theoretical orientations.

The publication of the DSM-III is a very important watershed because, for the first time, the clinical difference between schizophrenia and autism is officially sanctioned, free in its characteristics from psychotic symptoms, delusions and hallucinations. A few years later came the publication of the DSM-III-R (1987), which introduced further conceptual changes represented first of all by its definition: autism was no longer referred to as 'childhood autism' but as 'autistic disorder', making it possible to include within this condition not only very young children but also older ones. Furthermore, the DSM-III-R, much more clearly, divided the symptomatology of autism into three major areas: reciprocal social interaction, communication and area of interest. In order to obtain a diagnosis of autism, it was necessary to have behaviour that qualitatively impaired each of these areas.

In the years that followed, there was above all a revision of the categorisation aspects of autism. In this course of change, the work of Lorna Wing (1981) is of great significance.

Wing, before anyone else, used the term 'Asperger's syndrome' to refer to the condition observed by Asperger's years earlier. These conceptualisations would later be included in the DSM-IV (1994), which expanded the diagnostic category of generalised developmental disorders to include Asperger's syndrome and also Rett's syndrome. The publication of the DSM-IV allowed Asperger's syndrome to obtain an independent diagnostic categorisation. The publication of the DSM-IV-TR (2000) did not produce any radical changes.

Certainly, the most important transformations occurred in more recent times with the publication of the DSM-5 (2014).

The most important change relates to the inclusion of Autistic Disorder, Asperger's Syndrome and Pervasive Developmental Disorder

Not Otherwise Specified in the broader diagnostic category of Autism Spectrum Disorders.

Another change is related to the shift from a symptomatological triad (*qualitative impairment of social interaction, qualitative impairment of communication, restricted, repetitive and stereotyped behaviour, interests and activities*) to a dyad in which communication and social interaction aspects merge into a single diagnostic criterion.

A further change is inherent in the indication of possible peculiar reactivity in response to sensory stimuli. For the first time, a characteristic that has always been observed from a clinical point of view is listed among the diagnostic criteria.

Another novelty in the DSM-5 is the possibility of indicating the person's current level of severity, which is established by virtue of the need for support. A further possible specification relates to the concomitant presence of intellectual disability and communication disorder. Thus, the condition presented by the DSM-5 is highly variable, depending both on the level of severity and the way in which the symptoms are combined. We will go into these aspects in more detail in Chapter 7.

Autism in Other Cultures

It is always difficult to diagnose a mental disorder. It becomes particularly difficult in children and even more complicated in different cultural worlds. Diagnosis is in fact fundamentally an interpretation that, among other things, is only possible when people seek help, i.e. when they (or those close to them) perceive certain problems as symptoms of an illness. Although the latter is a universal experience, it will vary from culture to culture how people understand its causes, how they cope with suffering, the meaning they attach to what ails them, the therapeutic practices chosen and the actors involved in the treatment and healing process.

Like any other form of 'abnormality', autism does not exist outside of culture. There are communities around the world that do not even have a name for this disorder or do not consider it pathological. For example, the Navajo of the southwestern United States use the term 'state of unawareness' (*nidiniil geesh*), a definition that was recently coined based on the 1989 Arizona school system manual. Furthermore, Native Americans serenely accept a high degree of intellectual and physical disability in children. Only when they reach the age of six do the disabled begin to be educated by their mother and other authority figures, such as the maternal uncle. Since autistic persons are considered to be in a state of perpetual becoming, they are not solicited with superfluous demands. Adults only seek to 'repair', through rituals, what they believe to be underlying the problem: the manifestation of some kind of spiritual disharmony. The most common therapeutic ceremony is the blessing way, designed to ensure success and health at every stage of life. The ritual is considered successful even if the child does not appear changed, since harmony and order have been restored in the spirit world anyway. The Navajo also avoid institutionalising their children in order to avoid the loss of the ability to live in the community and to become true Navajos. Thus, only autistics who become too aggressive, physically or sexually, or who tend to run away, are placed in residences (Connors, 1995).

An opposite example to the Navajo is offered by the Wolof, Serer and Lebou of Senegal. These ethnic groups, who believe in reincarnation, call children with a problematic nature *Nit ku bon*. This nature endows them with specific abilities (e.g. the ability to decide whether to remain among humans or to 'depart'), determines specific risks (suicidal tendencies) and activates certain modes of interaction on the part of the parents and family entourage. Specifically, the *Nit ku bon* is a beautiful child, so much so that he attracts the gaze of others, who, however, shun him. He does not speak or express himself laconically. He isolates himself, hardly ever plays with peers and, if he does, he does not respond to teasing or the beatings he receives. He suddenly starts crying or laughing and often collapses suddenly, as if possessed by rabs (ancestral spirits). The surest symptom of recognising him is his reserved character and extreme sensitivity. Another decisive characteristic is the proximity of death: the *nit ku bon* child (Zempléni, 1965) can die suddenly without any previous, even apparent, illness and rarely reaches adolescence. The link with death can derive from three reasons: he is possessed by a r*ab* (or is a rab himself), is an incarnation of ancestors or is the child of mothers who have lost children.

As can be seen from these cultural examples, each society offers its own interpretations of autism, constructs its own diagnostic criteria and, above all, decides which interventions to implement (Grinker, 2021).

Notes

1. Miriam Gualtieri (2022) I disturbi dello spettro autistico tra storia e cultura—Nuova Rassegna di Studi Psichiatrici n.25.

   The distinction between 'first autism' and 'second autism' that I adopt in this book and that can be found in this chapter and in Chapter 8 were introduced by Bonnie Evans in 2013 [Evans B. (2013) How autism became autism: The radical transformation of a central concept of child development in Britain. Hist Human Sci. Jul;26(3):3-31.].
2. Challis N, Dewey HW. (1974) The Blessed Fools of Old Russia. Jahrbücher Für Geschichte Osteuropas.
3. Kumbier E, Haack K, Herpertz SC., (2008) Autism: exploring historical psychiatric and psychological concepts, in Fortschr Neurol Psychiatr. Vol. 76.
4. Tölle R., (2008) Eugen Bleuler and German psychiatry, in Nervenarzt. Vol. 79.
5. Lyons V, Fitzgerald M., (2007) Asperger (1906–1980) and Kanner (1894–1981), the two pioneers of autism., in J Autism Dev Disord., vol. 37.
6. Merrick J, Kandel I, Morad M., (2004) Trends in autism., in Int J Adolesc Med Health., vol. 16.
7. McKim, Jennifer B. (2006). "Some autistic kids' parents dispute criticism of nontraditional treatments". Knight Ridder Tribune News Service.

   Ratnayake, Hiran (2006). "Controversial autism therapy used in Delaware: Chelation blamed for PA death, but some parents say they have no choice". The News Journal.
8. Mims, Bob (1986). "Researchers seek autism Treatment". Orange County Register. Santa Ana, California.

   Lawton, Cathy (1989). "Life isn't much like the movies for OC autistics". Orange County Register. Santa Ana, California.

   Griffith, Kelly (2001). "Headline: Autism, ADHD in spotlight". Press & Sun-Bulletin. Binghamton, NY.
9. Kalk, Samara (1999). "Scrambling for secretin; questions remain on new treatment for autism...but parents want it now". Madison Capital Times. No. All Edition. Madison, Wisconsin.

   "Repligen against rights to potential autism treatment". PR Newswire. New York. March 10, 1999.

Maugh II, Thomas H. (1999). "In clinical trial, hormone shows no effect on autism; health: secretin, which parents had fervently sought as a cure, produces no more improvement than a placebo". Los Angeles Times. No. Home edition.

Mestel, Rose (2001). "Special Report: Autism. Parents look to unproven therapies to solve autism". Los Angeles Times. No. Home Edition.

10. For an examination of the work relevant to the autism discourse, reference is made to:

Bowlby J (1976) [1965]. Fry M (abridged &) (ed.). Child Care and the Growth of Love (Report, World Health Organisation, 1953 (above). Pelican books. Ainsworth MD (2 add. ch.) (2nd ed.). London: Penguin Books.

Bowlby J (1999) [1969]. Attachment. Attachment and Loss (vol. 1) (2nd ed.). New York: Basic Books.

Bowlby J (1973). Separation: Anxiety & Anger. Attachment and Loss (vol. 2); (International psycho-analytical library no.95). London: Hogarth Press.

Bowlby J (1980). Loss: Sadness & Depression. Attachment and Loss (vol. 3); (International psycho-analytical library no.109). London: Hogarth Press.

Bowlby J (1988). A Secure Base: Parent–Child Attachment and Healthy Human Development. Tavistock professional book. London: Routledge.

References

Arpinat, A.M., Giovanardi Rossi, P., Mariani Cerati, D., (2009) Tisca. Test di ingresso alla scuola per allievi con autismo, Roma, Armando

Asperger H. (1944) Autistic Psychopathy in Childhood. In: Frith U, ed. Autism and Asperger Syndrome Cambridge: Cambridge University Press

Bettleheim B. (1967). The empty fortress: Infantile autism and the birth of the self, New York, Free Press

Biever C, (2007) Web removes social barriers for those with autism, in New Sci,

Chess, S., Thomas, A. (2002) Conosci tuo figlio. Un'autorevole guida per i genitori di oggi, Firenze, Giunti Editore

Connors JL, Donnellan AM. Walk in Beauty: (1995) Western Perspectives on Disabilities and Navajo Family/Cultural Resilience. Resiliency in Native American and Immigrant Families. In: McCubbin HI, Thompson EA, Fromer JE, et al., eds. Thousand Oaks: Sage Publications

Evans B. (2013) How autism became autism: The radical transformation of a central concept of child development in Britain. Hist Human Sci. Jul;26(3)

Fendrik S., (2005) Autism and infantile psychosis. Theoretical and clinical contributions of the psychoanalysis, in Vertex., vol. 16

Freud, A., Dann, S., (1951) - An experiment in group upbringing, The psychoanalytic study of the child,

Frith, U. (2010) L'Autismo: Spiegazione di un enigma. Bari-Roma: Laterza

Grinker RR. (2021) Nobody's Normal. New York: W.W. Norton & Company

Harmon A, (2004) How about not 'curing' us, some autistics are pleading, in The New York Times

Kanner L. (1943) Autistic Disturbances of Affective Contact. Nervous Child.

Kendell, R. E., Brockington, I. F., & Leff, J. P. (1979). Prognostic implications of six alternative definitions of schizophrenia. Archives of General Psychiatry, 36

Kolvin I. (1972) Infantile Autism or Infantile Psychoses, British Medical Journal

Langdon Down, J., (1887) On some of the mental affections of childhood and youth. J & A Churchill

Maugh II, Thomas H. (2006). "Obituaries: Bernard Rimland, 78; Author was the father of modern autism research". Los Angeles Times. No. Home Edition. Los Angeles, California

Mistura, S. (2006) L'autismo come sintomo fondamentale, in Autismo. L'umanità nascosta. Torino, Einaudi

Mostert, M. P. (2002) Useless Eaters: Disability as Genocidal Mar-ker in Nazi Germany, "The Journal of Special Education", 36

Patel, R. (2009) The Value of Nothing. How to Reshape Market Society and Redefine Democracy—Portobello Books London

Powers, M.D. (1994) - Autismo. Guida per genitori ed educatori, Milano, Raffaello Cortina

Rimland B. (1964) Infantile Autism: The Syndrome and its implications for a neural theory of behavior, New York, Appleton-Century-Crofts

Rimland B. (1997) "When 'Alternative' medicine is prescient". Washington Post. No. Final edition. Washington, D.C.

Rosenhan, D. (1973) "On being sane in insane places". Science. 179

Rutter M. (1972) Childhood schizophrenia reconsidered, Journal of Autism and Childhood Schizophrenia

Rutter M. (1978) Diagnosis and Definitions of Childhood Autism, Journal of Autism and Developmental Disorders

Silverman C, (2008) Fieldwork on another planet: social science perspectives on the autism spectrum, in Biosocieties
Surian, L. (2005) L'autismo, Bologna, il Mulino
Stotz-Ingenlath, G., (2000) Epistemological aspects of Eugen Bleuler's conception of schizophrenia in 1911. in Med Health Care Philos.
Tredgold, A., (1908) Mental Deficiency: Amentia
Tustin, F., (1972) Autism and Childhood Psychosis
Venables, S. (2006) "Bernard Rimland; psychologist researcher in autism who overturned the theory that it was a reaction to bad parenting". The Independent. London
Waltz, M. (2023) Autism. A Social and Medical History Palgrave Macmillan—in particular: Autism Before and After the Enlightenment. In: Autism. Palgrave Macmillan
Wing, L. (1981). "Asperger's syndrome: a clinical account". Psychol Med. [for a critic lecture see Sheffer, E., (2018) Asperger's Children: The Origins of Autism in Nazi Vienna - N.Y: W W Norton & Co]
Zempléni A, Rabain J. (1965) L'enfant nit ku bon: un tableau psychopathologique traditionnel chez les Wolof et les Lébou. Psychopathologie africaine.

CHAPTER 3

An Initial Aetiology of Autism: From Early Psychological Theories to the Theory of Correlation with Vaccines

The Initial Aetiology

Aetiology is that part of scientific research that focuses on the causes of a phenomenon. Put in simple terms, the aetiology of a falling body is the force of gravity. We know this today because there has been an 'aetiological' study in physics to explain the cause of the phenomenon of falling bodies.

In the medical-clinical field, things get complicated.

Sometimes the cause of a pathological phenomenon is clear and unambiguous. And having found the cause, one has the direction in which to look for the cure and in the meantime the treatment of the pathological state.

This phenomenon became more pronounced in the twentieth century: as man's awareness of his capacity for research and knowledge increased, we looked for 'unique causes' in search of the 'miracle drug', everywhere. Cocaine for Freud (1885) was not far removed from the L-Dopa that Oliver Sacks (1973) talks about with extraordinary humanity.

Sometimes the phenomena are complex, have no single cause, let alone a miracle pill or a cure-all therapy.

Human omnipotence collapses, and with it often faith in science.

Autism is a 'complex state' and aetiological research has focused on numerous fronts in search of a cause, inexorably discovering new things and opening up new perspectives.

M. Di Salvo, *Autism Research between Psychology and Neuroscience*,
https://doi.org/10.1007/978-3-031-68338-1_3

The first hypothesis developed on the causes of autism was that of Leo Kanner, who first published a fairly comprehensive study on the syndrome in 1943. While concluding that it was an inborn disorder, Kanner, who had identified many parents, grandparents and relatives of high cultural status in families with autistic children, hypothesised that obsessiveness was a kind of fundamental characteristic of these families.

Leon Eisenberg, his close collaborator, emphasised how difficult it was to ignore the affective configuration of the family unit and hypothesised that the parents' behaviour did not help or stimulate the child to come out of its shell of 'affective refrigeration'. Kanner–Eisenberg's thesis has recently been taken up in the debate on the etiopathogenesis of the disorder following the discovery of the mirror neuron system (Iacoboni et al. 2006), which some authors believe could play a role in the genesis of autism (I will discuss this in more detail in Chapter 9).

Many of the early studies on autism after Kanner focused primarily on the role of parents. However, many different factors have been observed that may contribute to the development of the syndrome, including both hereditary and nonhereditary factors (Charles et al. 2008).

Since monozygotic twins (who have the same genetic make-up) are both affected in 60% of cases, it is likely that there is a genetic component, even if it is not the only trigger (otherwise 100% of monozygotic twins would develop the syndrome); a multifactorial cause with genetic and environmental elements is, therefore, assumed (Muhle, 2004).

The aetiology of autism spectrum disorder is not yet very clear. In most cases, however, the disorder is associated with a genetic abnormality, due to a mutation in part of the DNA; in other cases, the disorder results from a congenital lesion of one or more structures of the developing central nervous system (Galanti, Sales, 2017).

The Vaccination Correlation Fraud

An old hypothesis on the alleged cause of vaccination[1] has now been completely discredited: the hypothesis put forward by Andrew Wakefield turned out to be a scientific fraud, as his study, later retracted by the publisher, was based on a manipulation of experimental data.

As reported in the British Medical Journal[2] Wakefield et al. (1988) was paid to support the false evidence of a link between the disorder and the use of the trivalent vaccine (against measles, mumps and rubella).

Wakefield's publication prompted a series of other studies on a larger population to see if there really was a link.

None of these studies ever confirmed Wakefield's data.

For example, a well-known study of all children born in Denmark between 1991 and 1998 who received the vaccine (a sample of almost half a million children) found no difference in the incidence of autism compared to non-vaccinated children (cf. Madsen et al. 2002; Kaye et al. 2001).

There has been no single refutation of this claim: over time, many studies have been conducted with similar conclusions to those of the Danish study (crf. Farrinton et al. 2001).

Other studies had ruled out correlations between vaccines and autism, published in the same Lancet journal and on age- and gender-differentiated samples until 2008.[3]

The role of thimerosal as a risk factor has also been ruled out.[4]

The affair ended with the retraction of 10 of the 12 researchers who had published the manipulated 1998 study.[5]

In January 2011, the British Medical Journal published an extensive investigation on the subject, from which the fraudulent profile of the vaccination hypothesis definitively emerged, and on how some of the protagonists in the affair made false claims in exchange for financial compensation, thus carrying out a fraudulent fundraising campaign for personal gain.[6]

A 2014 meta-analysis (Taylor et al. 2014) reviewed ten major studies on autism and vaccines involving 1.25 million children worldwide; it concluded that neither the vaccine preservative thimerosal (mercury) nor the MMR vaccine, which never contained thimerosal, contributed to the development of autism.

Despite this, parental concern has led to lower rates of childhood vaccinations, epidemics of previously under-control childhood diseases in some countries, and the predictable death of many children.[7]

The Centres for Disease Control is publishing a thirty-four page guide entitled Guide to Vaccine Contraindications and Precautions.

"Contraindication", in medical terms, indicates a situation in which a certain intervention should not be practised, or should be practised after appropriate modifications, such as—to give a few examples—in the case of persons who have had strong reactions in the past to the vaccine or one of its components, or for persons with immune system abnormalities or certain types of infection, or for those who have had previous adverse reactions. There is a general consensus among doctors that vaccines may sometimes be harmful or at least contraindicated for certain categories of people in the presence of frailty.

The question in the case of autism is whether those children who might develop autism should not be included in the categories to which precautions and contraindications apply.

Martha Herbert (2013) gives a description of these children in her book.

Regarding children at risk for autism, Dr. David Amaral,[8] head of research at the Davis Mind Institute at the University of California, one of the world's leading institutes in the field of autism and inflammation research, recently stated that "in fact, for this category of children, vaccinations could be the environmental factor that pushes them over the autism threshold. And I think it is of paramount importance that we try to understand what vulnerabilities—if any—might make a small subset of children at risk'.

The new scientific branch of vaccinomics, which deals with the development of vaccines that are customised to the individual's genetic profile and medical history, recognises that the current generic and generalised approach to vaccines is not ideal and that for some individuals certain currently available vaccines may be ineffective, while for others they may be harmful (Moyer, 2010).

The Relevance of the Vaccination Correlation Fraud

The argument of vaccines as the cause of autism is relevant in many respects.

From a first point of view, it provided strong evidence of a very strong bias in scientific research, concentrated in the semantic frame of **confusion between causation, correlation and coincidence.**[9]

A related phenomenon is a phenomenon that intervenes, infers and contributes to another phenomenon, enhancing or detracting from its occurrence, but is not necessarily decisive.

A correlation is a statistical relationship between two variables such that each value of the former corresponds to a value of the latter, following a certain regularity. Correlation does not depend on a cause-effect relationship as much as on the tendency of one variable to change as a function of another.

The classic example is that of marriages and births. Generally, statistically, the increase in the number of marriages is correlated with the number of births in the following two years. But the two phenomena are independent, in the sense that one can easily generate children without necessarily contracting marriage, which is not a biologically necessary and determining requirement.

It is a typical statistical phenomenon, and is represented by curves that 'move together'.

A coincident phenomenon is one that occurs 'at the same time or in the same space' as another, but because of this coincidence alone it may also have no bearing on the phenomenon we are examining.

For example, if my team won the game while I was showering, it does not mean that they will always win when I shower or lose if I go to the stadium instead of showering.

It is an orientally intuitive phenomenon, in the sense that the relationship between the two is an individual and subjective intuition: it is I who connect the two phenomena and assign to them the value I prefer (to be clear, it is on this confusion that superstition is based).

Causation, on the other hand, is that (neither statistical nor intuitive but) ontological fact whereby one phenomenon necessarily makes another dependent on it.

If I put my hand on the fire I get burnt, and conversely fire is the cause of my burn, which also means that if I do not put my hand on the fire I will not get burnt.

We have here an example of three phenomenal directions: the identification of the theory of direct cause (fire is the cause of my burn), the proof of the property of causal action (I put my hand on the fire = I burn myself) and the 'reverse' verification (if I do not put my hand on the fire I will not burn myself). The hypothesis is false if the 'truth principle' is missing in any of these steps.

Quite simply, vaccines coincide with autism, in the broadest sense, that is: given that an autism spectrum disorder normally manifests itself between the ages of 9 months and two years, and given that these ages also correspond to those at which vaccination cycles normally take place, it was assumed that there was a correlation and then even a causation.

This must first of all warn those doing research to always check that an observed phenomenon does not fall into this cognitive bias. But it must also be a warning to all, because we are all observers of the world, not to fall into this bias that often makes us relate things that are not related.

Erwin Schrödinger (1929) has well traced the origin of causal thinking from a physics perspective:

"When a sense of higher consciousness is awakened in man, he is already part of an environment whose changes are of the utmost importance for the good or evil that can result from it. Experience—at first the unsystematic experience of the daily struggle for existence, later the systematic and organised experience of scientific research—shows him that the events that occur in his environment do not follow one another haphazardly [...] Of special biological importance is the case in which one group of characteristics precedes another in time. The circumstances preceding the unfolding of a certain phenomenon A repeatedly observed, typically split into two groups: the constant and the variable. [...] as the particular concatenations of a regular character are identified, the concept of the general and necessary connectability of phenomena originates, as an abstraction from the whole of them. [...] it is admitted that every natural phenomenon is quantitatively and absolutely determined at least by the totality of circumstances or initial physical conditions. [...] The confidence we have in this postulate, also called the 'principle of causality', is further consolidated through the progressive ascertainment of ever new particular determining causes. A natural law is thus for us nothing more than a regularity ascertained with sufficient certainty in the course of a phenomenon, insofar as it is considered a necessary regularity in the sense of the above postulate. [...] the overwhelming majority of phenomena, whose regular and invariable course has led to the establishment of the postulate of general causality, have chance as the common root of the strict regularity observed. [...] More than by many particular examples, our persuasion of the statistical character of the regularity of physical phenomena will be strengthened by the consideration of one of the most general empirical laws, the second principle of thermodynamics, or the law of entropy, which even takes part in every truly physical phenomenon, and which has proved to be the prototype of statistical laws. [...] Where, then, does the generally widespread belief in the absolute causal determinacy of molecular facts and the persuasion that the opposite is inconceivable come from? Simply from the habit, inherited from millennia, of thinking causally. It makes us see an indeterminate process, a primary, absolute accident, as a perfect absurdity, as something logically meaningless. Where does this habit of thinking come from? From the observation, protracted over centuries and millennia, of precisely those conformities to natural laws, of which we now know with certainty that they are not causal—at least not immediately—but immediately statistical".

From another point of view, the phenomenon showed **the power—both positive and negative—of the web**.

On the one hand, we have seen the exponential growth, in terms of diffusion, geographical distribution and speed, of misleading information where not directly false and conspiratorial. A phenomenon that, more than twenty-five years later, we are beginning to understand and counteract even normatively and with decidedly little effectiveness.

On the other hand, it showed an astonishing capacity—again, in size and speed—to bring together people with common interests and problems, to collect funds and information and to bring people together in a cooperative direction and with cooperative intent.

Depending on the point of view, each side has prevailed, but in a balance that has increasingly shifted towards awareness and cooperation, also contributing to the spread of denials.

In this sense, it also brought together and stimulated researchers, and allowed for a not new but all too rare phenomenon of cross-fertilisation of skills and comparison between research areas: psychologists, educators, doctors, physiotherapists, each with their own area of specialisation and personal experience and sensitivity, each with a background of experience and studies and focus of interests that were compared.

It was out of this confrontation that the first theories, which I call 'wide-ranging', were born.

From yet another point of view, the thesis of vaccine causation coincided with new diagnostic criteria, and through the over-information generated by the Internet, there was a perception of a real epidemic of autism (in the West).

As Steve Silberman (2015) put it "When, with the rise of the internet, fears of a vast conspiracy between large multinational drug companies and corrupt government officials to cover up the effects of a global wave of vaccine damage had been circulating, immunisation rates had begun to plummet worldwide, conjuring up the spectre of a resurgence of epidemics such as whooping cough, which had previously killed tens of thousands of children each year.

The official explanation for the surge in prevalence estimates was that the diagnostic criteria for autism had been gradually broadened over the years. But if that was the reason, why had they been so inadequately restrictive to begin with? How was it possible that a previously rare, obscure syndrome supposedly rooted in genetics suddenly seemed endowed with the gift of ubiquity?

On the wave of public outrage at the sudden rise in figures, autism research, long neglected by funding bodies such as the US NIH (National Institutes of Health) precisely because the condition was considered so rare, had found itself on the threshold of a golden age. Between 2000 and 2011, grants in the United States grew by an average of $51 million annually, in addition to the $1 billion budget for 2006 due to the passage of the Combating Autism Act by Congress. Private funding groups, such as the Simons Foundation, had done their part by pushing total investment in autism research to the highest level in history. In 2011, Autism Speaks, the world's largest autism fundraising organisation, announced a $50 million commitment, supported together by the Beijing Genomics Institute, to map the entire genome of ten thousand individuals from families with two or more autistic children. The organisation's Vice President for Scientific Affairs, Andy Shih, promised that the project would generate 'a transformative level of information'."

An Exponential Growth of Interest

Between 1946 and 1985, the average number of articles per year (mostly in English) stood at around 150.

In the 1990s, there was a growing interest in the subject on the part of popular publications (periodicals and books) and, above all, the film industry (the so-called 'Rain Man effect' for the 1988 film of the same name), which greatly increased interest in more in-depth studies. From 1991 to 2001, the average number of research articles was around 250 per year, and the peak of scientific articles on autism was 579 in 2001.

Every year since 2001, the number has practically doubled to 7833 in 2022.

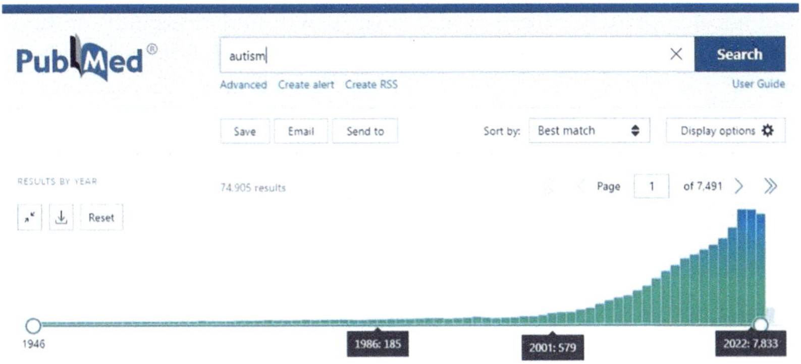

[author search - January 2024]

This was made possible by all the factors we have considered so far, but also by a new sensitivity. It was realised that the causes were not individual, that each field of research could offer analyses and data and contributions to be compared and put together with those of the others.

Approximately 10% of the publications are of a cross-sectional nature, uniting authors from different backgrounds and expertise. And in science this is no small thing.

More than 50 per cent of the published studies involve multiple authors from different research institutions and countries. A supremacy in medicine is only surpassed by genetic research, but in this case the figure is distorted by the (positive) fact that huge projects such as the 'Genome Project' are already born as global projects.

Before addressing the most recent and articulated aetiological theories of autism, it is good to dwell on 'what is good to save' from previous aetiological studies.

While it is true that Bettelheim's thesis is absolutely baseless, reasoning to the contrary has been disproved mainly by families, just as a behaviourist approach by Rimland has been disproved by the effectiveness of affective therapy, social integration and cognitive-behavioural approaches.

Sometimes it is the wrong approaches that help us understand which is the right way and also function as indirect evidence.

However, some of Rimland's insights into the links between 'pollutants' and environmental elements still form the basis of a number of aetiological approaches that—founded on new foundations and methodologies and technologies that were unimaginable in Rimland's time—open up completely new fronts of research, such as those examined at Harvard by Martha Herbert, for example.

If I have dwelled at length on the subject of the (false) correlation between vaccines and autism, it is because this idea has generated such a vast—this yes—scientific, worldwide, cross-sectional investigation, which has led to countless collateral discoveries, and has stimulated investments like never before, and new scenarios have opened up on this front.

By the end of the 1990s, the scientists had done exactly what they had studied for, specialised in and researched, and cynically speaking what they had 'been paid for'.

Molecular biologists had identified more than a thousand candidate genes and hundreds of de novo mutations associated with autism, and gained a better understanding of epigenetics, the scientific study of the factors that mediate interactions between genes and the environment.

The list of suspected environmental causes of autism grew daily, including dozens of commonly used chemicals and linking things that never were, again often wrongly.

Emily Willingham (2015) Forbes science columnist and mother of an autistic boy wrote on her blog: 'This Just In [...] Being Alive Linked to Autism'.

Today we know a lot more and, above all, we have some firm points to build on.

Autism is a complex phenomenon.

It is a state of the person that may (or may not) include a spectrum of syndromes and a symptom complex.

We know that it is neither a 'psychological problem' nor a psychiatric one, that it does not depend on parental behaviour of coldness or lack of affection.

At the same time, we know that it has profound psychological implications, we know that psychological support with targeted therapies can have surprising effects for cases of Asperger's and mild and intermediate autism and we know that love and affection and caregiving have extraordinary therapeutic power.

We can agree with at least two considerations of no small importance.

The first is by Simon Barhon-Cohen, according to whom it makes sense to diagnose autism if it serves the person.

The second is by Uta Frith according to whom we have to think of a long causal chain with many distinct links: 'the risk can be of many types, including genetic defects, chromosomal abnormalities, metabolic disorders, viral agents, immune intolerances, and anoxia resulting from perinatal problems. We can assume that each of these risks has the potential to produce very serious damage to neurodevelopment. Due to these severe interferences, permanent damage can be produced in the development of specific brain systems involved in higher mental processes. The damage may be slight or severe, but often involves the shutdown of one or more crucial systems at a critical stage. Our hypothesis is that only in such a case will autism occur".

Error, Bias and Noise

In the following chapters, I will explore and describe the major theories and scientific research on autism in the genetic, biogogical, psychological, psychiatric and neuroscientific fields.

It is good to address a relevant issue at the outset.

Scientific research is a form of judgement that tends towards a result.

Like all judgements, it is affected by bias, rumour and error, because researchers (individually understood) and research teams are affected by such bias, rumour and error.

Of course, this reasoning also applies to the reader.

We say that a **bias** exists when in a set of judgements most of the errors go in the same direction. The bias is the average error. Removing the bias from a set of judgements will not eliminate the error altogether. The errors that remain once the bias is removed are not shared: they are the unintended divergence in judgements, the unreliability of the measuring instrument we apply to reality; they are the noise, the variability of judgements that should be identical. Systemic **noise** is that which can be observed in organisations that use interchangeable professionals to make decisions, such as emergency room doctors, judges who impose criminal penalties and underwriters of an insurance company.

In science, bias is the difference, positive or negative, between the average of measurements. Noise, on the other hand, is the variability of results. In statistics, the most common measure of variability is the standard deviation, which is used to measure noise in judgements.[10]

Most judgements, especially predictive ones, can be likened to a measurement.

When we make a forecast, we try to get as close as possible to an actual value: one who makes economic forecasts, for example, aims to get as close as possible to, say, the actual value of the increase in gross domestic product in the following year, whereas a doctor aims to make the correct diagnosis (*the term forecast, in the technical sense used here, does not necessarily refer to the future: for our purposes, the diagnosis of an existing pathology is a forecast*).

Like any measuring instrument, the human mind is imperfect in that it is affected by bias and noise.

Many judgements are nothing more than predictions, and since verifiable predictions can be evaluated, by studying them we can learn a lot about noise and bias.

The study of forecasts in the economic and financial sector has shown, for example, that forecasts are accurate to the extent that employees whose potential was judged to be high at the time of recruitment receive high evaluations for their work.

One measure that exploits this insight is the **percentage of concordant pairs** (PC), which answers a precise question.[11]

PC has the advantage of being a very intuitive measure of covariation, but it is not the most widely used measure in the social sciences (Kamboj et al. 2018).

The standard measure is the **correlation** coefficient (r), which ranges from 0 to 1 when two variables have a positive relationship.

One can think of the correlation coefficient from different perspectives. The most intuitive is that the correlation between two variables corresponds to the percentage of determinants they have in common.

The loss of accuracy in cross-validation is more pronounced when the original sample is small, because then the coincidences stand out more.

The problem highlighted by Dawes (1974)—whom I have quoted in the introduction—is that the samples used in social research are generally so small that they negate the advantage of so-called optimal weighting. As the memorable subtitle of an article by statistician Howard Wainer (1976) on the estimation of optimal weighting puts it, *It Don't Make No Nevermind*. Or rather, in Dawes' words, 'we don't need more accurate models than our measurements' (Dana, 2008).

Fair weighting models work because they are not subject to sampling problems.

Dawes' study has a direct implication worth dwelling on: it is possible to make valid statistical predictions without possessing any prior data on the outcomes you are trying to predict: you just need a set of reliable predictors to correlate with the outcomes.

Low correlation coefficients will perhaps come as a surprise if one is used to reading about results that are defined as 'statistically significant', or even 'highly significant' (*we will see this with regard to psychological or genetic research*).

Statistical terms are often misleading for ordinary readers: the term 'significant' is the worst example. When a result is called 'significant', we should not draw the conclusion that the effect it describes is significant: it only means that it is unlikely to be a mere product of chance. With a sufficiently large sample, a correlation can be both 'significant' and too weak to merit attention.

The limited predictability of individual outcomes makes one reflect on the difference between understanding and prediction.

When we equate understanding with prediction (*or the absence of one with the absence of the other*), we are using the term understanding in a specific sense. In fact, this word also has other meanings: if we say that we understand a mathematical concept, or that we understand what love is, we are probably not referring to the ability to make any specific prediction.

However, in the social sciences, and very often in everyday conversation, to say that one understands something is to say that one understands its cause.

The sociologists who collected and studied the thousands of variables examined in the Fragile Families Study were trying to trace the causes of the observed outcomes. Doctors who understand what ails a patient claim that the condition they diagnosed is the cause of the observed symptoms.

Understanding, in this sense, means describing a causal chain?

The **ability to make a prediction** indicates whether and how far this chain has actually been identified. Similarly, **correlation**, the measure of predictive accuracy, indicates the extent to which we are able to explain a causation.

Perhaps it will come as a surprise if one is familiar with the fundamentals of statistics, and thus the classic caveat that 'correlation does not imply causation'.

However, it must be remembered that if correlation does not imply causation, the reverse is true: in the presence of a causal link, we should find correlation.

In summary: where there is causation, there is correlation. It follows that where there is causation, we should be able to make a prediction—and correlation, i.e. its accuracy, is a measure of how well we understand this causation.

Objective ignorance places a limit not only on our predictions, but also on our understanding.

What then do all the experts who claim to understand their field mean? How can they pronounce the causes of the phenomena they observe and arrive at convincing predictions to that effect? In short, why do experts, like the rest of us, seem to underestimate their own objective ignorance of the world?

Researchers are human beings.

One of the characteristics of human judgement is that when it is made, putting together data and sensations into a coherent story, at least apparently, internal feedback is received.

The **internal signal** is a self-recognition that one strives (sometimes not even that hard) for when coming to a conclusion about a certain judgement. And an emotionally satisfying experience, the pleasant feeling of having achieved coherence between the information examined and the judgement arrived at. All the pieces seem to fit together (often this impression of coherence is reinforced by hiding or neglecting information that one cannot integrate).

The internal signal is important—and misleading—because it is not interpreted as an impression, but as a conviction: an emotional experience ('I have the impression that this is right') is passed off as rational confidence in the validity of one's own judgement ('I don't know why, but I do').

Confidence, however, does not guarantee accuracy, and many confident predictions turn out to be wrong. If bias and noise contribute to errors in predictions, this does not depend so much on how good the predictive judgements are, but on the fact that there is a limit to how good they can be. In this chapter we will focus precisely on this limit, which we call objective ignorance.

The causal mode saves us a lot of energy by categorising events at the moment as normal or abnormal. Abnormal ones require a lot of effort from the outset to find relevant information, both in the surrounding environment and in our memory. Active expectation, i.e. remaining in diligent waiting for something to happen, also involves effort; in contrast, the flow of events in the valley of normality requires little mental work.

Causal thinking avoids unnecessary effort while maintaining the degree of alertness necessary to detect abnormal events.

As Daniel Kahneman (2011) has pointed out, on the contrary, statistical thinking is laborious, and requires attention that only System 2, the slow and deliberate mode of thinking, can mobilise. Beyond a certain basic threshold, it also requires specialised skills. This type of thinking starts with sets and considers individual cases as exemplifications of broader categories.

Relying on **causal thinking** for a single case leads to predictable errors; taking the statistical perspective, the outside view, allows one to avoid such errors.

Whenever causality is plausible, our mind gives a correlation, however low, explanatory causal force: the brilliance of a leader is accepted as a satisfactory explanation for a company's success, inexperience as an explanation for failure.

Relying on fallacious explanations is perhaps inevitable if the alternative is to give up understanding the world. Causal thinking and the illusion of understanding the past lead to overconfident predictions. The preference given to causal thinking also contributes to the failure to identify noise as a source of error, noise being an essentially statistical concept.

Causal thinking helps us make sense of a world that is much less predictable than we think and explains why we think it is much more predictable than it is.

The theory of judgement **heuristics** suggests that sometimes people resort to answering a simple question to answer a more difficult one.

Replacing one question with another is not only about similarity and probability: one runs into the same problem when one exchanges the impression of the ease with which examples of certain events come to mind with a judgement on their frequency.

Similarly, substituting likelihood for similarity leads to the neglect of base rates, which are completely irrelevant in judgements concerning the latter. And factors such as irrelevant variations in the outward form of the presentation document should have no weight in the assessment.

Then there is a different type of **bias, which** we call **prejudice**, or jumping to conclusions.

All of us, when we begin to make a judgement, often already have a bias towards a particular conclusion. In doing so, we let system 1, i.e. rapid, intuitive thinking, show the way. After that, we either jump to that conclusion, circumventing the process of information gathering and integration, or we mobilise system 2, the intentional thinking, to find arguments to support our bias. In that case, the evidence we find will be selective and biased: due to confirmation bias and desirability bias, we will tend to collect and interpret information selectively to favour, respectively, a judgement we already believe or wish were true.

People often develop plausible rationales for their judgements and come to think that they are the cause of their beliefs.

Prejudices are evident in any field, esp. they have an emotional component.

The psychologist Paul Slovic (2000) called it affection heuristics: people's thoughts are determined by their feelings.

Often professors notice that if at the end of the year, students give their teaching a positive rating, they also give the course materials a high rating; if, on the other hand, students do not like that professor very much, they give the same materials a lower rating. The same mechanism is activated even in the absence of an emotional factor: regardless of the true reasons behind a certain belief, you will be more inclined to accept arguments that seem to confirm it, even when the reasoning is wrong (*this type of reasoning has been called belief bias*—Evans et al. 1983).

A more subtle example of this tendency to jump to conclusions is **anchoring**, i.e. the effect an arbitrary number has on those who have to make a quantitative judgement.

Anchoring is an extremely robust effect, often used purposefully in bargaining (Galinsky, 2001). Making the first move will probably work to your advantage, because, by virtue of the anchoring effect, the other party will involuntarily be led to think about how your offer may be reasonable. People always look for logic in what is said: when faced with an implausible number, they almost unwittingly come to considerations to reduce its implausibility.

Directly related to anchoring is **over-consistency**: we quickly form consistent impressions and are slow to abandon them.

Confirmation bias—the tendency that leads us, in the presence of a bias, to ignore any contradictory data—can cause us to attach less importance than we should to subsequent data (*another term for this phenomenon is halo effect*) (Asch, 1946).

In general, we jump to conclusions and do not abandon them: we think our opinions are based on facts, but the facts we consider and our interpretation of them are likely to be distorted, at least to some extent, to fit our initial summary judgement. Consequently, on the whole, we remain consistent with the story that has taken over in our minds. In the event that the conclusions prove to be correct, this process is still acceptable, but when the initial assessment is wrong, the tendency not to break away from it against any evidence that proves it to be unfounded is very likely to amplify the error. And this effect is difficult to control, because the facts we have seen or heard are impossible to ignore, and almost always difficult to forget.

In courts and tribunals where juries are present, judges often ask jurors to disregard inadmissible evidence, but the request is unrealistic (*however useful it may be at the stage when the jury deliberates on the case, when arguments explicitly based on that evidence may be rejected*).

I have briefly presented three types of biases that operate in different ways: substitution bias, which leads to assigning wrong weight to facts; jumping to conclusions (or bias), a bias that leads us to circumvent facts or to consider them in a distorted way and over-consistency, which amplifies the effect of initial impressions and reduces the impact of contradictory information. All three of these types of bias, of course, can lead to statistical bias, but also create noise.

An example of an effect that amplifies this phenomenon is **matching.** If we said that judgement is an operation in which a value is assigned to a subjective impression (*or to an aspect of an impression*) within a scale, matching is a fundamental part of this process. It intervenes by replacing a judgement of likelihood with one of similarity.

Conflicting signals make it more difficult to arrive at a sense of coherence and arrive at a judgement that appears to correspond satisfactorily to reality. Since complex judgements are characterised by the presence of conflicting signals, we expect to find a lot of noise.

Some of the scales we rely on to make judgements are qualitative. They are distinguished by the fact that their values are not ordered: green is neither higher nor lower than red.

Many judgements, however, are made on the basis of quantitative intensity scales. Physical measurements of size, weight, brightness, temperature or sound volume and estimates of cost, value, probability or frequency are all examples of judgements using quantitative scales, as are those based on more abstract scales such as the degree of confidence, strength, beauty, anger, fear, immorality or severity of punishment.

The distinguishing feature of all these quantitative dimensions is that it is always possible to place any two values belonging to the same scale in ascending order.

People have great intuitive abilities when it comes to finding a correspondence between the intensity of unrelated dimensions by superimposing one intensity scale on another (Stevens, 1966).

Here I use the term 'matching' with a broader meaning than the one adopted by Stevens, which refers only to relative scales. The range of values within a scale varies depending on the context: even the concepts of large and small depend entirely on the reference system.

Although they lead to absurd results from a statistical point of view, it is difficult to resist the temptation of making predictions that seem to fit well with factual reality (*this speech, for example, may be useful in framing what I will say in Chapter 12 in a more objective manner*).

Predictions of this kind are bound to lead to great disappointment; on the other hand, if they are made at a particularly bad time, they are likely to be too pessimistic. Intuitive forecasts correlated with factual data are always too polarised, either positively or negatively (*technically, these prediction errors are called non-regressive, because they do not take into account the statistical phenomenon called regression to the mean*).

It should be noted, however, that substitution and matching are not always the basis of predictions. From the perspective of the two systems described by Kahneman, system 1, that of intuitive thoughts, proposes quick associative solutions to problems as they arise, but these intuitions must be accepted by system 2 of reflective thoughts before they become beliefs.

Predictions based on matching are sometimes rejected in favour of more complex answers: for example, people are more reluctant to correlate their predictions with unfavourable reality data.

When the available information forms a coherent picture, system 1 of fast thinking has no difficulty in finding meaning: simple judgement problems are easily solved and almost everyone will agree on the solution.

Multiple conflicting signals create the ambiguity that characterises complex judgement problems, which also explains why difficult problems are more prone to noise than simple ones. The rule is easy: if something can be seen in more than one way, people will see it differently. They will be able to select different elements to construct the core of their narrative and will, therefore, arrive at many possible conclusions.

In order to feel confident in one's judgement, two conditions must be fulfilled: the story must be consistent overall, and there must be no valid alternatives. Overall coherence is achieved when all the details of the interpretation adopted fit well into the story and reinforce each other.

Of course, one can also achieve the goal of coherence—albeit in a less elegant manner—by neglecting or dismissing everything that does not fit. The same applies to alternative interpretations: the expert who has succeeded in 'solving' a problem of judgement is able to say not only why his explanation is correct, but also why the others are not; however, it is also true that the same degree of certainty can be achieved with an incorrect interpretation, if the alternatives are disregarded or purposely concealed.

If one looks at self-confidence from this perspective, one can deduce that subjective confidence in one's own judgement is by no means a guarantee of accuracy.

Moreover, the concealment of alternative interpretations, a well-documented perceptual process, could lead to the phenomenon termed 'illusion of agreement': if people do not imagine possible alternatives to their own conclusions, they will tend to assume that other observers must also come to the same conclusions.[12]

Not all of us are always confident in our judgements, but we are almost always more confident than we should be (Moore et al. 2017).

As J. Michael Pearson (2015)—amazed that no one took Ida Noddack's proposal seriously at the time—and David Schwartz (2017) have pointed out, the discovery of fission is a typical case of **cognitive dissonance**.

"No one had seen the solution until it became impossible to ignore it.

In retrospect, the real cause was what psychologists call 'cognitive dissonance'. This phenomenon, first identified by academic psychologist Leon Festinger (1962), occurs when empirical data contradict what we 'know' about how the world works. To resolve the discrepancy, we ignore the data or try to force it into the structure of our preconceptions; sometimes, however, a crisis occurs and that structure falls apart. Hahn, Rutherford, Joliot-Curie, Bohr, Fermi and other great scientists involved in the analysis of the bombardment of uranium believed that the particles in the nucleus were like cemented bricks, rigid and incapable of assuming the necessary shape for fission to occur. Only when they were repeatedly confronted with evidence to the contrary did they realise that they were all wrong".

In Chapter 2 we saw the evolution of the various editions of the DSM.

Here I would like to focus on the **reliability of diagnoses in psychiatry.**

Numerous studies have documented that in terms of noise, psychiatry is an extreme case (among many, Rosenhan, 1973). When it comes to diagnosing the same patients using the same diagnostic criteria, psychiatrists often disagree with each other, which is why, at least since the 1940s, noise reduction has been a high priority in the field. And despite constant tweaking, the guidelines have not proved very helpful in this regard.

A 1964 study involving ninety-one patients and ten experienced psychiatrists found that the probability of finding agreement between two opinions did not exceed 57% (Kreitman 1961; Aboraya et al. 2006).

Another study on the subject, involving 426 hospital patients who had been seen by two psychiatrists, found agreement in only 50 per cent of cases for the two specialists' independent diagnoses of mental illness. A study involving 153 outpatients found 54% agreement. In these studies, the source of noise was not specified, but it is interesting to note that some psychiatrists were inclined to assign patients to specific diagnostic categories. Someone, for example, was particularly inclined to diagnose cases of depression, someone else cases of anxiety.

Often one of the reasons was 'clinician inconstancy': different schools of thought, different educational backgrounds, different clinical experiences and different interview styles. While one 'clinician with a background in developmental psychiatry could explain the hallucinatory experience in terms of a post-traumatic experience of an abuse suffered', another 'with a biomedical orientation could trace the same hallucinations back to the schizophrenic process' (Aboraya et al. 2006).

Such differences are typical examples of structural noise.

Even after the significant revision of the 2000 fourth edition, or DSM-IV (*originally published in 1994*), research showed that the noise level remained high (Cheniaux et al. 2009).

On the one hand, Ahmed Aboraya and his colleagues concluded that 'the use of diagnostic criteria for psychiatric disorders has led to a proven increase in the reliability of psychiatric diagnoses'. On the other hand, there was still a high risk that 'the admission records for a single patient may yield multiple diagnoses for the same individual' (Chmielewski et al. 2015).

The DSM-5 clinical trial revealed 'minimal agreement', which 'means that the senior psychiatric specialists involved in the study agreed on the patient's depression in between 4 and 15 per cent of cases'.

According to some field experiments, the DSM-5 only made things worse, showing an increased degree of noise 'in all major domains, with some diagnoses, such as mixed anxiety-depressive disorder, [...] so unreliable as to prove unreliable in clinical practice' (Lieblich et al. 2015).

These developments seem promising, but it remains to be ascertained to what extent they succeed in reducing the noise. To quote one observer 'the reliance on the patient's subjective symptoms, the clinician's interpretation of them, and the absence of objective measures (such as a blood test may be) sow the seeds of diagnostic unreliability in psychiatric disorders' (Aboraya et al. 2006).

In this sense, psychiatry may prove particularly resistant to attempts at noise reduction.

Before going into the recent aetiological and co-etiological theories (Chapter 11), it is good to at least briefly describe the general symptomatology of the autistic state complex and what it may entail, or with what characteristics it may be accompanied by from both a psychological and a symptomatic point of view.

Having definitively understood that autism has to do with neurodevelopment and the various factors involved at a causal level can generate permanent damage in the development of specific brain systems involved in higher mental processes, it is therefore essential, before going any further, to address and attempt to describe the developmental process of the system we call the brain.

Notes

1. Populism, Politics and Measles, New York Times, 2 May 2017 (Editorial Board). https://www.nytimes.com/2017/05/02/opinion/vaccination-populism-politics-and-measles.html
2. Brian Deer, (2011) How the vaccine crisis was meant to make money, British Medical Journal https://www.bmj.com/content/342/bmj.c5258
3. Smeeth L, Cook C, Fombonne E, Heavey L, Rodrigues LC, Smith PG, Hall AJ., (2004) MMR vaccination and pervasive developmental disorders: a case-control study, in Lancet, vol. 364. Baird G, Pickles A, Simonoff E, Charman T, Sullivan P, Chandler S, Loucas T, Meldrum D, Afzal M, Thomas B, Jin L, Brown D., (2008) Measles vaccination and antibody response in autism spectrum disorders, in Arch Dis Child., vol. 93.
4. Immunization Safety Review Committee. (2004) Immunization safety review: vaccines and autism. National Academies Press. Washington, DC, USA. p. 214.
5. Dan Lett, (2007) Vaccine-autism link discounted, but effect of 'study' is unknown, in CMAJ.
6. Brian Deer, (2011) How the vaccine crisis was meant to make money, British Medical Journal.
7. On the relationship between vaccines and autism see also as a reference Doja A, Roberts W (2006). 'Immunizations and autism: a review of the literature'. The Canadian Journal of Neurological Sciences. Le Journal Canadien des Sciences Neurologiques.

   Gerber JS, Offit PA (2009). 'Vaccines and autism: a tale of shifting hypotheses'. Clinical Infectious Diseases.

   Gross L (2009). 'A broken trust: lessons from the vaccine-autism wars'. PLOS Biology.

   Paul R (2009). "Parents ask: Am I risking autism if I vaccinate my children?". Journal of Autism and Developmental Disorders.

   Poland GA, Jacobson RM (2011). 'The age-old struggle against the antivaccinationists'. The New England Journal of Medicine.

   McBrien J, Murphy J, Gill D, Cronin M, O'Donovan C, Cafferkey MT (2003). 'Measles outbreak in Dublin, 2000'. The Pediatric Infectious Disease Journal.
8. Cited by Doidge, N.—in The Brain That Changes Itself, (2007)—Amaral's awards include the McKnight Foundation Scholars

Award, 1981, the Sloan Foundation Fellow, 1983 and the National Institute of Mental Health (NIMH) Merit Award, 1989 and 1993. He is a past president of the International Society for Autism Research and has been editor-in-chief of the society's journal Autism Research since 2015. In 2019, he was elected a member of the National Academy of Medicine.

9. See more extensively Pinker, S.—Rationality—2021—ch. 6 and 9
10. The standard deviation of a set of numbers is derived from another statistical measure, called variance. To calculate it, one must first obtain the distribution of the deviations from the mean and then calculate the square of each deviation. The variance is the mean of these squared deviations, while the standard deviation is the square root of the variance.
11. The percentage of concordant pairs (PC) is closely related to Kendall's W, also known as the coefficient of concordance.PC is calculated under the assumption that the joint distribution is bivariate normal, and the values shown in the table are approximations based on this assumption.
12. M.A. Gernsbacher, M.E. Faust, (1991) The Mechanism of Suppression: A Component of General Comprehension Skill, in Journal of Experimental Psychology: Learning, Memory, and Cognition, 17, pp. 245–262;
R. Blake, N.K. Logothetis, Visual competition, in Nature Reviews Neuroscience, 3(2002), pp. 13–21;
M.C. Stites, K.D. Federmeier, Subsequent to Suppression: Downstream Comprehension Consequences of Noun/Verb Ambiguity in Natural Reading, in Journal of Experimental Psychology: Learning, Memory, and Cognition, 41(2015), pp. 1497–1515.

References

Aboraya, A. et al. (2006) The Reliability of Psychiatric Diagnosis Revisited, in 'Psychiatry (Edgmont) 3

Asch, S. E. (1946) Forming Impressions of Personality, in Journal of Abnormal and Social Psychology, 41 No. 3, pp. 258-290.

Charles JM, Carpenter LA, Jenner W, Nicholas JS., (2008) *Recent advances in autism spectrum disorders*, in *Int J Psychiatry Med.*

Cheniaux E. et al. (2009) The Diagnoses of Schizopbrenia, Schizoaffective Disorder, Bipolar Disorder and Unipolar Depression: In-terrater Reliability and Congruence Between DSM-IV and ICD-10, in 'Psychopathology', 42

Chmielewski M. et al. (2015) Method Matters: Understanding Diagnostic Reliability in DSM-IV and DSM-5, in Journal of Abnormal Psychology, 124

Dana, J. (2008) What Makes Improper Linear Models Tick?, in J.I. Krueger (ed.), Rationality and Social Responsibility: Essays in Honor of Robyn M. Dawes, Psychology Press, New York pp. 71-89. Citation p.73.

Dawes, R. M., Corrigan, B. (1974) Linear Models in Decision Making, in Psychological Bulletin, 81, no. 2, pp. 95-106.

Evans, J.. Barson, J.L., Pollard, P. (1983) On the Conflict between Logic and Belief in Syllogistic Reasoning, in "Memory & Cognition", 11 no. 3

Farrington CP, Miller E, Taylor B., (2001) MMR and autism: further evidence against a causal association, in Vaccine.

Festinger, L. (1962) Cognitive Dissonance, in 'Scientific American', 207 No. 4 pp. 93-107.

Freud, S. (1885) Über Coca.

Galanti, M. A., Sales, B., Neurodevelopmental disorders and networks of care. Neuropsychiatric and pedagogical perspectives in dialogue., in Polifonica, Pisa, Edizioni ETS, 12 May 2017, pp. 151–183

Galinsky, A.D., Mussweiler, T. (2001) First Offers as Anchors: The Role of Perspective-Taking and Negotiator Focus, in Journal of Personality and Social Psychology, 81 No. 4, pp. 657-669

Herbert, M. (2013) The Autism Revolution: Whole-Body Strategies for Making Life All It Can Be. Harvard Health Publications

Iacoboni, M., Dapretto, M., Davies, M. S., Pfeifer, J. H., Scott, A. A., Sigman, M., Bookheimer, S. Y., (2006) *Understanding emotions in others: mirror neuron dysfunction in children with autism spectrum disorders*. Nature Neuroscience

Kahneman, D., (2011) Thinking, Fast and Slow. Farrar, Straus and Giroux.[also in Steven Pinker Rationality, Penguin Books Ltd, 2022].

Kamboj K. et al, (2018) A Study on the Correlation Between Foot Length and Height of an Individual and to Derive Regression Formulae to Estimate the Height from Foot Length of an Individual, in "International Journal of Research in Medical Sciences", 6

Kaye JA, del Mar Melero-Montes M, Jick H., (2001) Mumps, measles, and rubella vaccine and the incidence of autism recorded by general practitioners: a time trend analysis in BMJ

Kanner L. (1943) Autistic Disturbances of Affective Contact. Nervous Child.

Kreitman, N. (1961) The Reliability of Psychiatric Diagnosis, in 'Journal of Mental Science', 107, pp. 876-886

Lieblich, S. M. et al. (2015) High Heterogeneity and Low Remem in the Diagnosis of Major Depression Will Impair the Development of New Drugs, in "British Journal of Psychiatry Open"
Madsen KM, Hviid A, Vestergaard M, Schendel D, Wohlfahrt J, Thorsen P, Olsen J, Melbye M., (2002) A population-based study of measles, mumps, and rubella vaccination and autism, in N Engl J Med.
Moyer, M. W. (2010) 'Vaccinomics: Scientists Are Devising Your Personal Vaccine', Scientific American http://www.scientificamerican.com/article/vaccinomics-personal-vaccine/
Moore, D. A, Schatz, D. (2017) The three faces of overconfidence, in 'Social and Personality Psychology Compass', 11 no. 8
Muhle R, Trentacoste SV, Rapin I, (2004) The genetics of autism, in Pediatrics
Pearson, J. M. (2015) On the Belated Discovery of Fission, in Physics Today, 68, no. 6 pp. 40-45
Rosenhan, D. L. (1973). On being sane in insane places. Science, 179
Sacks, O. (1973) Awakenings
Schrödinger, E. (1929) Was ist ein Naturgesetz? in: Die Naturwissenschaften, vol. 17, No. 1.
Schwartz D (2017) The Last Man Who Knew Everything. Basic Books
Silberman, S. (2015) NeuroTribes: The Legacy of Autism and the Future of Neurodiversity. Avery
Slovic, P., Finucane, M. (2000) The affect heuristic in judgment of risks and benefits , in Journal of Behavioral Decision Making , 13
Stevens, S. S. (1966) On the Operation Known as Judgment, in American Scientist, 54 No. 4, pp. 385-401.
Taylor LE, Swerdfeger AL, Eslick GD (2014). "Vaccines are not associated with autism: an evidence-based meta-analysis of case-control and cohort studies". Vaccines. 3
Wainer, H. (1976) Estimating Coefficients in Linear Models: It Don't Make No Nevermind, in Psychological Bulletin, 83, no. 2.
Wakefield AJ, Murch SH, Anthony A, Linnell J, Casson DM, Malik M, Berelowitz M, Dhillon AP, Thomson MA, Harvey P, Valentine A, Davies SE, Walker-Smith JA., (1988) Ileallymphoid-nodular hyperplasia, non-specific colitis, and pervasive developmental disorder in children, in Lancet, vol. 351
Willingham, E., (2015) Forbes science columnist and mother of an autistic boy wrote on her blog: "This Just In […] Being Alive Linked to Autism" https://thinkingautismguide.com/2015/02/being-alive-linked-to-autism.html

CHAPTER 4

The Autism Characteristic and Symptomatic Picture

The purpose of this chapter is to give a common knowledge base of what are pathological symptoms, characteristics of autism and secondary and related pathologies.

In this picture, it will become clear that no symptom is single or isolated but the damage concerns a circuit, a higher function and a more complex system.

Once again, focusing on one aspect, one symptom and a single set of symptoms distracts from a general, overall picture.

In this chapter I will discuss characteristics, referring to autistic isolation, thinking and imagination, attention disorders, the search for immutability and order, social development, relations with other people, relations with objects, communication, repetitive behaviour and sleep disorders.

I will talk about symptoms by referring to verbal and non-verbal communication, social interaction, affectivity and sexuality, aggression, self-injury, imagination or repertoire of interests, the importance of order and obsessive–compulsive behaviour.

I will talk about sensory disorders, referring to hearing, sight, smell, taste, touch, pain and other symptoms. I will discuss the emotional picture, referring to anxiety, fear, sadness and joy, distrust and mistrust, as well as anger and rage, and the autistic child's defences against negative emotions. Finally, I will discuss co-symptomatic conditions, a definition

M. Di Salvo, *Autism Research between Psychology and Neuroscience*, https://doi.org/10.1007/978-3-031-68338-1_4

under which certain conditions of typical clinical pathology that are often associated and found concomitantly in persons diagnosed with autism are included.

Each of these descriptive blocks must be read in parallel and integrated with the other elements. It is clear that autistic isolation (in characteristics) is related to social interaction (between symptoms) and the entire emotional picture. I have chosen not to unify them in order to also allow an articulated understanding of the characteristic, symptomatic and psychological emotional picture.

I have chosen, methodologically, to describe, as far as and when possible, all these aspects in the words of autistic persons, leaving them to tell not so much the clinical symptom described from the outside, but how that symptom is experienced and described by the person.

I hope to succeed without oversimplification.

Features

Typically, symptoms begin to manifest slowly from the age of six months, becoming more explicit from the age of two or three years and continuing to increase into adulthood, although often in a less obvious form (Rogers 2009; Pinel 2011).

The autistic condition is distinguished not by a single symptom, but overall by a triad of characteristic symptoms: deficits in social interaction, deficits in communication and restricted and repetitive interests and behaviour.

Other aspects, such as an atypical diet, are also common, but are not essential for diagnosis.

The individual symptoms of autism can be found in the general population, but for one to be able to speak of autism, it is necessary to distinguish the situation by severity (Filipek et al. 1999; London 2007).

Autistic Isolation

Autistic closure or isolation, which is what gives this condition its name, is mainly implemented towards the outside world, in more severe cases this extreme defence can also be implemented towards a part of the stimuli coming from one's own mind or body.

This closure, which stems from the need to protect oneself from environmental stimuli that are too painful for one to be able to handle and endure, can be more or less severe and can, therefore, partially or totally exclude other human beings from one's world, but also animals and in the most severe cases even objects.

People with symptoms of autism describe this closure using various symbolism: Pier Carlo Morello (2016) describes it as enclosing oneself inside a glass dome, set above a frozen lagoon.

Temple Grandin (2006) instead uses the simile of the doors or glass panels inside which she felt trapped.

Donna Williams (2013) describes autistic closure as the search for a state of mind full of light, colour and enchantment, in order to escape the painful reality in which she lived, so as to find much-needed comfort.

This distancing and alienation from the reality that surrounds the child and losing oneself in an enchanted inner world can occur involuntarily and instinctively but can also be sought out through various stratagems. Donna Williams, for example, used coloured dots in the air, wallpaper patterns and the repetition of certain noises.

Pier Carlo Morello, on the other hand, shut himself away from the outside world in his room and used imagination and music.

People who are close to children or adults with symptoms of autism often try to bring them back to reality by using inappropriate ways such as scolding and reprimands, whereas it would be much more useful to be able to create an effective relationship, using the utmost gentleness and delicacy, for example, by playing with them, without constraints or conditioning (free self-managed play) (Tribulato 2013).

Thinking and Devising

The severe emotional disturbances from which people with autism symptoms suffer, modify and alter not only their sensory abilities, but also their ideational ones, generating confused and disorganised thoughts (Franciosi 2017; Sullivan 1962).

If these disorders are severe, these children or adults lose, at least in part, the ability to coherently order thoughts, ideas and feelings, resulting in difficulty, and in some cases inability, to understand words, gestures and situations in everyday life (De Rosa 2014).

Therefore, to observers, their behaviour will appear strange, unusual or deficient.

Autistic people who have been able to describe their mental status speak of a confused and chaotic, threatening and frightening inner world: "Imagine a state of hyperactivation in which one is pursued by a dangerous aggressor in a world of total chaos" (De Rosa 2014).

An inner world in which one can have the anguished feeling of falling apart at any moment (Frith 2019).

A world in which there is difficulty in understanding the logic behind things and events. A world that is completely incomprehensible and prevents thought control over the activities one wants to carry out. All these psychological alterations are, among other things, easily detectable in their drawings and stories.

They try to remedy this confused and chaotic inner world by using a series of defences: such as stereotyped behaviour, precise rituals and routine behaviour (Joliffe, cited by Grandin).

In other cases, in order to achieve a modicum of serenity and inner joy, these people use solitary games, such as working with numbers or constantly focusing their attention on certain, specific topics.

Fortunately, these psychic changes, especially in children, are not stable, so much so that they can improve, until they disappear, when their inner world regains a good degree of serenity (Tribulato 2020).

Attention Disorders

Attention is the ability to focus one's thoughts on a particular subject or object and to maintain this focus for as long as necessary. Good attention skills also occur when one is able to divide one's attention between various objects or topics, without confining it to a single object or topic (Silieri 1998).

This capacity depends to a large extent on the frontal cortex, which is able to facilitate or block all information concerning the task that, for that particular person, is prevalent at that moment (Oliverio 2013).

Moreover, attention is closely linked to the physiological phenomenon of the wandering mind, which consists of shifting attention from the activity being performed to internal feelings, thoughts and personal concerns. This phenomenon characterises 25% to 50% of our mind's activity during wakefulness (Zavagnini 2016).

Clearly, this involves being distracted from the task at hand. This occurs especially in times of stress and fatigue and manifests itself more in people who are sad, worried, anxious or depressed (Zavagnini 2016).

We find major attention disorders in persons with autism symptoms who often cannot follow the thought or activity of the moment, on which they are drawn or would like to place their concentration, just as they cannot follow the thought and reasoning of others due to the constant presence of a variety of involving internal emotions such as fear, distress and suffering (Tribulato 2020).

The Quest for Immutability and Order

One of the many characteristics attributed to autistic children is the presence of a lack of flexibility in thinking and a remarkable resistance to change.

They experience phobic terror when they are removed from their environment, if the location of objects or the appearance of rooms in their home is changed, or if their daily routine is altered.

Characteristic in these children is the ritualisation of certain habitual daily activities, such as eating, washing and leaving the house. Activities that need to take place according to rigid and possibly unchanging sequences (Militerni 2004).

They, therefore, do everything to ensure that situations, objects and schedules do not change and remain as they are. Says Grandin: 'Any disruption of routine causes panic attacks, anxiety and an escape response, unless the person is taught what to do when something goes wrong'.

The root cause of the need for immutability is certainly the anguish that pervades the mind, the anguish that they seek to continually fight and control, including through the pursuit of immutability and order.

Since every change and shift in objects or normal habits accentuates the instability of their psyche and aggravates their anxieties and fears to the point where they can no longer effectively control the acute suffering they constantly experience (Brauner et al. 2007).

However, autistic children, at other times or immediately after having sorted their toys absolutely perfectly, when they are left free to act and thus are not coerced by strict rules, like to 'go wild', throwing the same toys on the floor or in the air, thus creating moments of indescribable disorder.

For these particular children, it is much better to respect and follow their needs and their emotional needs of the moment, avoiding any pressure and forcing so that they feel the commitment of others not to neglect their needs and not to accentuate their discomfort.

Social Development

Social deficits distinguish autism from other developmental disorders (Rapin 2008). People diagnosed with autism present social difficulties and often do not have the same behaviours that many people take for granted. Temple Grandin explained that her inability to understand the social communication of neurotypicals, or people with normal neural development, makes her feel like "an anthropologist on Mars" (Sacks 1995).

Unusual social development becomes evident in early childhood. Autistic children show less attention to social stimuli, smile and look at others less often and respond less frequently to their own name. They also differ more markedly with regard to social norms; for example, they look others in the eye less often and do not often use simple movements to express themselves, such as pointing at things (Volkmar et al. 2014).

Autistic children aged three to five are less likely to understand social dynamics, approach others spontaneously, imitate and respond to emotions, communicate non-verbally and take turns in a discussion (Sigman et al. 2004).

Most autistic children show less secure attachment than neurotypical children, although this difference is not found in those with higher intellectual development or a less severe autistic condition (Rutgers et al. 2004).

Older children and adults with autism spectrum disorder perform worse on visual tests with regard to facial emotion recognition (Sigman et al. 2006), although this may be partly due to a lower ability to define their emotions (Bird Cook 2013).

High-functioning autistic children suffer from more intense and frequent loneliness than non-autistic peers, despite the common misconception that autistic children prefer to be alone.

Creating and nurturing friendships often proves difficult, but the quality of friendships, not the number of friends, is more influential: functional friendships, such as those resulting from invitations to parties or social activities, can have a greater influence on quality of life (Burgess 2007).

There are many anecdotal reports, but few systematic studies, concerning aggressive or violent attitudes on the part of persons with autism. Limited data indicate that, in children with mental retardation, autism may be correlated with aggression, harm and tantrums (Matson et al. 2007).

Relationships with Other People

The relationships of children with autism symptoms with other people are remarkably difficult. Their considerable social and relational problems include typical behaviours such as evading dialogue and the gaze of others, not accepting but resisting requests made to them, not showing proper attention to others (Franciosi 2017), not sharing the joys, interests and goals of loved ones, having a desire to be alone, not responding adequately to normal educational systems and so on.

People diagnosed with autism systematically feel that the people with whom they relate are often unable to understand their inner world, their deepest and truest problems, the reasons for their oppositional attitudes, the real causes of their unusual symptoms and the defences they put up to manage, diminish or ward off the most distressing and painful emotions from which they suffer.

Often the people with whom they communicate do not implement those behaviours and attitudes, desired by them, that can make the relationship effective and suitable to their needs.

Because they do not feel that their needs and desires are understood, they perceive both children and adults who are within the norm as bringing problems, anxiety, suffering and pain (Decety 2012).

Therefore, as Donna Williams states, the relationships they have with other human beings are often marked by mistrust, distrust, suspicion, if not considerable fear, so much so that they dislike and sometimes have impatience with the objects that represent them: dolls and dolllets, and they more readily accept cues from a tape recorder than from people (Brauner et al. 2007).

As far as relationships with other children are concerned, they often do not like to participate in the games of other children, so much so that they prefer to be alone, since, playing with peers is not pleasant and fun for them either because of the limitations they have or because of the difficulty they perceive in other children playing as they would like to (Vivanti 2009).

Perception of Faces and Eye Contact

The typical description of autism includes dysfunctions in social interaction: for example, autistic persons avoid eye contact, often do not respond when called by name and usually seem unaware of the feelings of others.

Regardless of the degree of autism, the person will manifest difficulties in social interaction and social communication.
This immediately noticeable characteristic prompted the question of whether lack of eye contact could explain the causes of autism and how this characteristic could be linked to typical social communication disorders: regardless of the severity of the disorder, aversion to looking at others was a constant symptom.
Since the periocular muscles express genuine emotions, this area of the face is crucial in social communication.
Richard Davidson's 1990 study shows that it is only through observation of the eyes that it is possible to clearly recognise whether someone is feeling a positive emotion.
Considering that most children and adults on the autism spectrum display an aversion to looking, Davidson concluded that they lacked important clues about the emotional states of others.
As Davidson (2012) stated, 'it is no wonder that autistic people experience so much difficulty in social interaction: they are unable to understand the feelings of others, nor the meaning of words or behaviour. I therefore suspected that, contrary to popular belief, this social and emotional blindness might not be the result of a specific deficit in the emotional processing taking place in the brain, but was instead the consequence of not looking others in the eye'.
Several studies (Schultz et al. 2003) had concluded that autistic children may have a fundamental abnormality in the fusiform gyrus, located in the visual cortex that sits at the back of the brain and is responsible for identifying faces.
In 1997, the discovery that a region of the brain specialised in face recognition seemed to make sense, as faces are of great importance in the social lives of humans and other primates (a 2009 study found that the fusiform gyrus is also present in the brains of chimpanzees).
Follow-up studies, however, found that the fusiform gyrus is not necessarily specialised in faces, but is activated whenever people perceive an object that ap: belongs to the category in which they are competent. In autistic people, the fusiform gyrus shows abnormal activation. When an autistic child undergoes MRI scans and is given face-recognition tasks (e.g. determining whether a face expresses happiness or anger) the fusiform gyrus is much less active than in neuronormal children.

Again in Davidson's words, 'I doubted the hypothesis of tracing autism to an innate deficit of the fusiform gyrus. I thought it more than plausible that these children would stare at an indefinite point in space in an attempt to calm themselves down, or simply close their eyes until the whole thing was over. If so, it was obvious that the fusiform gyrus was not activated. [...] those children may not have been looking at the faces projected on the ceiling or trying to distinguish the emotions they were expressing. I suspected that the inactivity in the fusiform gyrus did not reflect a real abnormality in that area of the brain, but was attributable to the children looking away from the faces the scientists were showing them. It would be like saying that the reason why a teenager does not answer when you call him to come to the table is the absence of activity in the auditory cortex, when in fact it is only because he has headphones stuck in his ears. An absence of activity does not necessarily imply functional defects: it may simply reflect a lack of input. [...] When we looked more systematically at the eye movement recordings of the thirty autistic children, after collecting all the data that re-watched them, we found that they had spent 20% less time looking at the eyes of the faces shown in the glasses, compared to children with normal development. Taking this fact into account, we could already explain the difference in activation of the fusiform region. There was nothing wrong with the fusiform area of autistic children: it remained dormant not because it was impaired, but because it did not receive any signal, and it did not receive any signal because the children looked away from people's faces, and in particular their eyes'.

A significant finding, which refuted the widely held view that autistic people suffer from a congenital neurological deficit in face recognition.

When performing the face-recognition task, the brain activity of autistic children also differed from that of typically developing children in another brain region: they showed increased activity in the amygdala.

The amygdala is an area of the brain essential for emotional learning and is a key structure in neuronal circuits related to fear and anxiety: the amygdala detects threats from the environment. Many autistic children who could hardly even look at the eyes of a face in a photograph (let alone a live face) showed hyperactivity of the amygdala.

The fact that this happens when autistic children look at faces—even if only for a fraction of a second, as in Davidson's experiment—suggests that this action causes them great discomfort or even fear: when they look into someone's eyes, their body and brain are flooded with messages

that are interpreted as threats. Only by looking away can they resist this attack.
In fact, when they looked away from the ocular region of the faces (as revealed by the tracking system) the activation of the amygdala ceased: avoiding eye contact is a strategy that calms and regulates emotions, settling anxiety and fear. By avoiding looking people in the eye, the autistic child is able to diminish the social stimuli that he or she finds so threatening.

If high levels of amygdala activation cause discomfort and are perceived by the brain as signals indicating the presence of a threat, as these results suggest, then it is likely that even harmless expressions on the face of a stranger or even a family member will be interpreted as intimidating. Autistic children are likely to discover the strategy of avoiding eye contact at an early age: they experience anxiety when they look at people's faces and find that they can alleviate it by looking away. However, that relief comes at a high cost. Avoiding eye contact deprives autistics of the important social information conveyed by faces, and in particular by eyes. Indeed, one of the first manifestations of social competence is the perception of faces. In adults, it takes very little to trigger the perception of a face: a caricature, a smiley face, a mask, etc. Surprisingly, this hypersensitivity is present from birth: an infant with only a few hours to live turns more quickly towards a baby's face than towards a similar image whose elements have been inverted (even when the experimenter is sure that the infant has never had the opportunity to see a face). One research team even managed the feat of using a light to project a stimulus through the wall of the uterus.The result: three dots arranged in the shape of a face (•.•) attract the foetus more than three dots arranged in a pyramid shape (.˙.). Face recognition appears to begin in utero (Reid et al 2017).
Many researchers believe that this almost magnetic attraction to faces plays a key role in the early development of attachment. By drawing attention to faces an innate propensity would compel us to learn to recognise them, from the earliest months of life—and indeed, from this age, a region of the visual cortex in the right hemisphere begins to respond more to faces than to other images[1].

The Relationship with Objects

The relationships autistic people have towards objects are certainly better and more intense than those they have towards human beings, with whom, on the other hand, they have considerable difficulty establishing strong and positive bonds (Grandin 2006).

This happens because the objects accept, without criticism, reprimand, punishment or protest, their need for order, their stereotypes, as well as their moments of aggression and destructiveness and all their other symptoms. Moreover, the objects accept their positive emotions as they are able to express them. They have a considerable attachment to some of these (pledge objects), so much so that they always carry them with them (Brauner et al. 2007).

Williams, for example, was particularly attached to a perfume and woollen objects because of the warm and positive experience she had had as a child with her grandmother, who, over time, had shown her respect, affection and understanding.

Since these 'pledge objects' help them to cope better with the anxieties, fears and moments of despondency from which they suffer, giving them some joy and security, it is good not to deprive them or criticise them for their attachment to them.

Communication

Approximately one-third to one-half of the autistic persons are unable to develop sufficiently natural language to meet their everyday communication needs (Noens et al. 2006). Communication deficits may present themselves as early as the first year of life and may include delayed onset of lalla, unusual gestures, decreased responsiveness and unsynchronised vocal patterns. In the second and third years, autistic children have less frequent and less diversified use of consonants, word combinations and lalla; their gestures are less frequently integrated with words. Children are less inclined to make requests or share experiences and are more likely to simply repeat the words of others (echolalia) or resort to pronoun reversal (Landa 2007; Tager-Flusberg, 2007; Kanner 1968).

In some studies, high-functioning autistic children between the ages of 8 and 15 performed as well as and better than adults on basic language checks involving vocabulary and spelling, both in combination and individually. However, it was seen that autistics performed worse on complex language tasks, such as figurative language, comprehension and inference. These studies, therefore, suggested that people who communicate with autistic people are more likely to overestimate what their interlocutor perceives (Williams et al. 2008).

Repetitive Behaviour

Autistic persons display many forms of repetitive or restricted behaviour, categorised as follows according to the Repetitive Behaviour Scale-Revised (RBS-R) (Lam, Aman 2007):

- Stereotypy is a repetitive movement, such as hand flapping or head rocking.
- Compulsive behaviour is expected and seems to follow rules, such as arranging objects in piles or lines.
- Monotony: it is resistance to change; for example, insisting that furniture should not be moved.
- Ritualistc behaviour entails an invariable pattern of daily activities, such as an unchanging diet and a ritual in dressing.
- Limited behaviour is focused on interests or activities, such as focusing on a single television programme, a single toy or a particular game.
- Self-injury includes movements that can harm or injure people (Johnson, Myers 2007).

No single repetitive or self-injurious behaviour appears, however, to be specific to autism, but autism appears to have a high onset and severity of these behaviours (Bodfish et al. 2000).

Sleep Disorders

Sleep disorders, often present in children with symptoms of autism, make life considerably more difficult for parents who are woken during the night by their children's cries, wails or needs for reassurance. Insomnia can be calm or agitated. In calm insomnia the child remains cross-eyed but does not cry or shout, in agitated insomnia the child shouts, whimpers or screams, in the grip of fears or anguish, without being able to calm down so as to be able to surrender to sleep (De Ajuriaguerra, Marcelli 1986).

Sleep is also disturbed by anxious dreams, nightmares and night terrors, during which the child wakes up, cross-eyed, screaming and distressed about something he sees or hears, without recognising who is around him.

Williams recounts: "Terror invaded me. Carping on the floor, I cried like a child. I felt the cold and hardness of the tiles and stared at my hands stretched out towards them. I felt I could not breathe. I felt the fear of the unknown lurking somewhere in the room. I groaned, terrified, lost and helpless. I cowered, trembling with fear and rocked like a child".

Visual and auditory hallucinations are also possible in this situation.

It is a cross-sectional observation that the efforts of parents and educators to make their child experience serene and joyful moments during the day, while at the same time avoiding any opportunities for anxiety and stress, markedly improves the child's nights.

Symptomatology

The severity and symptomatology of autism vary greatly from individual to individual and tend in most cases to improve with age, particularly if mental retardation is mild or absent, if verbal language is present and if valid therapeutic treatment is undertaken at an early age.

Autism can be associated with other disorders, but it is worth emphasising that there are different degrees of autism. Some autistic persons possess, for example, an extraordinary capacity for mathematical calculation, musical sensitivity, an exceptional audiovisual memory or other talents to an extent quite out of the ordinary, such as the ability to produce very faithful portraits or landscapes on canvas without having any technical knowledge of drawing or painting.

Normally, the symptoms, which only at first glance may seem similar to the characteristics of introversion, actually manifest themselves as a true autistic withdrawal (in the sense of noticeably abnormal and not always comprehensible behaviour, due to which the person is exposed to a high risk of social isolation), due to severe alterations in functional areas.

Verbal and Non-verbal Communication

'Many children with autism, a percentage that varies between 20 and 50%, do not acquire any verbal language. Another 25% acquire some words between 12 and 18 months and then go through a regression associated with the loss of verbal language' (Vivanti 2009).

Autistic persons who are able to use language express themselves on many occasions in a bizarre way, often repeating words, sounds or sentences they have heard pronounced.

Echolalia can be immediate (repetition of words or phrases immediately after hearing) (Pizzant 1981, pp. 241–249) or deferred (repetition of previously heard phrases or words at a distance of time) (Pizzant 1981, pp. 183–192).

Alongside echolalia, verbal stereotypies (the child repeats words or phrases unrelated to the situation and experiences of the moment) are often present.

Some children with autism invent new words (neolanguage) (De Ajuriaguerra, Marcelli 1986).

There may also be a disturbance in the melody of speech, which may appear sing-songy or excessively mannered.

Even if imitative skills are intact, these people often have considerable difficulty in applying new learning constructively to situations other than those that generated it in the first instance.

Thus: '*Communicating with a person with autistic disorder can be difficult or impossible for different and apparently opposing reasons. At the two extremes of the continuum are, on the one hand, subjects who have never acquired language and do not respond and do not initiate any communicative exchange, and, on the other hand, subjects who continuously initiate conversations using a rich and formally appropriate vocabulary, but who are not able to flexibly adapt their communication to the interactive context, to maintain reciprocity and alternating turns in the communicative exchange and to correctly interpret all the communicative exchanges expressed by the interlocutor*' (Vivanti, 2009).

It must be borne in mind that both verbal and non-verbal communication develops correctly when certain precise conditions are present: desire and pleasure in exchanging with the other and attention to the outside world, sufficient inner serenity, normal sensory capacities and appropriate age.

The autistic child unfortunately lacks some and sometimes all of these conditions.

Having no trust in the outside world, these children have no desire to communicate, but mainly feel the need to defend themselves against others. As their inner world is greatly disturbed by anxieties, fears and considerable restlessness, there is not the minimum of inner serenity that can enable them to hear and process sounds and thoughts correctly. Moreover, although children with autistic disorder hear perfectly well, their reaction to certain sounds produces alarm in them. They, therefore, try to defend themselves against this frustrating situation by alienating themselves from the outside world as much as possible.

However, when the child succeeds in acquiring a better inner serenity and greater trust in others and in himself, both verbal and gestural communication improve markedly; when this improvement occurs before the age of five to six, an age at which the centres for language are still well active. Unfortunately, when the child exceeds this age, the greatest acquisitions are mainly in the area of gestural communication.

Social Interaction

Autistic persons show an apparent lack of interest and relational reciprocity with others, a tendency towards isolation and social closure, apparent emotional indifference to stimuli or, on the contrary, hyperexcitability to them, and difficulty in making direct eye contact: the autistic child who continues to avoid the gaze of others around the age of two years shows, according to various studies, a greater social disability in the future (Jones 2008).

Autistic persons have difficulty initiating a conversation or taking 'turns', as well as having difficulty answering questions and participating in group life or games. It is not uncommon for children with autism to be initially checked for suspected deafness, as they show no apparent reaction (just as if they had hearing problems) when called by name.

Brief Note on the Concept of the Social Brain

Autism research has taught us a lot about social behaviour and the biology of social interactions and empathy. Some social interactions, for example, occur through biological movement, such as walking towards another person or extending a hand in greeting. In 2008, Kevin Pelphrey of Yale University discovered that autistic children have difficulty distinguishing biological movement.

In an experiment with autistic and non-autistic children, he monitored two regions of the brain while the children were observing biological or non-biological movement. One of the regions involved is a small visual area known as MT or V5, which is sensitive to any movement; the other is the superior temporal sulcus, which in neurotypical adults responds more strongly to biological movement. In both groups of children, the MT/V5 motion-sensitive brain region responded in much the same way to both types of movement. But in typically developing children the superior temporal sulcus responded more strongly to biological motion. In autistic children, the same brain area registered no difference between the two types of motion.

The ability to identify and integrate biological action with the context in which it occurs—for example, to integrate the observation of a person reaching for a glass of water with our assumption that person is thirsty—allows us to recognise intention, which is fundamental to a theory of mind. Thus, one of the reasons why autistic people have difficulty in social interactions is that they have a limited ability to read socially meaningful biological actions such as reaching out a hand.

People with autism also have a similar difficulty in reading faces. When looking at another person, they avoid the eyes and tend to stare at the mouth. Neurotypical people do the opposite: they mainly look at the eyes. Because a person's gaze—where they are looking - gives us important clues as to what that person desires, intends or believes. The words 'desire', 'intention' and 'belief' describe mental states. States of mind are not amenable to direct observation, but most of us act as if we can directly observe another person's states of mind, as if we can read minds.

In 1990, Leslie Brothers of the School of Medicine at the University of California at Los Angeles (UCLA) exploited insights into the theory of mind derived from autism studies to propose a theory of social interaction.

Leslie Brothers (2002) observed that social interaction requires a network of interconnected brain regions that process social information, giving rise to a theory of mind, and coined the term social brain to describe this network. The regions include the inferior temporal cortex (involved in face recognition), the amygdala (emotion), the superior temporal sulcus (biological motion), the mirror neuron system (empathy) and the temporo-parietal junction areas involved in the theory of mind.

Only today is neuroscience beginning to decipher how the social brain regions identified by cognitive psychology are connected and how they interact to influence behaviour. Stephen Gotts (2012) and his colleagues at the National Institute of Mental Health have used functional brain imaging to confirm that the neuronal circuitry of the social brain is indeed altered in people with autism spectrum disorders. Specifically, the disrupted connections are found in three regions of the social brain: those involved in the emotional aspects of social behaviour, those involved in language and communication, and those involving the interaction between visual perception and movement. Normally, the patterns of activity in these three regions are coordinated with each other, but in autistic people, they are not: they are out of sync with each other and with the rest of the social brain.

Of particular interest are the anatomical findings concerning the timing of brain development and growth in autistic children.

As we have seen in detail, before the age of two, the circumference of an autistic child's head (Schumann et al 2010) is often larger than that of a typically developing child. In addition, certain regions of the autistic child's brain may develop prematurely during the first years of life, particularly the frontal lobe, which is involved in attention and decision-making, and the amygdala, which is involved in emotions. This is important because when one or more brain regions develop out of sequence, they can severely disrupt growth patterns in other brain regions with which they are connected.

Affectivity and Sexuality

When autism is not excessively severe, affective feelings towards those who understand, help and manage to relate well to children, adolescents and adults diagnosed with autism are by no means the exception. They are capable of loving and bonding, showing clear signs of friendship towards those who care for them and managing to respect their needs, difficulties and problems (Tribulato 2020).

Similarly, adolescents and adults with non-severe symptoms of autism are also able to experience intense love feelings towards the opposite sex. Just as they have sexual thoughts and desires, to the extent of practising masturbation (De Rosa 2014; Morello 2016).

In spite of this, we cannot deny the difficulties they experience when they would like to establish, and then maintain, an affective, loving and sexual relationship that is also lasting, pleasant and gratifying. These difficulties are present both when their interests are directed towards people who fall within the norm, and when they try to establish a loving relationship with adolescents and adults who present symptoms of autism. And this causes them disappointment and frustration (Williams 2013; Morello 2016).

These difficulties are due to the complexity present in every stable and lasting love relationship. In this type of relationship, communication skills, psychological and emotional well-being, dialogue and acceptance skills, the correct management of physical contacts and the ability to harmonise one's own needs with those of the other are indispensable. And this, if it is arduous for people who fall within the neuronormal range, is much more so for adolescents and adults with autism because of the difficulties they present, both in the relational and communicative sphere and due to the presence of emotional disorders: anxieties, fears and insecurities (Williams 2013).

Before helping them to enter into emotional bonds, it is, therefore, necessary to work to diminish, and if possible solve psychological problems, so that these experiences, of fundamental importance for every human being, are not only possible but also enjoyable and gratifying.

Aggression

If aggression frequently arises as a response to the suffering endured, it is no wonder that aggressive behaviour is present in the child or young person with autism, who is immersed in a state of emotional dysregulation, (Franciosi 2017) with a continuous state of anxiety and fear towards a world that they perceive as hostile and threatening. And since the subjective element is important in aggressive reactions, whereby the reaction depends on the personality characteristics of the subject and his or her experiences at the time, these reactive behaviours can be present even without a clear and immediate provocation (Bonino 2005).

However, aggressive feelings are not always manifested. In the most severe cases, these are as if frozen and sterilised, to prevent destructive reactions from the surrounding world (Williams 2013).

Typically, aggressive manifestations are directed at objects that are slammed on the ground or walls, torn or destroyed or towards people whose behaviour does not respect their needs for serenity, tranquillity and their fear of being approached or touched (Williams 2013).

Self-Harm

One of the symptoms that is most disconcerting and upsetting to family members and caregivers who care for a child or young person diagnosed with autism is self-injury. It is certainly traumatic to witness one's own child or pupil hurting himself: banging his head on the wall or on some piece of furniture, biting his arms or tongue, slapping or scratching himself and injuring himself with his nails or some sharp or pointed object.

The causes of self-injury may be different and may coexist in the same person.

While at times this behaviour stems from a fear of turning aggression outwards, so as not to suffer the destructive consequences of their thoughts or behaviour towards others (Williams 2013), in other cases it can also be a means of challenging those who, through behaviour not in keeping with their needs and feelings, have provoked their anxiety and fears (Brauer A., F.; 2007).

Self-aggressiveness may also manifest itself as guilt for having done, said or thought something that should not have been said, said or done to a person who had been good and kind (Williams 2013).

Or it may simply be a way of sensing something in the vacuum of the autistic condition: one's body, one's emotions, oneself, using a painful sensation (Williams 2013).

Finally, self-aggression can be an extreme and desperate act, when they notice that those around them have no care and attention towards their suffering, their fears and their needs.

Fortunately, when parents and caregivers respect their need for non-interference and their constant search for moments of tranquillity and peace, both hetero-aggressiveness and self-aggressiveness diminish to the point of ceasing altogether. Just as they decrease and cease when they accept that aggressive feelings can be vented on objects or through non-injurious play with adults (Tribulato 2020).

Imagination or Repertoire of Interests

Usually a limited repertoire of behaviours is repeated obsessively; stereotyped postures and sequences of movements (*e.g. twisting or biting hands, waving them in the air, rocking, making complex head movements,* etc.) known as stereotypies can be observed.

Autistic persons may show excessive interest in objects or parts thereof, particularly if they have round shapes or can rotate (*oval balls, marbles, spinning tops, propellers,* etc.).

At times, the autistic person tends to withdraw from reality to isolate himself in a kind of 'virtual world', in which he feels he lives for all intents and purposes (*sometimes conversing with invented characters*).

While in many cases maintaining an awareness of his own fantasising, it is with difficulty and only with external prompting (*sudden sounds, reminders from other people*) that he manages to participate to varying degrees in group life.

Importance of the Order

In some autistic persons, there is a marked resistance to change, which for some can take on the characteristics of a true phobic terror. This may occur if the person is removed from their environment, or if the location of objects, furniture or otherwise the appearance of the room is inadvertently changed.

The same can occur if objects are left in disorder (*chairs moved, windows opened, newspapers in disarray*): the spontaneous reaction of the autistic person will be to immediately restore things to their order or, if unable to do so, to manifest restlessness anyway. The person may then explode into fits of crying or laughter, or even become self-destructive and aggressive towards others or objects. Others, on the contrary, show excessive passivity, motor apraxia and hypotonia, which seems to make them impervious to any stimulus (Ming et al. 2007).

Sensory Disorders

Sensory disturbances are very frequent in people with autism symptoms. Sometimes they occur with an increased sensitivity to stimuli from the outside world, but also from one's own body (*sensory over-response*) while, in other cases, at other times or due to other sensations, they may manifest themselves with a decreased sensory response (*sensory hypo-response*).

Altered interpretations coming from the senses are also very common. In these cases, even a trivial stimulus can be interpreted as something aggressive or damaging, resulting in reactions of fear, anxiety, if not outright panic attacks.

Therefore, there can sometimes be an excessive and abnormal search for particular stimuli or, on the contrary, a clear rejection of, and therefore a distancing from, specific sensory experiences (Franciosi 2017).

The consequences of these altered perceptions are reflected in their behaviour, e.g. in their ability to relate and socialise with other peers, but also in their learning abilities (De Rosa 2014; Notbohm 2015).

These sensory alterations, caused by the severe psychological disorder of these people, tend to worsen their already greatly disturbed inner world, making them even more anxious, unstable, irritable and confused.

Hearing

As far as hearing is concerned, persons with autism symptoms, due to their irritability, fragility and altered inner reality, manifest fears and phobias more frequently and with more emotional involvement, resulting in nervous breakdowns or escapes, due to numerous types of sounds: Loud noises, ambulance sirens, loud music, bangs, shouting and hubbub present at parties or in classrooms, the echoes created in gyms and school toilets and so on (Grandin 2006; Morello 2016).

It is important what meaning a particular sound or noise takes on in their mind. Thus the same noise may be perceived as pleasant by one person while it may terrify another.

In other cases and in other children, it is as if a condition of deafness were present, as they do not respond in any way and are absolutely indifferent to even loud and persistent sounds, when they are immersed and estranged in their magical world, to the extent that they exclude the world outside of them. On the other hand, the same children may become agitated and disturbed by a soft, gentle sound. This suggests that there is no specific anatomical alteration of the receptors that amplifies or reduces the sounds heard, but that the way in which these people, in a certain psychic situation, experience, interpret and perceive certain sounds or noises is fundamental.

The View

Visual perceptions can also create fear, anxiety and explosive behaviour.

For example, fluorescent lights, reflective surfaces, objects moving rapidly or at irregular speed, flashing sirens, (Notbohm 2015) things that are constantly moving, like eyes (Grandin 2006).

In addition, people with autistic disorders may have difficulty seeing the object in its entirety (Frith 2019) to the extent that some, in order to recognise the object, pass it in front of their eyes, as on a scanner.

Moreover, experiencing suspicion and anxiety when dealing with humans, they may have little recollection of the faces of the people they encounter while perfectly recalling animals, objects or sights that are pleasant and interesting to them (Morello 2016).

The Sense of Smell

There are also sensory alterations in smell and taste that accentuate their suffering and modify or alter their behaviour.

For example, the pleasure of smelling people (Grandin 2006), being attracted to bad smells and repelling scents (Brauer 2007).

Being attracted to smells that refer to a safe environment such as one's home, or to people they value and love (Williams 2013).

The variability and particularity with which these disorders present themselves, with unusual preferences and exclusions, points to a particularly troubled and disturbed inner psychological reality.

Taste

Also with regard to taste, people diagnosed with autism are generally very selective and picky, so they generally do not like to taste crispy, jelly-like, too hot, too cold food.

Like many young or psychologically disturbed children, they tend to associate favourite foods with pleasant and pleasant people, animals and situations, while, conversely, they reject foods they associate with unpleasant or unpleasant people, animals and situations (Williams 2013).

Touch

Touch is connected to very intimate and primitive sensations.

Autistic children, who are effectively and psychologically very immature and disturbed, experience these feelings in particular ways, whereby they are often afraid of being touched and hugged not only by strangers but also by their parents (De Rosa 2014; Williams 2013) to the extent that they react aggressively towards those who do not respect their needs and fears.

However, if they hate some types of touching, they may love other, strange and unusual types of touching, such as touching curtains, furniture and other special objects, such as the stringer (also known as the 'hugging machine') per Grandin (2006).

Also in this area, there are no constant characteristics, whereby other children diagnosed with autism desire and like to be touched by their parents and also by strangers (De Rosa 2014).

The Pain

Even for painful sensations there can be considerable differences, so that some children with autistic disorders will not tolerate and will scream over a small scratch, while others, or the same ones, at other times and on other occasions, may not mind very intense painful sensations (sensory anaesthesia), either when the pain is caused by themselves (self-injury) or when it is caused by others (Williams 2013).

Ultimately, all sensory input is felt according to their age, their specific psychological characteristics, their experiences and their inner experiences at the time.

Other Symptoms

Autistic persons may present certain symptoms that are independent of the diagnosis, but which may affect their life or family sphere, and which may interact and infer with other typical characteristics (Filipek et al. 1999).

It is estimated that from about 0.5 to 10 per cent manifest unusual abilities, ranging from great skills in specific tasks, such as an extraordinary ability to memorise irrelevant details, to the development of conditions known as '*savant syndrome*' (Treffert 2009).

Many people with autism spectrum disorder show superior abilities to the general population in perception and attention (Plaisted Grant, Davis 2009).

Sensory abnormalities are found in over 90 per cent of cases (Geschwind 2009) although there is no evidence that sensory symptoms differentiate autism from other developmental disorders (Ben-Sasson et al. 2009; Rogers 2005).

It is estimated that approximately 60–80 per cent of people diagnosed with autism have motor signs that include poor muscle tone, apraxia and predominantly toe walking.

Deficits in motor coordination are widespread (Fournier et al. 2010).

In approximately three-quarters of autistic children, unusual eating behaviour is found. Selectivity is the most common problem in which both ritual eating and refusal of food can occur (Dominick et al. 2007).

This does not appear to cause episodes of malnutrition. Although some children with autism experience gastrointestinal symptoms, there is no adequate quantitative analysis of published rigorous data to support that this occurs more than the average for peers (Erickson et al. 2005; Buie et al. 2009).

The Emotions of the Autistic Child

Many prejudices accompany the autistic syndrome.

One of the most widespread claims is that these children feel no or only modest emotions. This is absolutely untrue, and would, moreover, be incompatible with the findings of high levels of anxiety, numerous distressing fears, often combined with manifestations of anger and rage.

The presence of an emotionally disturbed inner world is already evident in the stories and drawings that these children sometimes manage to construct. Tales and drawings in which distressing, gory, gruesome or coprolalic themes predominate.

Temple Grandin describes her emotions as follows: "*Some people believe that people with autism have no emotions. I do have them, but they are more like the emotions of a child than those of an adult*" (Grandin 2006).

Anxiety

As far as anxiety is concerned, this emotion, in the mild forms of autism, expresses itself mainly with symptoms such as lability of attention, hyperactivity, hyperkinesis and considerable reactivity even to small frustrations.

In these forms, when the child wishes to make friends with peers, anxiety and inner excitement severely impair his or her interpersonal skills, so that, in relationships with peers, because the child lacks the serenity to listen to others, accepting their needs and desires, he or she is often rejected and rejected.

In severe forms of autism, despite the fact that anxiety is masked by more severe symptoms such as stereotypies, apparent apathy and indifference, it can easily be detected in the unpredictable, sudden and frequent mood swings and acute crises of anguish, provoked by the slightest frustration. Moreover, in many cases, this distressing emotion manages to disrupt the structural organisation of thought with alterations in language that can become disconnected and incoherent.

Fears

When the child is confronted with particular situations, objects and tactile, visual or auditory stimuli, or when confronted with slight changes in the world around him, fears may also manifest themselves dramatically, with screams and disarrayed attitudes.

Temple Grandin describes her fears as follows: "*The problems of such a person are further complicated by a nervous system that is often in a state of heightened fear and panic. Since fear was my main emotion, it spilled over into all events that had any emotional significance'. "Since puberty I had experienced constant fears and anxieties, accompanied by strong panic attacks, occurring at varying intervals from a few weeks to several months. My life was based on avoiding situations that could trigger a panic attack. With puberty, fear became my main emotion*" (Grandin 2006).

Sadness and Joy

It is not always possible to highlight these two emotions as sometimes, and in some children, they are presented in an excessive and abnormal manner, while in others or at other times they are not always evident, as they are masked by non-congruent facial expressions (Militerni 2004). Thus, a facial expression that is always the same or attitudes with excessive and foul-mouthed displays of laughter may conceal great sadness and distress or, on the contrary, moments of true serenity and joy.

Nevertheless, when adults, be they parents, teachers or operators, manage to listen to the child's deepest emotions, without being distracted by his or her more superficial or extreme behaviour and emotional manifestations, it is not so difficult to grasp his or her true emotions so as to behave accordingly.

Mistrust and Distrust

The inner world of children diagnosed with autism is not only greatly disturbed by anxiety, sadness, phobias and fears and an abnormal state of arousal, but it is also altered due to the considerable distrust and mistrust of the world around them, which is frequently perceived as bad, treacherous, inconsistent and bringing constant anguish and frustration. They, therefore, often find themselves alone in an environment in which they do not feel understood and accepted, and this increasingly leads them to closure.

The Autistic Child's Defences Against Negative Emotions

From what has been said, it is easy to understand that a large part of the symptoms can be traced back to defences, often archaic and therefore not very functional, that children put in place to avoid, alleviate or overcome their suffering, caused by intense negative emotions such as anxiety, fear, depression and considerable distrust of others and themselves.

Autistic children, for example, try in every way to avoid, through closure, people, places, objects and situations in which they are uncomfortable or which may accentuate their discomfort. Since every change accentuates their anxieties and fears, they have an aversion to every new experience, whether it is a new food, a different object, place or time. To diminish their sadness they sometimes resort to nervous laughter, since by laughing their sadness and anxiety diminish, while, at the same time, this mimic expression not only does not offend or hurt anyone, but is frequently accepted by others, as it is mistaken for a manifestation of joy.

Another way to diminish anxiety and inner malaise is to implement repetitive behaviour, such as stereotypies. Self-injury can also be used to decrease confusion and reduce inner tension, as the pain that is provoked serves to distract them for a few moments from distressing experiences, while simultaneously allowing them to be more present (Schmahl 2013).

Co-symptomatic and Co-pathological Conditions

In this definition, we discuss typical clinical pathology conditions that are associated with concomitant symptoms in persons diagnosed with autism.

While this is a statistical list, which does not necessarily have an ecological correlation, in some cases it is almost the consequent pathological effects of other biological changes.

Abnormal Folate Metabolism

There is a variety of evidence of abnormalities in folate metabolism. These abnormalities can lead to decreased production of 5-methyltetrahydrofolate, alter the production of folate metabolites and reduce the transport of folate across the blood–brain barrier to neurons. The most significant abnormalities of folate metabolism associated with autism spectrum disorders may be autoantibodies against the folate receptor alpha (FRα). These autoantibodies have been associated with brain folate deficiency. Autoantibodies can bind to FRα and significantly impair its function.

Furthermore, significant improvements in the general clinical picture were noted in autistic children with brain folate deficiency when taking folinic acid.

In a series of five children with cerebral folate deficiency and low-functioning autism with neurological deficits, an overall reduction in typical symptoms was found with folinic acid use in one child and substantial improvements in communication in two others (Frye 2014).

Abnormal Redox Metabolism

An imbalance in glutathione-dependent redox metabolism has been shown to be associated with autism spectrum disorders. Glutathione synthesis and intracellular redox balance are related to folate metabolism and methylation, metabolic pathways that have been shown to be abnormal in persons diagnosed with autism. Together, these metabolic abnormalities define a distinct endophenotype of TSA that is closely associated with genetic, epigenetic and mitochondrial abnormalities, as well as autism-related environmental factors. Glutathione is involved in neuroprotection against oxidative stress and neuroinflammation by enhancing the antioxidant stress system.

Interestingly, recent studies have shown that N-acetyl-1-cysteine, a glutathione precursor supplement, is effective in improving symptoms and behaviour associated with autism (Lee et al 2021).

However, glutathione was not measured in these studies.

Small, medium and large open DPBC studies and clinical trials show that new treatments for children with autism spectrum disorders for oxidative stress are associated with improvements in baseline symptoms, sleep, gastrointestinal symptoms, hyperactivity, seizures and parental impressions, sensory disturbances and motor symptoms. These new treatments include N-acetyl-l-cysteine, methylcobalamin with and without oral folinic acid, vitamin C and a vitamin and mineral supplement including antioxidants, Q10 enzyme and B vitamins.

Several other treatments that have antioxidant properties, including carnosine, have been reported to significantly improve behaviours associated with autism spectrum disorders, suggesting that treatment of oxidative stress may be beneficial.

Many antioxidants may also help improve mitochondrial function, suggesting that clinical improvements with antioxidants could occur through a reduction in oxidative stress and/or an improvement in mitochondrial function (Ghanizadeh et al. 2012).

Intestinal Disease

Gastrointestinal symptoms are a common co-symptom in persons with autism spectrum disorders, although the underlying mechanisms are largely unknown. The most common gastrointestinal symptoms reported by a proprietary tool developed and administered by Mayer, Padua and Tillisch (2014) are abdominal pain, constipation, diarrhoea and bloating, reported in at least 25% of participants.

Carbohydrate digestion and transport are impaired in autistic persons and are thought to be attributed to functional disorders that cause increased intestinal permeability, a deficiency in disaccharide enzyme activity, a secretin-induced increase in pancreatico-biliary secretion and an abnormal faecal flora of Clostridia taxa (Williams et al. 2011).

Developmental Coordination Disorder (Dyspraxia)

Early reports of Asperger's syndrome and other diagnostic schemes (Ehlers et al. 1993) include descriptions of developmental coordination disorders. Children diagnosed with autism may have a delay in acquiring motor skills that require dexterity, such as riding a bicycle or opening a can, and may appear clumsy. They may be poorly coordinated, have an awkward or bouncy gait or posture, poor handwriting, other hand/dexterity disorders or problems with visual-motor integration, visual-perceptual skills and conceptual learning (Klin 2006).

They may show problems with proprioception (sensation of body position) with disturbance of coordination, balance, tandem gait and finger-thumb apposition (McPartland et Klin 2006).

Fetal Alcohol Spectrum Disorder

Fetal alcohol spectrum disorder or FASD is a common disorder that can mimic signs of autism (Bruer-Thompson 2019).

Although study results are mixed, it is estimated that 2.6 per cent of children with FASD also have an autism spectrum disorder, a rate almost twice as high as that reported in the general US population (Lange et al. 2018).

Fragile X Syndrome

Fragile X syndrome is the most commonly inherited form of intellectual disability. It was named so because a part of the X chromosome has a 'defective piece' that appears squashed and fragile when observed under a microscope. Fragile X syndrome affects approximately 2–5% of people diagnosed with autism (National Fragile X Foundation 2013).

If a child has Fragile X, there is a 50% chance that children born to the same parents will have Fragile X.

Hypermobility Spectrum Disorder and Ehlers-Danlos Syndromes

Studies have confirmed a link between inherited connective tissue disorders such as Ehlers-Danlos syndrome (EDS) and hypermobility spectrum disorder (HSD) with autism, as comorbidity and co-occurrence within the same families (Casanova et al 2020; Cederlöf et al. 2016).

Mitochondrial Diseases

The central actor in bioenergetics is the mitochondrion. Mitochondria produce about 90 per cent of cellular energy, regulate the cellular redox state, produce ROS, maintain Ca2 + homeostasis, synthesise and degrade high-energy biochemical intermediates and regulate cell death through activation of the mitochondrial permeability transition pore (mtPTP). When they do not function or function poorly, less and less energy is generated within the cell. Damage and even cell death follow. If this process is repeated, entire organ systems begin to fail.

Mitochondrial diseases are a heterogeneous group of disorders that can affect several organs with varying severity. Symptoms may be acute or chronic with intermittent decompensation. Neurological manifestations include encephalopathy, stroke, cognitive regression, seizures, heart disease (*cardiac conduction defects, hypertensive heart disease, cardiomyopathy,* etc.…) (Siasos et al. 2018; El-Hattab et Scaglia 2016) diabetes, vision and hearing loss, organ failure, neuropathic pain and peripheral neuropathy.

Prevalence estimates of disease and mitochondrial dysfunction in studies range from about 5 to 80%. This could be, in part, due to the unclear distinction between disease and mitochondrial dysfunction. The studies indicating the highest rates of mitochondrial diagnosis are usually the most recent.

Some Drugs are Toxic to Mitochondria

They can trigger or aggravate mitochondrial dysfunction or disease.

Antiepileptics: valproic acid (*also used in various other indications*) and phenytoin are the most toxic. Phenobarbital, carbamazepine, oxcarbazepine, ethosuximide, zonisamide, topiramate, gabapentin and vigabatrin are also included (Finsterer et al. 2012, Mithal et al. 2017).

Corticosteroids (*such as cortisone*), isotretinoin (*Accutane*) and other vitamin A derivatives, barbiturates, some antibiotics, propofol, volatile anaesthetics, non-depolarising muscle relaxants, some local anaesthetics, statins, fibrates, glitazones, beta-blockers, biguanides, amiodarone, some chemotherapeutics, some neuroleptics, nucleoside reverse transcriptase inhibitors and various other drugs (Finsterer et al. 2010; de Oliveira 2015).

Neuroinflammation and Immune Disorders

The role of the immune system and neuroinflammation in the development of autism is controversial. Until recently, there was little evidence to support the immune hypothesis, but research into the role of the immune response and neuroinflammation may have important clinical and therapeutic implications. The exact role of the heightened immune response in the central nervous system in people diagnosed with autism is uncertain, but it may be a primary factor in triggering and sustaining many of the comorbid conditions associated with autism. Recent studies indicate the presence of increased neuroimmune activity in both the brain tissue and cerebrospinal fluid of parsons diagnosed with autism, supporting the hypothesis that an increased immune response may be an essential factor in the onset of autistic symptoms (Pardo et al. 2005).

A 2013 review also found evidence of microglial activation and increased cytokine production in post-mortem brain samples from autistic people (Gesundheit et al. 2013).

Reduced NMDA Receptor Function

Reduced NMDA receptor function has been linked to reduced social interactions, locomotor hyperactivity, self-injury, pre-pulse inhibition (PPI) deficits and sensory hypersensitivity. The results suggest that NMDA dysregulation may contribute to the core symptoms of autism spectrum disorders (Gandal et al. 2012).

Strabismus

According to several studies, there is a high prevalence of strabismus in persons diagnosed with autism with rates 3–10 times higher than in the general population (Williams 2021).

Tinnitus

According to one study, 35% of people with autism are reported to suffer from tinnitus, a much higher percentage than in the general population (Danesh et al. 2015).

Tourette Syndrome

The prevalence of Tourette syndrome among autistics is estimated to be 6.5%, higher than the prevalence of 2–3% in the general population. Several hypotheses have been put forward for this association, including common genetic factors and dopamine, glutamate or serotonin abnormalities (Zafeiriou et al. 2007).

Tuberous Sclerosis

Tuberous sclerosis is a rare genetic disease that causes the growth of benign tumours in the brain and other vital organs. It has a consistently strong association with the autistic spectrum. One to four per cent of autistic persons also suffer from tuberous sclerosis (Smalley 1998).

Studies have reported that between 25 and 61% of people with tuberous sclerosis fit the diagnostic criteria for autism with an even higher percentage showing characteristics of a broader pervasive developmental disorder (Harrison et Bolton 1997).

Inference of Metal Ions

Several recent studies have revealed that biometallic homeostasis plays a crucial role in the pathogenesis of neurological disorders, including autism. Substantial evidence indicates that disturbed neuronal homeostasis of several metal ions such as iron, copper, lead, mercury, selenium and zinc (Fe, Cu, Pb, Hg, Se, Zn) may be at the origin of synaptic dysfunction and impaired synapse formation and maturation. In vitro studies were conducted on the consequences of an imbalance of transition metals on glutamatergic synapses of hippocampal neurons, and it was evaluated whether a biometric profile characteristic of autistic persons influences synapse formation, maturation and composition with regard to NMDA receptor subunits and Shank proteins. The results show that an autism-like biometric profile leads to a reduction in NMDAR receptors (NR/Grin/GluN) subunits 1 and 2a and Shank gene expression, as well as a reduction in synapse density. In addition, synaptic protein levels of GluN2a and Shank are reduced. Although Zn supplementation is able to rescue the aforementioned alterations, Zn deficiency is not the only causative factor. According to Simone Hagmeyer (et al 2015), balancing Zn levels in autism could be a first target for normalising synaptic alterations caused by biomeostasis.

Vitamin Deficiencies

Vitamin deficiencies are more common in autism spectrum disorders than in the general population.

Vitamin D: In a German study, vitamin D deficiency affected 78% of the hospitalised autistic population. Of the entire group in the study, 52% were severely deficient, which is much higher than in the general population. Other studies also show a higher rate of vitamin D deficiency (Endres et al. 2016).

Vitamin B12: Researchers found that, overall, vitamin B12 levels in the brain tissue of autistic children were three times lower than in the brain tissue of children without a diagnosis of autism. This lower-than-normal vitamin B12 profile persisted throughout life in the brain tissue of patients with autism. These deficiencies are not visible by conventional blood sampling (Zhang et al. 2016).

As for the classic vitamin B12 deficiency, it affects up to 40% of the population; its prevalence has not yet been studied in autism spectrum disorders. Vitamin B12 deficiency is one of the most serious.

Vitamin B9 (folic acid): Studies were conducted on folic acid supplementation in autistic children. "Results showed that folic acid supplementation significantly improved certain symptoms such as sociability, verbal/preverbal cognitive language, receptive language, emotional expression and communication. In addition, this treatment improved the concentrations of folic acid, homocysteine and redox metabolism of standardised glutathione" (Sun et al. 2016, Frye et al. 2018).

Vitamin A: Vitamin A may induce mitochondrial dysfunction. According to a non-autism-specific study, "*Vitamin A and its derivatives, the retinoids, are micronutrients required in the human diet to maintain several cellular functions of human development in adulthood as well as during ageing (...) Although an essential micronutrient, vitamin A has several toxic effects on the redox environment and mitochondrial function. The exact mechanism by which vitamin A causes its deleterious effects remains unclear (...) Vitamin A and its derivatives, the retinoids, disrupt mitochondrial function by a mechanism that is not fully understood*" (de Oliveira 2015).

Zinc: The incidence rates of zinc deficiency in children aged 0–3, 4–9 and 10–15 years were estimated at 43.5%, 28.1% and 3.3% for boys and 52.5%, 28.7% and 3.5% among girls (Yasuda et Tsutsui 2013).

Magnesium: The incidence rates of magnesium deficiency in children aged 0–3, 4–9 and 10–15 years were estimated at 27%, 17.1% and 4.2% for boys and 22.9%, 12.7% and 4.3% for girls.

Calcium: The incidence rates of calcium deficiency in children aged 0–3, 4–9 and 10–15 years were estimated at 10.4%, 6.1% and 0.4% for boys and 3.4%, 1.7% and 0.9% for girls.

It has been found that inappropriate special diets for children diagnosed with autism usually result in excessive amounts of certain nutrients and persistent vitamin deficiencies (Stewart et al. 2015).

Often one focuses on a specific trait or category of typical autistic traits.

What I have briefly (and certainly not exhaustively) tried to portray is a general picture of the symptomatology and the overall characteristic picture.

Through this schematic synthesis, a picture emerges clearly, both physical and psychological, that has clear origins in the neurodevelopmental stages, and not at the level of specific localisation, but in circuit and intra-circuit dynamics.

Not all of the characteristics and symptoms described, however, must necessarily be traced back to a general diagnosis of autosmo. It is still possible for one or more characteristics to be due to different aetiologies, and only coincidentally to be found in the same person.

This brings us back to the centrality of an absolutely personalised and individualised diagnosis, which takes into account all the factors and moments of development, and which does not merely sum up in a necessarily unitary frame of reference characteristics that may be randomly correlated.

This breakdown of the picture into a more articulated and differential analysis is not a wasteful exercise devoid of meaning: through this procedure, one or more personalised and, therefore, more effective interventions can be identified.

Having definitively understood that autism has to do with neurodevelopment and the various causally implicated factors can generate permanent damage in the development of specific brain systems involved in higher mental processes, it is essential at this point to address and attempt to describe the developmental process of the system we call the brain.

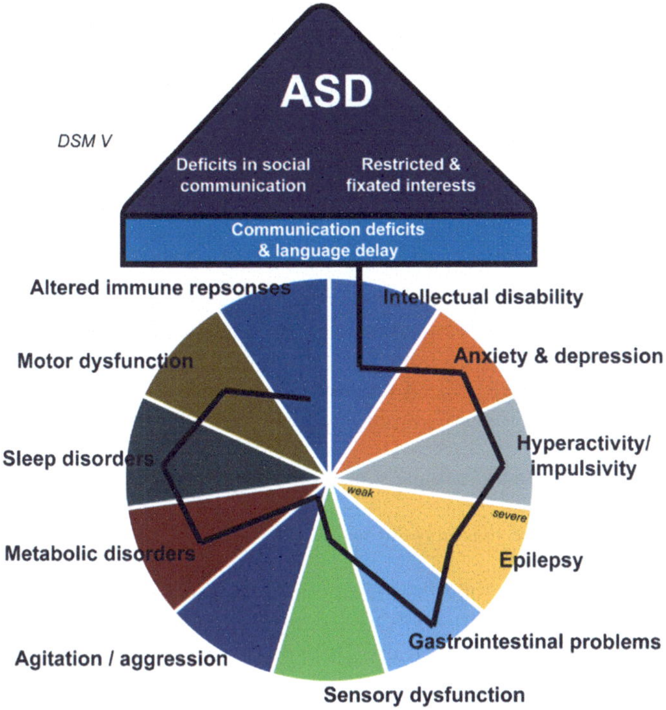

Clinical Features of ASD. The three core features of ASD are deficits in social communication, restricted and fixated interests together with speech deficits and language delays. The first two features are used to diagnose ASD according to the Diagnostic and Statistical Manual of Mental Disorders, 5th Edition. In addition, individuals with ASD may present further symptoms and comorbidities such as cognitive deficits (intellectual disability), anxiety, depression, attention deficits and hyperactivity, impulsivity, seizures, gastrointestinal problems, sensory dysfunction, aggression, metabolic disorders, sleep disorders, motor dysfunction and altered immune responses. However, not all symptoms and comorbidities are present in individuals with ASD, and those that are, vary in severity. This results in a considerable heterogeneity of clinical features of individuals with ASD.The image is published courtesy of Andreas Grabrucker, who processed it, and originally published it in *Grabrucker AM. Sauer AK, Stanton JE, Hans S,* (2021) *Autism Spectrum Disorders: Etiology and Pathology. In: Autism Spectrum Disorders. Grabrucker AM (Editor). Exon Publications, Brisbane, Australia.*

NOTE

1. On face recognition in children and development of the specialised brain area see: Adibpour, Dubois, Dehaene-Lambertz (2018); Deen et al. (2017); Livingstone et al. (2017).

REFERENCES

Ben-Sasson A, Hen L, Fluss R, Cermak SA, Engel-Yeger B, Gal E, (2009) A meta-analysis of sensory modulation symptoms in individuals with autism spectrum disorders, in J Autism Dev Disord, vol. 39

Bird G., Cook R., (2013) Mixed emotions: the contribution of alexithymia to the emotional symptoms of autism, in Translational Psychiatry

Bodfish JW, Symons FJ, Parker DE, Lewis MH, (2000) Varieties of repetitive behavior in autism: comparisons to mental retardation, in J Autism Dev Disord

Bonino S., (2005) Il sé massimo e il no dell'altro, in Psicologia contemporanea, n. 182

Brauner A., Brauner F., (2007) Vivere con un bambino autistico, Firenze, Giunti

Bruer-Thompson, C. (2019) "Overlapping Behavioral Characteristics of FASD's & Related Mental Health Diagnosis". www.fasdfamilies.com. Cathy Bruer-Thompson

Brothers, L.A. (2002), The social brain: A project for integrating primate behavior and neurophysiology in a new domain, In Concepts in Neuroscience, 1, pp. 27–51.

Buie T, Campbell DB, Fuchs GJ, Furuta GT, Levy J, Vandewater J, Whitaker AH, Atkins D, Bauman ML, Beaudet AL, Carr EG, Gershon MD, Hyman SL, Jirapinyo P, Jyonouchi H, Kooros K, Kushak R, Levitt P, Levy SE, Lewis JD, Murray KF, Natowicz MR, Sabra A, Wershil BK, Weston SC, Zeltzer L, Winter H, (2009) Evaluation, diagnosis, and treatment of gastrointestinal disorders in individuals with ASDs: a consensus re-port, in Pediatrics, vol. 125

Burgess AF, Gutstein SE, (2007) Quality of life for people with autism: raising the standard for evaluating successful outcomes, in Child Adolesc Ment Health, vol. 12

Casanova EL, Baeza-Velasco C, Buchanan CB, Casanova MF (2020) "The Relationship between Autism and Ehlers-Danlos Syndromes/Hypermobility Spectrum Disorders". Journal of Personalized Medicine

Cederlöf M, Larsson H, Lichtenstein P, Almqvist C, Serlachius E, Ludvigsson JF (2016). Nationwide population-based cohort study of psychiatric disorders in individuals with Ehlers-Danlos syndrome or hypermobility syndrome and their siblings. BMC Psychiatry

Danesh AA, Lang D, Kaf W, Andreassen WD, Scott J, Eshraghi AA (2015) Tinnitus and hyperacusis in autism spectrum disorders with emphasis on high functioning individuals diagnosed with Asperger's Syndrome. International Journal of Pediatric Otorhinolaryngology

Davidson, R.J., Friesen, W.V. (1990) The Duchenne Smile: Emotional Expression and Brain Physiology I, Journal of Personality and Social Psychology, 58 pp. 342–353.

Davidson, R.J., Begley, S. (2012) The Emotional Life of Your Brain Reid VM, Dunn K, Young RJ, Amu J, Donovan T, Reissland N. The Human Fetus Preferentially Engages with Face-like Visual Stimuli. Curr Biol. 2017 Jun 19; 27(12): 1825–1828.e3. https://doi:10.1016/j.cub.2017.05.044.

Decety J. (2012) La forza dell'empatia, in Mente e cervello, n. 89

De Ajuriaguerra J., Marcelli D. (1986) Psicopatologia del bambino, Milano, Masson Italia Editori

de Oliveira MR (2015) "Vitamin A and Retinoids as Mitochondrial Toxicants". Oxidative Medicine and Cellular Longevity

De Rosa F., (2014) Quello che non ho mai detto. Cinisello Balsamo. San Paolo

Dominick KC, Davis NO, Lainhart J, Tager-Flusberg H, Folstein S, (2007) Atypical behaviors in children with autism and children with a history of language impairment, in Res Dev Disabil

Ehlers S, Gillberg C (1993) "The epidemiology of Asperger syndrome. A total population study". Journal of Child Psychology and Psychiatry, and Allied Disciplines

El-Hattab AW, Scaglia F (2016). "Mitochondrial Cardiomyopathies". Frontiers in Cardiovascular Medicine

Endres D, Dersch R, Stich O, Buchwald A, Perlov E, Feige B, et al. (2016) "Vitamin D Deficiency in Adult Patients with Schizophreniform and Autism Spectrum Syndromes: A One-Year Cohort Study at a German Tertiary Care Hospital". Frontiers in Psychiatry

Erickson CA, Stigler KA, Corkins MR, Posey DJ, Fitzgerald JF, McDougle CJ, (2005) Gastrointestinal factors in autistic disorder: a critical review, in J Autism Dev Disord, vol. 35

Filipek PA, Accardo PJ, Baranek GT, Cook EH, Dawson G, Gordon B, Gravel JS, Johnson CP, Kallen RJ, Levy SE, Minshew NJ, Ozonoff S, Prizant BM, Rapin I, Rogers SJ, Stone WL, Teplin S, Tuchman RF, Volkmar FR, (1999) The screening and diagnosis of autistic spectrum disorders, in J Autism Dev Disord, vol. 29 [This paper represents a consensus of representatives from nine professional and four parent organizations in the US]

Finsterer J, Segall L (2010) "Drugs interfering with mitochondrial disorders". Drug and Chemical Toxicology

Finsterer J, Zarrouk Mahjoub S (2012). "Mitochondrial toxicity of antiepileptic drugs and their tolerability in mitochondrial disorders". Expert Opinion on Drug Metabolism & Toxicology

Fournier KA, Hass CJ, Naik SK, Lodha N, Cauraugh JH, (2010) Motor coordination in autism spectrum disorders: a synthesis and meta-analysis, in J Autism Dev Disord v.40

Franciosi F., (2017) La regolazione emotiva nei disturbi dello spettro autistico, Pisa, Edizioni ETS

Frith U., (2019) L'autismo - Spiegazione di un enigma, Milano, Economica La Terza

Frye RE, Rossignol DA (2014). "Treatments for biomedical abnormalities associated with autism spectrum disorder". Frontiers in Pediatrics

Frye RE, Slattery J, Delhey L, Furgerson B, Strickland T, Tippett M, et al. (2018). "Folinic acid improves verbal communication in children with autism and language impairment: a randomized double-blind placebo-controlled trial". Molecular Psychiatry

Gandal MJ, Anderson RL, Billingslea EN, Carlson GC, Roberts TP, Siegel SJ (2012). "Mice with reduced NMDA receptor expression: more consistent with autism than schizophrenia?". Genes, Brain and Behavior.

Geschwind DH, (2009) Advances in autism, in Annu Rev Med, vol. 60

Gesundheit B, Rosenzweig JP, Naor D, Lerer B, Zachor DA, Procházka V, et al. (2013) "Immunological and autoimmune considerations of Autism Spectrum Disorders". Journal of Autoimmunity

Gotts, S.J., Simmons, W.K., Milbury, L.A., Wallace, G.L., Cox, R. W., Martin, A. (2012), Fractionation of social brain circuits in autism spectrum disorders. In Brain, 135, 9, pp. 2711–2725.

Ghanizadeh A, Akhondzadeh S, Hormozi M, Makarem A, Abotorabi-Zarchi M, Firoozabadi A (2012) "Glutathione-related factors and oxidative stress in autism, a review". Current Medicinal Chemistry

Grandin T. (2006) Pensare in immagini, Erickson

Grabrucker AM., Sauer AK, Stanton JE, Hans S, (2021) Autism Spectrum Disorders: Etiology and Pathology. In: Autism Spectrum Disorders. Grabrucker AM (Editor). Exon Publications, Brisbane, Australia.

Hagmeyer S, Mangus K, Boeckers TM, Grabrucker AM (2015) Effects of trace metal profiles characteristic for autism on synapses in cultured neurons. Neural Plast. https://doi:10.1155/2015/985083

Harrison JE, Bolton PF (1997). "Annotation: tuberous sclerosis". Journal of Child Psychology and Psychiatry, and Allied Disciplines

Jones W, Carr K, Klin A. (2008) Absence of preferential looking to the eyes of approaching adults predicts level of social disability in 2-year-old toddlers with autism spectrum disorder., in Arch Gen Psychiatry, vol. 65

Johnson CP, Myers SM, (2007) Identification and evaluation of children with autism spectrum disorders, in Pediatrics, vol. 120

Kanner L. (1968) Autistic disturbances of affective contact, in Acta Paedopsychiatr, vol. 35,

Klin A (2006) Autism and Asperger syndrome: an overview. Revista Brasileira de Psiquiatria

Lam KS, Aman MG, (2007) The Repetitive Behavior Scale-Revised: independent validation in individuals with autism spectrum disorders. J Autism Dev Disord, v.37

Landa R, (2007) Early communication development and intervention for children with autism, in Ment Retard Dev Dis-abil Res Rev, vol. 13

Lange, Shannon; Rehm, Jürgen; Anagnostou, Evdokia; Popova, Svetlana (2018) Prevalence of externalizing disorders and Autism Spectrum Disorders among children with Fetal Alcohol Spectrum Disorder: systematic review and meta-analysis. Biochemistry and Cell Biology

Lee TM, Lee KM, Lee CY, Lee HC, Tam KW, Loh EW (2021) Effectiveness of N-acetylcysteine in autism spectrum disorders: A meta-analysis of randomized controlled trials. The Australian and New Zealand Journal of Psychiatry

London E, (2007) The role of the neurobiologist in redefining the diagnosis of autism, in Brain Pathol

Mayer EA, Padua D, Tillisch K (2014) Altered brain-gut axis in autism: comorbidity or causative mechanisms? BioEssays

Matson JL, Nebel-Schwalm M, (2007) Assessing challenging behaviors in children with autism spectrum disorders: A review, in Research in Developmental Disabilities, vol. 28

McPartland J, Klin A (2006) Asperger's syndrome. Adolescent Medicine Clinics

Militerni R., (2004) Neuropsichiatria infantile. Napoli, Idelson Gnocchi

Ming X, Brimacombe M, Wagner GC. (2007) Prevalence of motor impairment in autism spectrum disorders. in Brain Dev. vol. 29

Mithal DS, Kurz JE (2017) Anticonvulsant Medications in Mitochondrial Disease. Pediatric Neurology Briefs

Morello P. C. (2016) Macchia, autobiografia di un autistico, Milano, Salani

National Fragile X Foundation. "Autism and Fragile X Syndrome". 2013

Noens I, van Berckelaer-Onnes I, Verpoorten R, van Duijn G, (2006) The ComFor: an instrument for the indication of augmentative communication in people with autism and intellectual disability, in J Intellect Disabil Res, vol. 50

Notbohm E., (2015) 10 cose che ogni bambino con autismo vorrebbe che tu sapessi, Trento, Erikson

Oliverio A., Effetto cocktail party, in Mente e cervello, novembre, 2013

Pardo CA, Vargas DL, Zimmerman AW (2005) Immunity, neuroglia and neuroinflammation in autism. International Review of Psychiatry

Pelphrey, K.A., Carter, E.J. (2008) Brain mechanisms for social perception: Lessons from autism and typical development, in Annals of the New York Academy of Sciences, 1145, pp. 283–299.

Pinel JPG, (2011) Biopsychology, Boston, Massachusetts, Pearson

Plaisted Grant K, Davis G, (2009) Perception and apperception in autism: rejecting the inverse assumption, in Philosophical Transactions of the Royal Society B, vol. 364

Prizant BM, Duchan JF., (1981) The functions of immediate echolalia in autistic children., in J Speech Hear Disord., vol. 46

Rapin I, Tuchman RF, (2008) Autism: definition, neurobiology, screening, diagnosis, in Pediatr Clin North Am, vol. 55

Reid, V. M., Dunn, K., Young, R. J., Amu, J., Donovan, T., & Reissland, N. (2017). The Human Fetus Preferentially Engages with Face-like Visual Stimuli. Current biology : CB, 27(12), 1825–1828.e3. https://doi.org/10.1016/j.cub.2017.05.044

Rogers SJ, (2009) What are infant siblings teaching us about autism in infancy?, in Autism Res, vol. 2

Rogers SJ, Ozonoff S, (2005) Annotation: what do we know about sensory dysfunction in autism? A critical review of the empirical evidence, J Child Psychol Psychiatry, v. 46

Rutgers AH, Bakermans-Kranenburg MJ, van Ijzendoorn MH, van Berckelaer-Onnes IA, (2004) Autism and attachment: a meta-analytic review, J Child Psychol Psychiatry, v.45

Sacks O. (1995) An Anthropologist on Mars: Seven Paradoxical Tales, Knopf

Schmahl, C., (2013) Farsi male per farsi bene, in Mente e cervello, n. 98.

Schultz RT, Grelotti DJ, Klin A, Kleinman J, Van der Gaag C, Marois R, Skudlarski P. 2003 The role of the fusiform face area in social cognition: implications for the pathobiology of autism. Philos Trans R Soc Lond B Biol Sci. Feb 28; 358 (1430): 415–27.https://doi:10.1098/rstb.2002

Schumann, C.M., Bloss, C.S., Barnes, C.C., Wideman, G.M., Et Al. (2010), Longitudinal magnetic resonance imaging study of cortical development through early childhood in autism. In The Journal of Neuroscience, 30, 12, pp. 4419–4427.

Siasos G, Tsigkou V, Kosmopoulos M, Theodosiadis D, Simantiris S, Tagkou NM, et al. (2018) Mitochondria and cardiovascular diseases-from pathophysiology to treatment. Annals of Translational Medicine

Sigman M, Dijamco A, Gratier M, Rozga A, (2004) Early detection of core deficits in autism, in Ment Retard Dev Disabil Res Rev, vol. 10

Sigman M, Spence SJ, Wang AT, (2006) Autism from developmental and neuropsychological perspectives, in Annu Rev Clin Psychol, vol. 2

Silieri L., Lorenzoni L. e Tasso D., (1998) Il problema dell'attenzione nella scuola media, in Psicologia e scuola

Smalley SL (1998) Autism and tuberous sclerosis. Journal of Autism and Developmental Disorders

Stewart PA, Hyman SL, Schmidt BL, Macklin EA, Reynolds A, Johnson CR, et al. (2015) Dietary Supplementation in Children with Autism Spectrum Disorders: Common, Insufficient, and Excessive. Journal of the Academy of Nutrition and Dietetics

Sullivan H.S. (1962) Teoria interpersonale della psichiatria, Milano, Feltrinelli Editore

Sun C, Zou M, Zhao D, Xia W, Wu L (2016) Efficacy of Folic Acid Supplementation in Autistic Children Participating in Structured Teaching: An Open-Label Trial. Nutrients

Tager-Flusberg H, Caronna E, (2007) Language disorders: autism and other pervasive developmental disorders, in Pediatr Clin North Am, vol. 54

Treffert DA, (2009) The savant syndrome: an extraordinary condition. A synopsis: past, present, future, in Philosophical Transactions of the Royal Society B, vol. 364

Tribulato E. (2013) Autismo e gioco libero autogestito, Milano, Franco Angeli

Tribulato E. (2020) Bambini da liberare - Una sfida all'autismo, Messina, Centro studi logos ODV

Vivanti G., Conghi S. (2009) La comprensione del linguaggio nell'autismo, in Psichiatria dell'infanzia e dell'adolescenza, vol. 76

Volkmar FR, Paul R, Rogers SJ, Pelphrey KA, (2014) Handbook of Autism and Pervasive Developmental Disorders, Assessment, Interventions, and Policy, John Wiley & Sons

Williams BL, Hornig M, Buie T, Bauman ML, Cho Paik M, Wick I, et al. (2011) Impaired carbohydrate digestion and transport and mucosal dysbiosis in the intestines of children with autism and gastrointestinal disturbances". PLOS ONE

Williams D. (2013) Nessuno in nessun luogo, Roma, Armando

Williams E, Thomas K, Sidebotham H, Emond A. (2008) Prevalence and characteristics of autistic spectrum disorders in the ALSPAC cohort, in Dev Med Child Neurol., v. 50

Williams ZJ (2021) Prevalence of Strabismus in Individuals on the Autism Spectrum: A Meta-analysis. medRxiv

Yasuda H, Tsutsui T (2013) Assessment of infantile mineral imbalances in autism spectrum disorders (ASDs). International Journal of Environmental Research and Public Health

Zafeiriou DI, Ververi A, Vargiami E (2007). Childhood autism and associated comorbidities. Brain & Development

Zhang Y, Hodgson NW, Trivedi MS, Abdolmaleky HM, Fournier M, Cuenod M, et al. (2016) Decreased Brain Levels of Vitamin B12 in Aging, Autism and Schizophrenia. PLOS ONE

Zavagnini M., De Beni R., (2016) La mente che vaga, in Psicologia contemporanea

CHAPTER 5

Brain Construction and Neurogenesis: A Basis for the Neuroscientific Understanding of Autism

For a geneticist and neuroscientist, this chapter is probably superfluous.

Looking for omissions there are many, I am aware of that.

But the spirit of this book is to offer an overall picture that can reach, and thus actively involved, the 'differently specialised'.

I have tried to make things simple, without losing scientificity and reliability of information.

I think it is necessary, for what I will say next, that certain concepts are clear, and to do this, this chapter offers a common basis for deepening and reasoning.

Remember when I quoted my uncle in the preface?

During the broadcast in which he had recounted the discovery of his laboratory, he had spoken of chromosomes. In the evening, when he got home, a lady called him (*I cannot imagine how he would have got her number today, but at the time there were telephone directories and the explanation can be as simple as that*). She had a son with a genetic disease and asked him if he could 'give her the chromosome her son was missing'.

When my uncle told us about this episode we all had tears in our eyes, including him, because of the sense of impotence of science even at the moment of great discoveries.

I believe that this little anecdote shows forcefully, perhaps even violently, the need for science not to remain closed in on itself. I believe there must be a middle way between trivialising popularisation and

M. Di Salvo, *Autism Research between Psychology and Neuroscience*,
https://doi.org/10.1007/978-3-031-68338-1_5

specialist academic analysis, which at least succeeds in providing elements to be properly informed. So that 'too difficult to explain' does not become the new 'it is too difficult, you cannot understand'.

Einstein is said to have said 'if you can't explain it to a six-year-old, it means you don't understand it either'.

Apart from the fact that if one is a great physicist, that does not mean one has to be a phenomenon even in epistemology, not everything can be explained to a six-year-old without making such radical simplifications that the meaning of what one is trying to communicate is completely distorted.

On the contrary, this 'simplification anxiety' risks becoming a global problem.

That is why a way must be found to explain without trivialising or oversimplifying.

To go back to Einstein, the correct phrase is quite different and is 'Everything should be made as simple as possible, but not simpler', which, on closer inspection, has the opposite meaning (Robinson 2018).

Those involved in research are first and foremost in the service of the community. I have tried to 'made as simple as possible, but not simpler'. I hope I have succeeded.

How a Brain is Born

Every child represents an adventure towards a better life, an opportunity to change the old system and renew it (Hubert H. Humphrey).

With this quote begins a chapter of a splendid book in which Joseph LeDoux (2002) describes the process of brain construction.

Billions of neurons in the brain are interconnected in a complex way. Because the various steps of growth and development take time, and because different circuits run through these steps according to different planes, our behavioural and mental repertoire develops gradually, unevenly, during childhood.

Today, no one would seriously argue that the brain is a blank slate at birth, a tabula rasa waiting to be written by experience, or, conversely, that it is genetically predetermined, a kind of unchangeable repository of dispositions to act, think and feel in a particular way. It is widely believed that brain circuits are realised through a combination of genetic and non-genetic influences, and the discussion currently focuses less on the dichotomy and more on how nature and culture contribute to brain development.

The nature–culture dichotomy begins to blur as one realises that, with regard to questions of mind and behaviour, nature and culture are really two ways of doing the same thing—connecting synapses—and both are necessary for development to take place.

We commonly believe that experiences leave their mark on the brain through the memory register and that memory is a product of synapses. It is less common to think that genes also exert their influence on us in the form of memory. In this case, however, synaptic memory intervenes as a result of hereditary rather than personal history.

The earliest events in embryonic brain development are mainly controlled by genes, their products and the particular chemical environment in which they are found. The function of genes is to synthesise proteins, which regulate many aspects of brain development. Some proteins are enzymes that trigger chemical reactions, others induce other genes to produce additional proteins, some form barriers that direct and restrict multiple cell movements, and still others provide cells with adhesive outer walls to which other cells attach themselves as they make their way to their final locations. When we speak of genetic influences on brain development, we are essentially describing the effects of proteins and their secondary chemical effects.

The embryo, lacking sensory systems, is largely insulated from direct perceptual contact with the external environment, but even at these early stages of development, genes do not operate independently of the outside world.

The embryo's chemical environment is, by necessity, in direct contact with the chemistry of the maternal body. The embryo cannot itself produce the amino acids used to assemble the proteins necessary for brain and body development. These must be acquired from the mother, who obtains them from food.

In this sense, food—and in general the environment in the sense of the maternal ecosystem—can also be the source of toxins and chemical additives (*food, air, water, drugs, alcohol, tobacco*...). Moreover, the level of maternal stress will affect its hormonal state, which can affect the embryo, not unlike the antibodies that the mother produces to fight infections. Although the main characteristics of the brain are dictated by a genetic programme (*which ensures that all human brains look and function virtually the same*), this programme imposes certain conditions on the internal chemical environment in which neurons will develop. If this gene–environment interaction is disturbed, so will disturbed normal brain development. At this stage nature, culture and environment have already met long before birth, and at perhaps the most delicate moment.

Brain development begins in the ectoderm, which, together with the mesoderm and endoderm, form the three main parts of the embryo. These give rise to the various districts and organs of the body. From the ectoderm, through the inductive action of the notochord and the prechordal plate, the neural plate is formed. On the 18th day, the neural plate forms the neural shower, from the sides of which cells are formed that migrate to the periphery and form the neural crests, then fold to form the neural tube, outside of which the brain begins to structure itself.

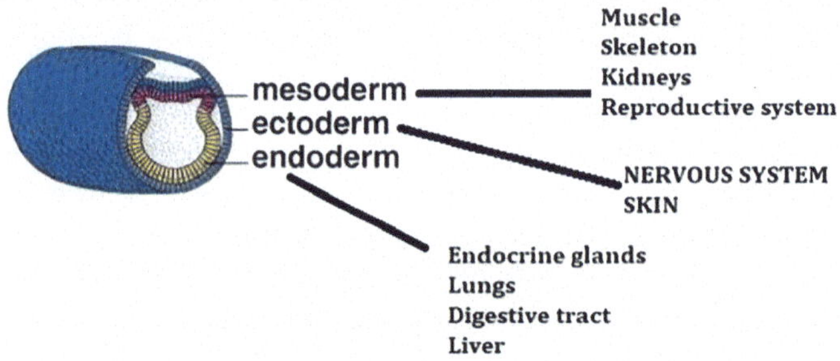

[graphic processing by the author]

In the neural tube, non-neural cell precursors divide and give rise to neurons. In humans, the vast majority of neurons are created in the months immediately before birth.

At peak production about 250,000 neurons are created per minute. This process is controlled by hormones that diffuse into the neural tube from underlying tissues and activate protein-producing genes, which, in turn, regulate neuronal production from cell precursors (*the hormones themselves are gene products created in non-neural portions of the embryo*). If the production of neurons is interfered with during this initial phase (*e.g. by alcohol consumption or drug taking*), congenital abnormalities may occur. Of course, not all abnormalities can be traced back to the intake of the wrong substances, just as the intake of such substances does not always induce congenital abnormalities.

Until a few years ago, it was believed that new brain cells did not develop in adult mammals. Recent studies by Fernando Nottebohm (1989) and Bruce McEwen (1994), Elizabeth Gould (1997 1998 1999a 1999b) and other researchers have challenged the traditional view, showing that neurons continue to be generated in certain areas of the adult brain. This discovery opens up new frontiers that neuroscientists may one day be able to develop therapies that promote the growth of new neurons to remedy today's largely incurable neurological disorders, such as spinal cord damage, apoplexy and Alzheimer's disease.

It is important to make a distinction at this point between neurogenesis—the birth of new neurons—and synaptogenesis—the creation of new synapses between pre-existing neurons. Synaptogenesis is a common phenomenon, which probably continues to occur until death. Again, the discovery is recent (Diamond 2001) although not so recent, and has led to considerable studies on neuroplasticity (Di Salvo 2024).

Synapses change considerably during the early stages of existence.

Soon after their birth, neurons begin to separate, dividing into cells destined to be part of the rhombencephalon, midbrain and forebrain, engaging different regions of the developing neural tube. Proliferation is under the direct control of a particular set of genes known as homeotic genes, which produce proteins that control the arrangement of cells, providing barriers that guide and contain cell movement and adhesive outer surfaces that allow cells to aggregate.

Some scientists believe that autism may be due to a mutation in homeotic genes that induces defective brain organisation and interconnection (Rodier 2000).

The function of homeotic genes has been discovered in studies on fruit flies and has been further investigated in studies on flies and worms, particularly the C-elegans worm (Caenorhabditis elegans). It may be hard to believe that we have much in common with flies and worms, yet the reality is that these genes, as well as many others, have been preserved through many levels of evolutionary history (Chan 1999; Rechert, Simeone 1999).

Eventually, the divided cells differentiate, which means that they take on different shapes and sizes and end up with different neurotransmitters and neuromodulators. This is how, for example, neural projections and interneurons come to differentiate. Differentiation occurs under genetic control, but it is not a strictly genetically determined process. Studies have shown that if cells from a particular brain area are implanted in a different region in another animal prior to differentiation, they take on the expected characteristics of the host organism's cells instead of those of the donor (Schlaggar 1991).

Chemical factors in the local environment appear to determine the final cell type that will then be expressed. However, transformation can only occur in very young cells, which suggests that once a cell type is determined, its fate is decided. This indicates that a cell type is not rigidly imposed by genes and is greatly influenced by the environment, although in this case it is the local chemical environment surrounding the cell, rather than the environment outside the organism as a whole. The local signals involved, however, are themselves genetically encoded proteins, so the non-genetic contributions to cell differentiation are not too dissociated from the genetic ones.

In the course of development, the neural tube expands and folds, eventually assuming a shape that begins to resemble the brain. While initially segregation concentrates cells destined to be in the same brain areas within the neural tube, as the tube expands cells must migrate from their separate resting place to reach their final destinations in the developing brain. Migration, for example, ensures that the different regions that make up the adult neocortex receive from the neural tube the correct amount and type of cells needed for those areas to perform their functions.

Pasko Rakic's (1995) work on cortex development indicates that the process involves the formation of temporary structures or chemical pathways that migrating cells follow.

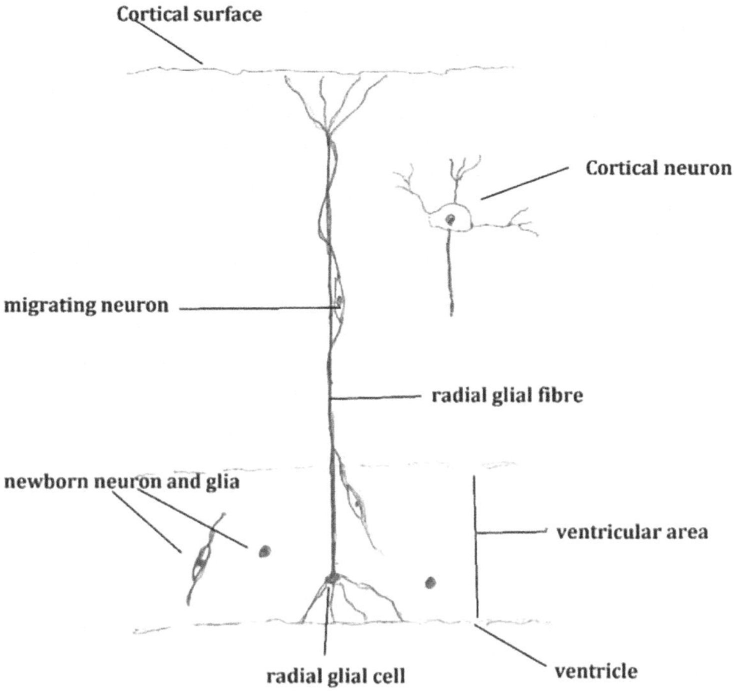

[Simplified diagram of cell migration through the glial scaffold. Author's drawing reconstructed on the basis of Rakic 1995].

These tracks consist of glial cells (*non-neural cells that contribute in important ways to development and other aspects of brain functioning*). Glial cells are, in turn, guided by local chemical tracks, produced by genes and their by-products, which act as molecular signalling, creating barriers that restrict movement and providing external adhesive surfaces.

As the developing brain increases in size, the migrating glia produce fibres that extend beyond the proliferation zone (*the place where cell precursors are generating neurons and neuroglia*) towards the brain surface (*where the neurons are to go to form the cortex*). Following the glial network, the young neurons find their way to their final destination.

The glial network causes some of the neighbouring neurons within the isolated neural tube to end up paired in the cortex.

This explains, in part, how cortical areas come to have the cells they have: adjacent ones in the tube travel together along the same glial fibres and tend to end up at the same cortical site (how cortical or other brain areas are assembled is not yet fully understood).

One possibility is that the function of a cell (*whether it will be involved in vision rather than tactile sensations, for example*) is genetically determined, so that it is destined to end up in specific areas dedicated to a specific function before it leaves the neural tube. Another possibility is that cells learn what to do from the context in which they go.

Supporting the second possibility is the fact that if a portion of the visual cortex is removed and implanted in the region of the somatosensory (touch) cortex, it will accept somatosensory axons from a lower station (nucleus) of the brain (the somatosensory thalamus) and function in the same way as the somatosensory cortex. This indicates how cells adapt to the local situation (the chemical environment and the axons afferent to it) in the process of assuming their own functionality (Shatz 1992; Rakic 1992).

Support for the first possibility (*that a cell's functional fate is genetically determined*) comes from a recent study showing that under particular conditions where axons are prevented from developing within the cortex, cortical cells will still possess the chemical molecules that characterise that area. This suggests that the chemical marking of the cell is determined by the cell itself, presumably by its genes (Miyashita-Lin et al. 1999).

From the point of view of brain functioning, the activity performed by a cortical area depends more on its synaptic connections than on the molecules present. Consequently, although the chemical marking of an area is determined by genes, the area's function will most likely be predominantly determined by its synaptic connections. Cells moving along the glial reticulum prove to be functionally ambivalent in this respect, at least within certain limits.

Once they reach their destinations, neurons begin to develop axons.

These newly formed fibres must then find their way to their targets, the neurons with which they will form synapses. The targets may be in neighbouring or distant areas, or even in both dislocations. Their so-called pioneering reconnaissance ability depends on growth cones, Japanese fan-like structures that extend from the ends of growing axons (Raper et al. 1998).

The ends of growth cones have chemical affinities for specific proteins in the surrounding space: they only bind to some proteins and are repelled by others.

When one of the ends of a growth cone finds the right substance, it binds to it. If the other ends are not bound, the growth cone rotates in the direction of the binding, and the axon follows.

The same protein can be both an attractant and a repellent for an axon, depending on the chemical state of the axon (Terman et al. 1999).

Slowly, gradually, the growth cone proceeds through the chemical pathway, projecting the axon forward, until it reaches its destination. Then certain signals (the nature of which is not yet fully understood) cause the growth to stop.

The first axon along the pathway, the 'pioneer', acts as an outrider, marking a pathway that other nerve fibres from the same area follow. The reconnaissance capacity of the fibres is not dissimilar to cell migration along the glial pathway.

As the axons reach their targets, and receive the signal to stop growing, the growth cones give way to axon terminals, which begin to form synaptic connections with postsynaptic cells.

Initial connections, like other early events in brain development, are essentially determined by genetic and other intrinsic factors. But the brain is not entirely constructed from the inside out. The final stages of its structuring, particularly the phases during which the connections are perfectly tuned in a way that enables it to perform its activities, require something more.

The human genome contains the instructions for physical development, including, as we have seen, the development of a general brain structure. It is believed that some 50–70 per cent of all genes in the human body are involved in the construction of the brain. Nevertheless, we probably do not have enough genes to explain the wiring of every particular synaptic connection of the three trillion present in the brain.

Almost all researchers agree that the transition from the initial, immature neural connections of the young brain to the mature, highly specific connectivity that characterises the adult brain involves neural activity, i.e. transmission across synapses. The exact role of neural activity, however, is still hotly debated; the main bone of contention is whether activity, particularly that triggered by environmental stimulation, contributes to mature connections or merely selects from the initial set of intrinsically established connections that will be maintained. Since neural activity triggered by environmental factors is implicated in both the management and selection of the system of connections, this is not so much a genes/contextual experience dispute as it is a dispute over the precise contribution of experience.

Neural Darwinism

Neo-Darwinian theories arising from evolutionary biology were adopted and adapted from the field of immunology and were then applied to brain functioning.

The most influential interpreter of neo-Darwinism is probably Gerald Edelman (1987), who received the Nobel Prize for his work on the immune system.

Edelman argued that synapses in the brain, like animals in their environments, compete for survival. Used synapses compete successfully and survive, while unused ones die. For Edelman, the structure of neural circuits is neither determined nor systematically modified in response to external influences. External influences select synapses through the activation and consolidation of the particular patterns of neural activity that comprise them.

Neo-Darwinians assume that genetic and non-genetic factors interact at every stage of brain development. Selection acts on genetically predetermined pre-existing connections (*which produce the proteins that help orient axons to the appropriate areas*), working in concert with non-genetic factors (*chemicals of maternal origin, for example*). However, genes and the chemical environment are not totally responsible for defining the initial connections. Neo-Darwinians also assume that some degree of randomness is involved: terminals and dendrites that are randomly located in close proximity realise the opportunity to form synaptic connections, independent of the overall directing pattern established by genes.

Consequently, in spite of a genetically programmed overall plan, even the pre-existing connections on which selection eventually intervenes possess a peculiar, individualistic nature, on which experience then operates a selection. Since the experiences of each individual are different, different patterns of connections are selected. Therefore, genes dictate that we will have a human brain type with roughly the same types of circuits; however, there will be random individual differences, and the connectivity of circuits, selected by synaptic activity, will shape the individual's brain.

Three of the fundamental principles of neo-Darwinism are illustrated in the conditions of redundancy (*more synapses are produced than are actually retained*), utilisation (*retained synapses are the active ones*) and subtraction (*unused connections are eliminated*). It was thought that structuring the nervous system in this way provided a method for coping with the lack of information available to the brain from alternative sources, such as the outside world, during early development. If the neo-Darwinians are right, the connections we end up with as adults are those that have not been eliminated: the self, they would say, is not constructed, it is selected from pre-existing possibilities.

The pioneering studies of Victor Hamburger and Rita Levi Montalcini (1949) in the 1930s showed that subtractive or degenerative events occur in the nervous system, a view that is now widely accepted (Oppenheim 1998; O'Leary 1992).

During development, cells are produced and eliminated, chemicals come and go and functions change. Our main interest here turns to synaptic degeneration, the pruning of redundant projections that were unused during the early stages of development.

A control of synaptic degeneration was proposed in a widely cited study by Pasko Rakic (1986) and his collaborators, who found that the number of synapses in different brain areas of the primate cortex increases and then decreases during the first year of life. A study involving a more detailed analysis of a single cortical area indicated that the decrease did not occur before puberty (Bourgeois et al. 1994).

Since cognitive development takes place almost entirely during puberty, a decrease in synapses that does not occur before puberty cannot explain the maturation of the mind.[1]

Although few such investigations have been carried out on humans, one widely cited study would suggest that throughout the cortex the greatest number of synapses are present at around twenty-four months of age (*evidently well before puberty*), but with different areas reaching their maximum value at different times (O'Leary 1992).

Undoubtedly, much more research work is needed to determine the extent to which synaptic suppression occurs, when it intervenes in specific neural circuits, whether there are significant species-related differences, and how these data should be interpreted. The crux of the matter is not whether synapses are suppressed during development (*undoubtedly they are*), but whether the contractions that have occurred provide decisive confirmation for the strong version of neo-Darwinism, the version that maintains that activity is limited to the suppression of synapses.

The Origin of Neurons

From an evolutionary point of view, neurons probably derived from eukaryotic cells that used to change their shape and produce tube-like extensions as they moved around perceiving their surroundings, incorporating nourishment. The pseudopods of an amoeba give an idea of such activity. The tubular extensions of this organism, created at the moment through an internal rearrangement of microtubules, are dismantled once the cell has performed its task. However, when they evolved from temporary to permanent, these extensions probably gave rise to what we now know as axons and dendrites, the tubular components that make neurons distinctive. After all, the process of axonisation, orientation and dendrisation follows this same path in neurogenesis: a stable system of cables and antennae was born, ideal for emitting and receiving signals.

Although they have a very characteristic functioning that paved the way for complex behaviour and the mind, neurons have retained a close affinity with other cells in the organism. If we limit ourselves to considering neurons and the brain as radically different structures, leaving aside their origins, we risk over-separating the brain from the body (Damasio 2010).

Mirror neurons, which I will only introduce here, are a particular type of neuron, not so much in their form as in the peculiarity of the activity they perform.

Attempts to explain the existence of mirror neurons have emphasised their possible role in enabling us to understand the actions of others by putting ourselves in a similar body state. When we witness the action performed by another person, our brain adopts the same configuration as it would if we were the one moving, and in all likelihood it does so not by using passive sensory configurations, but through the pre-activation of motor structures (*ready for action but not yet allowed to act*) and, in some cases, through actual motor activation.

How did such a complex physiological system evolve? The suggestive idea proposed by Damasio (1983) is that it evolved from a pre-existing 'as if' circuit, long used by complex brains to simulate their body states.

This would entail a clear and immediate advantage: the rapid and energetically cheap activation of maps of certain body states, which were, in turn, associated with past knowledge and relevant cognitive strategies. Eventually, the 'as if' system was applied to others, and it became widespread because of the equally obvious social benefits to be gained from knowing the body states of other individuals, an expression of their mental states. In short, Damasio sees the 'as if' body circuit within each organism as the precursor to the function of mirror neurons.

Modulating neurons distribute neurotransmitters (***dopamine, norepinephrine, serotonin, acetylcholine***) in diffuse regions of the cerebral cortex and subcortical nuclei. One can describe this set-up as follows: (1) innate regulatory circuits are involved in promoting the survival of the organism, and for this reason 'know' what is going on in the most modern areas of the brain; (2) these circuits are regularly signalled as to whether things are going well or badly; (3) their intrinsic reaction is to influence the way the rest of the brain is shaped, so that it can contribute to survival in the most effective way.

Thus, as we move from infancy to adulthood, the design of the brain circuits that represent our evolving body and its interactions with the world seems to depend on the activities in which the organism is engaged and the action of innate bioregulatory circuits as they react to those activities. We can, therefore, see how it is inappropriate to conceive of brain, mind and behaviour in terms of the opposition between nature and culture, between genes and experience. At birth, our brains are not tabula rasa, nor is the mind. Neither are they completely genetically determined: genetics casts a large but not total shadow. Genes provide brain components with a definite structure and others whose exact structure must be determined. But the latter can only be arrived at under the influence of three elements: (1) the exact structure; (2) individual circumstances and activities (and here the last word comes both from the human and physical environment and from chance); (3) self-organising drives that arise from the very complexity of the system. The unpredictable trajectory of the individual's experiences makes itself felt, directly and indirectly, in the circuit design, through the reaction it induces in the innate circuits and through the consequences these reactions provoke in the overall process of circuit formation (Posner, Petersen 1990).

Neuronal Activity During Development

Neurons that are apparently identical and thus derived from the same embryonic neuroblasts—the embryological precursors of the same neurons—acquire different functional specialisation.

There is functional specialisation in an apparently morphologically homogeneous population. In fact, each Purkinje cell possesses a peculiarity that enables it to be distinguished from its neighbours.[2]

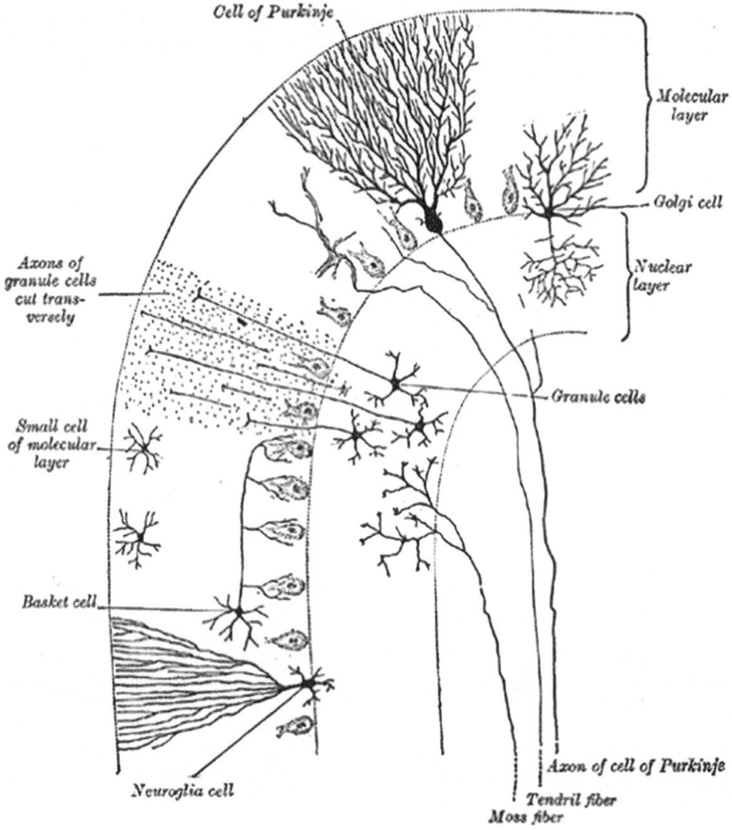

Henry Gray (1918) Anatomy of the Human Body.
[Opensource Image].

The singularisation of a neuron thus consists of an increase in its functional anisotropy, which allows it to be distinguished from its neighbours. Singularisation could give rise, for example, to differential localisation of neurotransmitter receptor molecules, which are uniformly distributed on the surface of an embryonic cell but will later cluster under specific nerve endings. When the first synaptic contacts are formed, they are much more numerous than in adults. This synaptic exuberance is transient, being followed by the synaptic stabilisation of certain privileged contacts that give the neuron its functional specificity, i.e. its singularity.

The 'singularity' of a given neuron within a class would manifest itself both at the connective level and in the configuration of the genes expressed.

Jean-Pierre Changeux's (2008) hypothesis is that the activity of a neuronal network during development intervenes in this singularity. The term 'activity' refers to the set of processes that directly or indirectly produce changes in the electrical properties of the neuronal membrane. This includes the propagation of an action potential, transmission and chemical modulation of the synapse and electrical coupling.

There are several reasons in favour of this position. Firstly, all electrical membrane phenomena can produce cellular integration. Furthermore, from the perspective of the systemic model, activity in a developing neuronal network represents a form of interaction between elements—including cells, nerve centres and sensor and effector organs—that ensures both integration and diversification. This activity introduces a new combination of signals through convergence and divergence properties specific to the neural network. Finally, spontaneous activity, present very early in embryonic development, can be modulated or even replaced by evoked activity resulting from the interaction between the developing organism and the outside world. The range of endogenous signals is enriched by the evoked signals that can participate in singularisation, which in mammals and especially in humans continues long after birth, particularly in the cerebral cortex. The hypothesis that nerve activity participates in the regulation of embryonic development allows a considerable economy of genes in the formation of the nervous system as a whole.

Various electrical phenomena have been observed experimentally at the early stages of embryonic development and even in the unfertilised egg.[3]

The electrical activity of the embryonic nervous system has mainly been studied in the chick. In 1885 William Preyer noted that at three and a half days of incubation the embryo manifests spontaneous movements. Rita

Levi Montalcini and Fabio Visintini (1938) found that these spontaneous movements are interrupted by curare and coincide with electrical activity of a similar frequency to that in the spinal cord, leaving little doubt that the origin of the embryo's motor activity is neuronal. Electrophysiological data from the embryonic chick spinal cord on day five demonstrate units with regular periodic discharges and units with burst discharges.

The thermodynamic conditions for oscillations to appear have been defined: for Ilya Prigogine (1982) and co-workers, oscillations can only take place in an open thermodynamic system that continuously exchanges energy with the outside world. Furthermore, this system should be stable, but not in equilibrium. Such dissipative structures can emerge if non-linear relationships exist between flow and forces, due, for example, to the existence of cooperative interactions and feedback effects between the elements of the system.

We must distinguish two types of oscillation in the nervous system. Some oscillations appear in neuronal chains and depend directly on their organisation, such as the circuits that control respiratory movements in vertebrates or walking in some insects. Others appear in isolated cells, such as the oscillations responsible for spontaneous embryonic activity. Among these cellular oscillations, cytoplasmic and membrane oscillators are distinguished.

The enzymatic chain of glycolysis is an example of a cytoplasmic oscillator.

Mollusc neurons that discharge in bursts, such as the R-cell of the abdominal ganglion of Aplysia (Strumwasser 1965; Meech 1979), constitute one of the best-known examples of the membrane oscillator. One channel is selective for calcium and its potential-sensitive opening is accompanied by a depolarising current of either calcium or sodium. The other channel is selective for potassium and its opening is activated by the calcium present within the cell and is accompanied by a hyperpolarising outgoing current of potassium. The regenerative depolarisation (*positive cooperative effect*) associated with the opening of the slow calcium channel causes an increase in the intracellular calcium concentration which, by activating the slow potassium channel (*positive feedback*), causes a change in potential of the opposite sign to that which had allowed its entry. These regulatory signals lead to slow oscillations.

When the membrane potential crosses the threshold for an action potential, burst discharge signals appear at the crest of the slow waves. These discharges open fast ion channels that cause ionic changes that modulate the basic oscillator. The opening of these four types of channels leads to a passive loss of ions that is continuously compensated for by an intense pumping activity by the membrane ATPase.

This very simple membrane system can be categorised as an open thermodynamic system.

The pattern described with the basic oscillator of burst neurons in the molluscum applies without major modifications to the potential oscillations of Purkinje fibres in the heart or the secretory β-cells of the pancreas. This probably also accounts for spontaneous burst discharges in the spinal cord. Clearly, the genetic cost of such spontaneous activity is no higher than that of a few structural genes.

Whenever one has looked for it, spontaneous electrical activity has always been found in vertebrate embryos and foetuses. In humans, in particular, spontaneous movements appear during the second month, when the embryo is still only four centimetres long (Fransson et al. 2007).

They involve the head and trunk and resemble type I movements in the chicken embryo. During the third and fourth months, the movements become generalised to the entire foetus and the first reflexes appear along with facial expressions and sucking and swallowing movements; the foetus actually drinks amniotic fluid. The movements are perceived by the mother and have been shown to originally involve the spinal cord and brainstem (*they are spinobulbaric*). They present similarities to the type I activity of the chicken embryo. From the fourth to the sixth month, more specialised activities develop, such as stable and regular respiratory movements that have a supra-bulbar component, i.e. they involve the cerebral hemispheres, including the cortex, and are analogous to the type I activity of the chicken embryo. As happens in the chicken embryo, during foetal development first an acceleration phase and then a deceleration phase takes place. Around the eighth month, the foetus makes two to a maximum of ten movements per minute. Naturally, spontaneous electrical activity accompanies these foetal movements.

The first recordings were made in 1969 by Staffan Bergström on the brainstem of a seventy-day-old foetus. The activity continued and diversified over the following months.

Electroencephalography on humans allows a global appreciation of the brain's electrical activity and, therefore, its most obvious manifestations such as wakefulness and sleep. Sleep studies in adults have revealed the existence of command centres with periodic oscillatory activity, consisting of groups of neurons located in various nuclei of the Varolius Bridge region in the brainstem.

Through their strongly divergent axonal branches, they distribute signals to a large number of regions, even to the entire brain. The locus coeruleus thus controls wakefulness, and the anterior raphe nuclei regulate slow-wave sleep.

In the embryo, nerve activity is essentially endogenous.

As the sense organs become functional, evoked responses replace spontaneous activity at least during wakefulness. In humans, touch develops very early. As early as the forty-ninth day of gestation, the foetus responds to tactile stimulation of the lips, and between the fifth and seventh month the sensitive innervation of the fingers is practically complete. The vestibular apparatus becomes functional between ninety and one hundred and twenty days, and auditory sensations, as far as could be assessed from changes in the heartbeat, between one hundred and eighty and two hundred and ten days. A flash evokes potentials in the premature infant at twenty-nine weeks. These sensitive functions of the foetus are much less efficient than those of the adult.

The maturation of the sense organs, characterised by a reduced sensitive threshold and a decreasing latency time, continues long after birth.

At birth, the human brain weighs about 300 grams. It increases by about five times to reach its adult weight at around seven years of age. To have a yardstick, the weight of the chimpanzee's brain at birth is already about 60 per cent of its adult weight. In the cerebellum and cerebral cortex, dendritic arborisation and synaptogenesis are mostly postnatal. In 1975 Brian Cragg estimated that in the primary visual cortex of the cat, the average number of synapses per neuron increases from a few hundred to 12,000 between the 10th and 35th day after birth. This number then decreases by 20 or 30 per cent to reach its adult value. It is, therefore, legitimate to think that activity evoked by interaction with the environment replaces spontaneous activity and governs synaptic development from then on.

One might believe that from an empirical point of view, interaction with the outside world causes activity in the nervous system whose characteristics would be directly related to the types of physical signals received by the sense organs.

In reality, this is not the case. Action potentials in the auditory nerve have the same nature as those in the optic nerve. A cellular analysis of the functioning of the sense organs allows us to go even further. In most cases, the effect of the physical signal causes a modulation of spontaneous activity that obviously precedes interaction with the environment.

For example, in 1968 Horace Barlow showed that in the cat, illumination of the retina produces an increase in ON ganglion cell activity of forty to seventy impulses per second. The first signs of spontaneous activity appear eight days after birth, i.e. before any response to light.

Another example is the vestibular nerve, where there is considerable spontaneous activity in the adult cat: when the position of the head changes, this activity ranges from five to thirty-five impulses per second. In 1979, Ian Curthoys showed that in mice, spontaneous activity in the vestibular nerve is present from birth and increases almost sixfold during the first twenty postnatal days. Signal sensitivity evolves in parallel.

Traces of this activity can be recorded down to the level of the neurons of the vestibular nuclei and also in the cerebellar cortex. The sensory receptors thus form peripheral oscillators whose activity largely contributes to the spontaneous activity recorded in the nervous system: interaction with the environment modulates the rhythm of the peripheral sensory oscillators which then regulate the central oscillators.

One can, therefore, imagine the molecular mechanism by which the activity evoked in the peripheral oscillators could modulate the activity of the central oscillators, more or less in the long term. Only a small number of macromolecules, receptors and cyclases would be required to make this 'trace' available to neurons, and these could moreover be used for different tasks in other tissues. The cost in terms of structural genes would again be particularly limited.

Contrary to some biochemical theories of memory, no more synthesis of new types of molecules is required to produce a lasting modification of neuronal properties following an organism's interaction with its environment.[4]

The early theory of selective synapse stabilisation defined the term 'genetic envelope' as the set of genetic determinants, or genetic constraints, involved in axonal guidance, including target recognition, synaptic adhesion and stabilisation or elimination of initial contacts.

One of the most important findings in studies in recent years has been the demonstration of pathological changes in connectional epigenesis brought about by disorders of the genetic envelope. These could be at the origin of serious childhood diseases. One particularly interesting gene, discovered by Katariina Hannula-Jouppi and co-workers (2005), is homologous to the axon guidance gene called ROBO, identified in Drosophila and which, when found in mice, disrupts midline axons in the brain.

What Happens at a 'Somewhat Larger' Level?

It is only with the advent of magnetic resonance imaging that we have been able to visualise the early organisation of the human brain and discover that from birth, practically all the circuits of the adult brain are already present in that of the child.

At the age of two months, when a child hears sentences in its native language, it already activates the same brain regions that are activated in an adult.

The first area to activate is the primary auditory cortex, the entry point for all auditory information in the cortex. It is activated as soon as a sentence begins. Until recently, some researchers estimated that the sensory areas were so unorganised at birth that the senses were confused: during the first weeks of life, the baby's brain mixes hearing, vision and touch. Today, we know that this theory is wrong: from birth, hearing activates the auditory areas, vision the visual areas and touch the tactile areas, without the need to learn it. This organisation into distinct regions for each sensory modality is due to genes, is common to all mammals and its origin is lost in the mists of time of our evolution (Dehaene-Lambertz et al., 2002, 2006).

From the age of two months, the child activates the same hierarchy of phonological, lexical, syntactic and semantic brain areas as an adult. Also in his case, the further one advances in the cortex hierarchy, the slower the responses are and the more they integrate higher-level information (Elman et al., 1996; Quartz, Sejnowski 1997).

In his brain, linguistic information is already channelled into highly specialised circuits, as in adults. If a child learns so early to understand and speak, whereas all other primates are not able to, it is undoubtedly because its left hemisphere already houses a structure of regions specialised in the detection of statistical regularities at all levels: sound, word, sentence and text.

Since all these brain areas are activated in a chain, it follows that they must be linked together. In adults, we begin to understand which bundles of axons connect the various language areas. There is, in particular, a large cord made up of millions of nerve fibres, called the arcuate fasciculus, which connects the temporal and parietal lobes with the frontal lobes and, in particular, with the famous 'Broca's area'. It is an indicator of the evolution of language. In 96% of humans, it is much larger in the left hemisphere, which is dedicated to language.

This asymmetry only exists in the human species, not in other apes, not even in chimpanzees. Once again, this anatomical feature is not the result of learning: it is present from the very beginning.

When we examine the connections of an infant's brain, we find that not only the arcuate fascicle, but also all the large fascicles of connections between the cortex and subcortical regions are already present at birth.

These 'brain highways' are built during the third trimester of pregnancy. Without the intervention of the outside world, a network of nerve connections specific to the human species is established and this mechanism will later be perfected through learning, but the initial scaffolding is innate and is built in utero.

Only twenty years ago, many researchers thought it extremely unlikely that the brain was nothing more than a messy mass of random connections. They could not imagine that our DNA, which contains only a small number of genes, could contain the detailed blueprint of different specialised circuits such as those of vision, language or motor skills.

From the first months of gravi-dance, many genes are selectively expressed at different points of the cortex. Around 28 weeks of gestation, the human brain begins to structure itself in a series of folds and the main grooves appear. In a 35-week-old foetus, all the main folds of the cortex are already well formed and one can already see the characteristic asymmetry of the temporal region, which houses the language areas.[5]

The formation of these cortex folds and the growth of connections in the brain continue throughout pregnancy. Subsequently, secondary and tertiary folds appear: folds upon folds. Their epigenesis becomes increasingly dependent on the activity of the nervous system.

Depending on the feedback signal that the brain receives from the sense organs, some circuits will be stabilised, while others, unnecessary, will degenerate. Today, it is believed that the appearance of folds in the cortex, which allow the development of different functions, is a process of self-organisation that depends on both genes and the chemical environment of the cells, but requires very little genetic information to occur.

The cortex is full of columns, stripes or boundaries that constitute specialised modules for processing a particular piece of information. In the visual cortex, for example, one can see an alternation of strips specialised in information coming from the left eye and the right eye: the ocular dominance columns. But the same mechanism also occurs at a higher level, not necessarily on the surface of the cortex, but in a more abstract space. One of the most spectacular examples, which earned its discoverers the Nobel Prize, is the existence of grid cells whose discharges form almost perfect hexagons. These neurons were discovered by recording the activity of a precise region of the rat's brain, the entorhinal cortex, as the animal moved around a large room.[6]

It was already known that there were 'position cells' in a neighbouring region (the hippocampus) that map space, i.e. they are only activated if the animal is in a specific place in the room. The new discovery was that the grid cells do not just respond to a position in the room, but to a whole series of positions, which are distributed according to a network of equilateral triangles that group together to form a hexagon. The Nobel Committee called this system the 'GPS of the brain', and it works like a spatial coordinate system, a grid that divides the room into triangles, and allows the animal to constantly know where it is because these neurons activate even in the dark. Neurons probably self-organise during development and those hexagonal structures appear in the most natural way as soon as a physical system stabilises. Several mathematical models correctly explain the emergence of grid cells: this organisation into triangles and hexagons is a spontaneous attractor of entorhinal cortex dynamics (Nieder, Dehaene, 2009; Hannagan et al. 2018).

We know, albeit indirectly, that the adult brain also contains a neural map with hexagonal symmetry. We also know that a child, even a very young one, has a sense of space. Even a blind child at an early age can find his way around a room. We know that his visual cortex contains, from a few months of age (though not from birth), a region that responds preferentially to faces, rather than to images of houses. This is certainly, in part, the result of learning, but strictly channelled, guided and forced by cerebral connectivity to always flow into the same area, into a specific module of the cortex. We also know that the infant's parietal cortex is already sensitive to the number of objects, the same region that is activated when an adult performs a calculation or when a monkey memorises a number of objects.

The view that emerges is that of the power of genes and self-organisation in the creation of the human brain. From birth, the child's cortex has almost the same folds as an adult's and is subdivided into specialised sensory and cognitive areas, which are already interconnected by precise and reproducible fibre fascicles.

It is a collection of modules and each of these projects a particular type of representation of the external world. The entorhinal cortex, with its grid cells, draws two-dimensional planes (perfect for encoding and navigating space). Other regions, such as the parietal cortex, draw lines (excellent for encoding linear quantities such as number, size or time passing), while Broca's region projects trees embedded in each other (ideal for encoding language structures). We inherit, from our evolution, a set of primitives from which we will be able to select those that are best suited to represent the situations and concepts we are to learn (Dehaene, 2019).

* * *

Autism is known to be associated with accelerated axonal growth in the brain between six and fourteen months of age.

Bourgeron (2016) and collaborators identified two loci on the X chromosome linked to autism, which corresponds to neuroligins 3 and 4. These are postsynaptically present adhesion proteins that associate with neuressin, proteins that can have very different shapes along alternative splicing. Changes in neuroligins that interfere with cortical synaptogenesis mainly affect long-axon neurons, which are directly involved in the conscious neuronal workspace.

These neurons should be particularly vulnerable due to the very high ratio of axon volume to somatic volume.

Fragile X syndrome is the most common form of mental retardation, caused by an abnormality at the end of the long arm of the X chromosome. The gene responsible was identified by Jean-Louis Mandel and his group of researchers in 1991, and is called fragile X mental retardation (Vincent et al. 1991).

The coding region is 1900 base pairs, and the mutations responsible for fragile X result from the expansion of CGG triplets hundreds of times, until loss of function occurs. The end result is the disruption of the neuronal cytoskeleton required for synaptic stabilisation.

There are, therefore, genetic constraints on synaptic epigenesis.

Notes

1. Quartz, Sejnowski, 1997, and Karz, Shatz, 1996. First, it is very difficult to accurately measure synaptic density in a given brain region while that region is changing its size with the passage of time. Furthermore, if synaptic changes are not related to specific cell types, it is complicated to know what the implications would be. Finally, the relationship between structural measurements (such as the number of synapses) and functional measurements (the synapse is in activity) is difficult to assess. During early development, synapses are functional before they take on the appearance of synapses, and may not be counted.
2. Purkinje cells or Purkinje neurons are a class of GABAergic neurons located in the cortex, more specifically in the ganglionic layer. They are inhibitory neurons that regulate complex and coordinated movements, for example, by preventing too abrupt a movement. They are named after their discoverer, the Czech anatomist Jan Evangelista Purkyně, who identified them in 1837.
3. For example, see Kusano, K., Miledi, R., Stinnakre, J. (1977) Acetylcholine receptors in the oocyte membrane. In Nature, 270, pp. 739–741.
4. On the whole subject see Jean-Pierre Changeux (2008) cit, trad it. Il bello, il buono, il vero. Un nuovo approccio neuronale ed Raffaello Cortina. Cap 4 pp. 215–252.
5. For a discussion of these recent findings see: Early expression of specific genes in certain cortical areas: Kwan et al. (2012); Sun et al. (2005). Early origin of brain asymmetry: Dubois et al. (2009); Leroy et al. (2015). Brain asymmetry in right- and left-handed men: Sun et al. (2012). Model of self-organising cortex folds: Lefevre, Mangin (2010).
6. Grid cells in rats: Banino et al. (2018); Brun et al. (2008); Fyhn et al. (2004); Hafting et al. (2005). Assumption of grid cell selforganisation: Kropff Treves (2008); Shipston-Sharman, Solanka, Nolan (2016); Widloski, Fiete (2014); Yoon et al. (2013). Rapid emergence of position, grid and head cells: Langston et al. (2010); Wills et al. (2010). Grid cells in the human brain: Doeller, Barry, Burgess (2010); Nau et al. (2018).

References

Bourgeois, J. P. et al. (1994). Cereb. Cortex, 4

Bourgeron T. (2016) Current knowledge on the genetics of autism and propositions for future research. C R Biol.

Chan, Y. M., Jan, Y. N. (1999). Curr. Opin. Neurobiol

Changeux, J. P., (2008) Du vrai, du beau, du bien. Une nouvelle approche neuronale. Odile Jacob

Damasio, A. (2010) Self comes to mind: Constructing the conscious brain. Pantheon

Damasio, A.R., van Hoesen, G.W. (1983), Emotional disturbances associated with focal lesions of the limbic frontal lobe, in Neuropsychology of Human Emotion, a cura di K.M. Heilman e P. Satz, The Guilford Press, New York

Dehaene-Lambertz, G. et al. (2002) Functional Neuroimaging of Speech Perception in Infants.Science298 https://DOI:10.1126/science.1077066

Dehaene-Lambertz, G., Dehaene, S., Anton, J.-L., Campagne, A., Ciuciu, P., Dehaene, G.P., Denghien, I., Jobert, A., LeBihan, D., Sigman, M., Pallier, C. and Poline, J.-B. (2006) Functional segregation of cortical language areas by sentence repetition. Hum. Brain Mapp., 27: 360–371. https://doi.org/10.1002/hbm.20250

Dehaene, S. (2019) How We Learn: the New Science of Education and the Brain – Penguin

Diamond, M. C. (2001) Response of the Brain in Enrichment, in Anais da Academia Brasilera de Ciencias

Di Salvo, M. (2024) Psicoterapia e plasticità cognitiva e neuronale. Cervello sociale e disturbo post traumatico da stress. - Ricerca Psicoanalitica Journal of Relations in Psychoanalysis

Edelman, G. (1987). Darwinismo neurale. Tr.it. Einaudi, Torino 1995.

Elman, J. L., Bates, E. A., Johnson, M. H., Karmiloff-Smith, A., et al. (1996) Rethinking innateness: A connectionist perspective on development. The MIT Press.

Fransson, P., Skiöld, B., Horsch, S., Nordell, A., Blennow, M., La-Gercrantz, H., Aden, U. (2007) Resting-state networks in the infant brain. In Proceedings of the National Academy of Sciences

Gould, E. et al. (1997). J. Neurosci.

Gould, E. et al. (1998). Proc. Natl. Acad. Sci. USA, 95

Gould, E. et al. (1999a). Nat. Neurosci.

Gould, E. et al. (1999b). Science, 286

Hamburger V, Levi Montalcini R. (1949) Proliferation, differentiation and degeneration in the spinal ganglia of the chick embryo under normal and experimental conditions. J Exp Zool.

Hannagan T., Nieder A., Viswanathan P., Dehaene S. (2018) A random-matrix theory of the number sensePhil. Trans. R. Soc. B37320170253 http://doi.org/10.1098/rstb.2017.0253

Hannula-Jouppi K, Kaminen-Ahola N, Taipale M, Eklund R, Nopola-Hemmi J, Kääriäinen H, Kere J. (2005) The axon guidance receptor gene ROBO1 is a candidate gene for developmental dyslexia. PLoS Genet

LeDoux, J. (2002) Synaptic Self: How Our Brains Become Who We Are - Penguin, NY

Levi Montalcini R, Visentini, R. F. (1938) Boll. Soc. ital. Biol. Sper.

McEwen, B. S., Schmeck, H. M. (1994). The Hostage Brain. Stati Uniti: Rockefeller University Press

Meech, R. W. (1979) “Membrane potential oscillations in molluscan ‘burster’ neurones”. In Journal of Experimental Biology

Miyashita-Lin, E. M. et al. (1999). Science, 285

Nieder A, Dehaene S. (2009) Representation of number in the brain. Annu Rev Neurosci. 2009; 32: 185–208. https://doi:10.1146/annurev.neuro.051508.

Nottebohm F. (1989) From bird song to neurogenesis. Sci Am.

O’Leary, D. D. (1992). Curr. Opin. Neurobiol.

Oppenheim, R. W. (1998). In Zigmond, M. (a cura di), Fundamental Neuroscience. Academic Press, San Diego

Posner, M.I. e Petersen, S.E. (1990), The attention system of the human brain, in Annual Review of Neuroscience

Prigogine, I. (1982) Le strutture dissipative. Auto organizzazione dei sistemi termodinamici di non equilibrio, Sansoni, Firenze, consultato in italiano.

Quartz SR, Sejnowski TJ. The neural basis of cognitive development: a constructivist manifesto. Behav Brain Sci. 1997 Dec; 20(4): 537–56; discussion 556-96. https://doi:10.1017/s0140525x97001581

Rakic, P. (1992). Science, 258

Rakic, P. (1995). Trends Neurosci

Rakic, P. et al. (1986). Science, 232

Raper, J. A., Tessier-Lavigne, M. (1998). In Zigmond, M. (a cura di), Fundamental Neuroscience. Academic Press, San Diego

Robinson, A. (2018) Did Einstein really say that? - Nature. https://www.nature.com/articles/d41586-018-05004-4

Rodier, P. M. (2000) The early origins of autism. Sci Am.

Rechert, H., Simeone, A. (1999). Cur. Opin. Neurobiol.

Schlaggar, B. L., O’Leary, D. D. (1991). Science, 252

Shatz, C. J. (1992). Science, 258

Strumwasser, F. (1965), “The demonstration and manipulation of a circadian rhythm in a single neuron” . In Circadian Clocks. North-Holland Publ., Amsterdam

Terman, J.R., Kolodkin, A. L. (1999). Neuron

Vincent, A., Hertz, D., Petit, C., Kretz, C., Oberlé, I., & Mandel, J. L. (1991). Abnormal pattern detected in fragile-X patients by pulsed-field gel electrophoresis. Nature, 349

CHAPTER 6

Cranio-Cerebellar Features Typically Found in Persons Diagnosed with Autism

Before delving into the most well-established and accredited aetiological theories of the last 30 years, it is necessary to further investigate the cranio-cerebellar characteristics of persons diagnosed with autism.

Several studies (i.e. DiCicco-Bloom et al. 2006) have shown that the cranial circumference of autistic children of preschool age is significantly greater due to an increased amount of white matter and grey matter (Keller, 2003).

In particular, it was found that marked brain growth occurs in the first year of life (Courchesne et al. 2003).

In recent decades, the challenge of research into the brain basis of autism has been met with enthusiasm. The difference between artificial intelligence and human intelligence can ultimately be summarised as follows. If artificial intelligence is exponentially efficient in finding answers in the imponderable set of available data, human intelligence is the only one capable of creativity, i.e. it is the only one capable of imagining questions. This should be taken into account well beyond the discourse we are dealing with, because the ability to ask questions is ultimately the very engine of research and discovery.

The range of factors to consider is vast. Where should one expect to find abnormalities? At the level of individual neurons? At the level of a specialised brain system, such as that for movement, vision, hearing or language? Or should we look at the interconnections between these

M. Di Salvo, *Autism Research between Psychology and Neuroscience*,
https://doi.org/10.1007/978-3-031-68338-1_6

systems, rather than the systems themselves? Should we focus on specific neurochemical transmission systems?

For a long time, the only way to obtain information on brain abnormalities was to observe the brain under a microscope after death. Pioneering work in this field was done by Margaret Bauman and Thomas Kemper (1994) at Harvard Medical School.

The number of brains available for autopsy is small, and among them, the number of cases that cannot be confused with other diseases, such as epilepsy or severe mental retardation, is very small.

At the same time, each brain is a continent to be explored under the microscope, an almost impossible task when it is not guided by prior knowledge and assumptions about what and where to look for (which is why I said in the preface that we need 'big theories').

Nevertheless, detailed observations of cell structure have shown anomalies in certain regions. The most striking ones concerned increased cell density. These anomalies can be dated according to their time of origin and their consequences for development. In this way, Margaret Bauman has dated the brain abnormalities of autism to a gestation stage of up to 5 weeks.

If there is any deviation at an early stage of a developing system, this will lead to 'cascading' effects that may be invisible even under the microscope.

Since autism is a developmental disorder, we can assume that underlying it is not some sudden damage, but an abnormality in brain development that can only manifest itself gradually. While formulating the hypothesis as to what kind of abnormality can cause autism, one must, therefore, keep in mind the normal development of the brain. Since nerve cells follow growth instructions written in genes, abnormalities appear when the programme of a gene fails. Obviously, such failure can arise for many reasons. Compared to the mature brain, the immature brain has a higher cell density and a greater number of synapses per cell. In this sense, the autistic brain resembles an immature brain. Developmental problems may be caused by a failure to stop, rather than start, the growth of connections. If this is the case, it could explain the enigmatic finding that autistic people have larger brains (Hardan et al. 2001).

To date, the most robust recent finding of brain abnormality in autistic persons is that the circumference of the head is larger and the brain is heavier (Bailey et al. 1998).

Generally, an increase in overall brain volume is detected, and in about 30% of cases, the brain is abnormally large. This finding was overshadowed by another subgroup of autistic people who presented with a very small brain (microcephaly). Microcephaly is known to be associated with certain clinical conditions and severe mental retardation. If this group of people, who also suffer from other severe problems, is not taken into account, then the average brain size of autistic people will be well above the normal average (Fombonne et al. 1999).

What implications does the presence of a larger brain have for autism? An important clue is given by the fact that the larger size is not evident at birth, but becomes so in early or later childhood (Lainhart et al. 1997; Courchesne et al. 2001).

The fact that these brains are larger may be due to an overgrowth of connections between nerve cells and a lack of 'pruning'.

During childhood and adolescence, the normal brain goes through phases of growth and successive 'pruning' that lead to reorganisation. This process is controlled by genes, but environmental factors also play a part. At first, synapses proliferate. Then a large number of connections that have just formed are shed (Huttenlocher, 1999).

This two-stage reorganisation suggests the image of a well-maintained garden where plants need to be pruned vigorously after an accelerated regurgitation. Just as in a garden, where rapid growth occurs at different times for different plants, so the time for synaptic growth and pruning is different for different brain regions. It is, therefore, possible that a failure to eliminate overproduction only affects certain brain systems, and not others (Huttenlocher, 1997).

Since we have no direct evidence of the failure of pruning in autism, all we can do is to reason about this possibility and its implications. It is plausible that cognitive processes that are served by brain regions containing too many synapses may be inefficient. One possible scenario is that in the autistic brain, the lack of synaptic thinning may mainly affect top-down processing systems, which depend on efficiently organised retroactive neuronal connections. Bottom-up processing systems depend on afferent neural connections, which are probably drawn earlier and remain more stable during development (Burkhalter, 1993).

Retroactive neuronal connections, on the other hand, may go through various stages of reorganisation, each involving growth followed by thinning. It is plausible that each reorganisation in which overproduction of cells is not followed by elimination makes the system less efficient.

Epilepsy, for example, which is due to an increase in neuronal discharges, may reasonably be the consequence of an increased number of inefficient connections.

An inefficient top-down control system could lead as a consequence to a poor ability to modulate bottom-up systems. This could be a possible cause of many social and non-social features of autism.

One of the main stages of growth and subsequent synaptic 'thinning' may certainly coincide with the 18-month revolution, which marks the end of very early childhood. It may also mark the beginning of new ways of learning by gathering information from others. This form of learning is based on the ability to grasp the intentions of others, and is particularly evident in the way children learn the meaning of words; it is also evident in the way they communicate with peers during social play. The 18-month revolution appears to be absent or delayed in autism, and this may be caused by a lack of reorganisation of specific brain regions that support the ability to 'read minds'.

Since reorganisation phases may involve other brain regions at different times during development, other hypotheses are also possible. The prefrontal cortex is one of the last brain regions to mature. This may imply, rather than the formation of new synapses, a gradual loss and reinforcement of remaining synaptic connections (Casey et al. 2000).

This maturation process may be delayed or disturbed in autism.

If this is the case, this would explain the deficits in executive functions, which we know to be more evident at older ages rather than earlier ages. Bruce Pennington and his colleagues, who had shown much evidence of executive dysfunction in autism, found no particular signs of this dysfunction in 4-year-olds (Griffith et al. 1999).

This result was confirmed by Geraldine Dawson (2002).

These studies suggest that developmental abnormalities should be followed over time in the same person.

The possibility of observing the living brain while it thinks has revolutionised the study of the mind and brain. The decisive step came with PET.

Positron emission tomography uses radioactive tracers.

When small quantities of radioactive substances, such as Oxygen15 in an aqueous solution, are injected intravenously into the bloodstream, it is possible to follow the traces in the bloodstream and see where this stream goes in the brain by detecting the level of radioactivity. The entire brain is scanned continuously until, after two minutes, the radioactivity decays. The idea is that the neurons need a supply of energy. The blood carries precisely this energy in the form of oxygen and glucose. By detecting where the blood flow is greatest, we can identify where the brain is most active.

Magnetic resonance imaging, or MRI, is newer and has further advantages: it always makes the blood flow visible, but for shorter periods of time, and does so without using radioactivity.

It does, however, have disadvantages. The head is placed inside a powerful magnetic coil, which produces such loud echoes that earplugs are necessary. In addition, the person being scanned must lie absolutely still in a confined space for a period of several minutes to an hour.

Stimuli can be shown on a monitor, and the person is asked to just look at it, or press a button. Due to very small changes in its magnetic properties, oxygen-carrying blood can be detected and it is possible to see into which parts of the brain oxygenated blood is flowing, thus deducing which regions are working most intensively. MRI can also be used to obtain three-dimensional photographic images of the structure of the brain at rest. It is possible to see the shape of the cerebral cortex's circumvolutions, grooves (depressions) and gyri, as well as the structures underneath the cortex.

Radioactivity prevents PET from being used freely, while the magnetism of MRI makes scanning safe, but involves very loud noise. In all scanning experiments, subjects have to remain still in the same position for a long time; when it comes to children, this means that they are usually scanned while sleeping, which makes it difficult to observe brain activity during a particular task.

The development of scanners is only one part of the technical revolution that has made it possible to observe the living brain. The next step was to make sense of the thousands of images that the scanner captured. This is an enormous statistical analysis problem.

The smallest unit is called a voxel and consists of a tiny cube of a few millimetres. The analysis not only has to cover the entire brain, but also has to take into account the fact that people's brains, just like bodies, have different shapes and sizes, even though they have roughly the same structure.

By examining the brains as a whole, so that each brain is statistically transformed to correspond to a standard image, averages of the activations of an entire group of subjects can be obtained. Structural or functional abnormalities can be detected by means of a suitable comparison group, but this too obviously has to be statistically processed so as to match the standard image (Abell et al. 1999; Salmond et al. 2003).

This technique, when applied to blood flow measurements, is called 'statistical parametric mapping'. In the technique known as 'voxel-based morphometry', an average brain, consisting of thousands of these small cubic images, is processed for one group of people and then compared with the average brain of another group. Any anatomical differences obtained are independent of brain size (Ashburner, 2000).

Beyond these technical and practical difficulties, the results of neuroimaging studies are still difficult to interpret because we still know very little about the functions of most brain regions. A major limitation is the fact that the increased blood flow occurs in large areas of the brain and affects millions of neurons.

Studying the activity of individual neurons with special probes, which is generally only possible with animals, provides essential information on how brain cells actually work. Despite these difficulties, the use of neuroimaging has led to new insights into cognitive processes. For example, it has been possible to confirm that the brain's executive functions are distributed throughout the frontal cortex; it has also been possible to identify certain brain areas that process specific information, such as faces. Most of this knowledge is still preliminary, but the techniques have already been applied to autism.

There are essentially two approaches for choosing relevant tasks for a study of autism using functional neuroimaging techniques. One approach is to identify a brain region of interest and then select a task that is already known to be able to elicit activity in that region. The other approach is to identify a cognitive process of interest and then select tasks that have positively correlated this process with the activation of circumscribed regions.

Obviously, it is problematic to compare the results of neuroimaging studies when some subjects do not perform the task very well.

The differences found may simply be related to the level of performance and tell us nothing we do not already know. It is difficult to distinguish cause and effect when a group has difficulty performing the task: the resulting activation may be as much the cause as a simple reflection of the poor performance.

In the studies described above, this problem was avoided by using simple tasks and people diagnosed with autism with an IQ in the normal to high range. Therefore, we still do not know what brain activation we might find in people with more severe forms of autism. Another still unanswered question is how children differ from adults, as research into the neurophysiological effects of development on cognitive functions is still in its infancy.

In all studies to date on mentalisation in persons with autism or Asperger's syndrome, the three regions of the 'mind reading' network were less active. A coincidence is that one of the three components of mentalisation, the one located in the medial prefrontal cortex, the so-called paracingulate sulcus, in studies with structural magnetic resonance imaging (Gusnard et al. 2001; Aylward et al. 1999; Pierce et al. 2001), shows an abnormally reduced grey matter density in autistics. The amygdaloid region, the second component, has also been found to be abnormal in the autistic group, as it shows an increased grey matter density (Baron-Cohen et al. 2000; Zilbovicius et al. 2000; Aznedar et al. 2000; Howard et al. 2000).

The study using animations provided a possible clue as to why 'mind reading' is deficient in autism. The new clue concerns one of the two additional visual regions, which is highly activated in all subjects while watching sequences that cause attribution of mental states. This region is located in the extra-striate visual cortex. In this area, known to be associated with the earliest (bottom-up) stages of perceptual processing, the autistic group in the critical condition showed a completely normal pattern of increased activation. This is in stark contrast to the other regions of the mentalisation network (probably involved in top-down processing), which show reduced activation. In other words, relevant visual information fails to pass through regions that should be able to extract social meaning (Fletcher et al. 1995; Happé et al. 1996; Baron-Cohen et al. 1999; Frith, 2001; Castelli et al. 2002).

As we have seen, correlation measures in the autistic group showed significantly less connection between the extra-striate visual regions and the superior temporal sulcus (STS), one of the three main components of the mentalisation system. There is, therefore, less information connection between these regions than there is in a normal brain.[1]

The classical psychological explanation of information processing was bottom-up. For example, a visual scene would first be analysed in terms of visual features, then these would be combined together into objects. Recent evidence from anatomy, neurophysiology and neural network models suggests that the brain does not work in this way. In addition to neuronal connections forward (feedforward) and corresponding to bottom-up processing, there are also retroactive (feedback) connections corresponding to top-down processing, and these constitute the majority (Douglas et al. 1995).

Incoming information is modified by these retroactive connections (Lamme, 2000).

This requires that there are strong and probably long-range connections between neurons that cross different regions of the brain. Through these connection pathways, signals can be sent that modify neurons located in distant areas. For example, when a person deliberately pays attention to a moving stimulus, neural connectivity is elevated between two areas of the visual cortex, one related to basic visual processing (V2) and the other related to motion processing (V5).

This increase in connectivity is generated by top-down signals from an attention system located in the parietal and prefrontal cortex (Friston, 2000).

Neuroimaging studies during task performance have all shown a reduction in brain activation in areas known to be the most relevant for that task. One way to explain the reduction in activation is to consider the possibility of a reduction in connectivity between neuronal connections.

There is also the possibility—argued, for example, by Uta Frith—that in autism lower-level perceptual processes function well, but are not modulated by higher-level processes. Perhaps the blockage in information processing in autistic individuals can be linked to their increased brain size.

The increase in brain size may be caused by an overabundance of synapses that under normal conditions undergo vigorous pruning. If this is the case, retroactive connections may be more affected by this defect than 'forward' connections, which develop earlier and are more stable. A failure in synaptic pruning could cause abnormalities in the flow of information, and this could then be associated with an inadequate top-down control system.

The research by Robert Emerson and colleagues (2017) also revealed a diagnostic use of fMRI, which would make it possible to identify children at high risk of developing autism as early as the first few months of life, i.e. long before symptoms appear.

The study published in 'Science Translational Medicine' opens up a new diagnostic perspective, given that there is currently no way to detect signs of the disorder before the first symptoms appear around the age of two.

"Some studies have shown that the neuroanatomical abnormalities typical of autism are structured before symptoms emerge", Diana Bianchi explained in Science "If future studies confirm our findings, instrumental detection of these differences will allow doctors to diagnose and treat autism much earlier than is possible now".

Emerson and colleagues focused, in particular, on the connections between the brain regions most directly affected by autism spectrum disorders, i.e. those that oversee language and social relationships, by performing functional magnetic resonance imaging scans on 59 6-month-old infants while they slept. The babies were considered to be at high risk for autism because they all had an older autistic brother or sister. Of these, 11 were later diagnosed with autism at the age of 2 years.

The authors then used a machine learning technology that could autonomously learn to observe differences in brain scan images of autistic people compared to another non-autistic group to predict future diagnoses. The method correctly identified 9 of the 11 children with autism in Bianchi's study, demonstrating 82% sensitivity, and correctly identified all those who did not subsequently develop the disorder. In a second analysis conducted on other cases, the sensitivity of the test was 93%.

In total, 974 abnormalities in neural connections associated with autistic behaviour were identified. The results are now expected to be repeated and validated on much larger samples before the method can enter clinical practice.

Although these are undoubtedly preliminary results, the study suggests that in the future, neuroimaging could be useful in diagnosing autism or helping healthcare professionals assess children's risk of developing the disorder.

Note

1. G.R. Fink, P.W. Halligan, J.C. Marshall, C.D. Frith, R.S.J. Frackowiak, R.J. Dolan, (1997) Neural mechanism involved in the processing of global and local aspect of hierarchically organized visual stimuli, in «Brain», 120.

H. Ring, S. Baron-Cohen, S. Wheelwright et al., (1999) Cerebral correlates of preserved cognitive skills in autism - a functional MRI study of Embedded Figures Task performance, in «Brain», 122.

R.J.R. Blair, N. Smith, F. Abell, L. Cipolotti, (2002) Fractionation of visual memory: Agency detection and its impairment in autism, in «Neuropsychologia», 40.

A. Puce, T. Allison, M. Asgari, J.C. Gore, G. McCarthy, (1996) Differential sensitivity of human visual cortex to faces, letter strings and textures: A functional magnetic resonance imaging study, in «Journal of Neuroscience», 16.

N. Kanwisher, J. McDermott, M.M. Chun, (1997) The fusiform face area: A module in human extrastriate cortex specialized for face perception, in «Journal of Neuroscience», 17.

R.T. Schultz, I. Gauthier, A. Klin et al. (2000) Abnormal ventral temporal cortical activity during face discrimination among individuals with autism and Asperger's syndrome, in «Archives of General Psychiatry», 57.

H. Critchley, E. Daly, E. Bullmore et al. (2000) The functional neuroanatomy of social behaviour: Changes in cerebral blood flow when people with autistic disorder process facial expressions, in «Brain», 124.

References

Abell, F., Krams, K., Ashburner, J. et al. (1999) The neuroanatomy of autism: A voxel-based wbole brain analysis of structural scans, in «Neuroreports, 10

Ashburner, J., Friston, K.J. (2000) Voxel-based morphometry - the methods, in «Neuroimage», 11

Aylward, E.H., Minshew, N.J., Goldstein, G., et al. (1999) MRI volume of amygdala and hyppocampus in non-mentally retarded autistic adolescents and adults, in «Neurology», 53

Aznedar, M.M., Buchsbaum, M.S., Wei, T.C. et al. (2000) Limbic circuitry in patients with autism spectrum disorders studied with positron emission tomography and magnetic resonance imaging, in «American Journal of Psychiatry», 157

Bailey, A., Luthert, P., Dean, A. et al. (1998) A clinicopathological study of autism, in «Brain», 121

Baron-Cohen, S., Ring, H.A., Wheelwright, S. et al. (1999) Social intelligence in the normal and autistic brain: An fMRI study, in «European Journal of Neuroscience», 11

Baron-Cohen, S., Ring, H., Bullmore, E., Wheelwright, S., Ashwin, C., Williams, S., (2000) The amygdala theory of autism, in «Neuroscience and Biobehavioural-Reviews», 24

Bauman, M.L., Kemper, T.L. (1994) Neuroanatomic observations of the brain of autism, in T.L. Kemper (a cura di), The Neurobiology of Autism, The Johns Hopkins University Press, Baltimore

Burkhalter, A. (1993) Development of forward and feedback connections between areas V1 and V2 of human visual cortex, in «Cerebral Cortex», 3

Casey, B.J., Giedd, J.N., Thomas, K.M., (2000) Structural and functional brain development and its relation to cognitive development, in «Biological Psychology», 54

Castelli, F., Frith, C.D., Happe, F., Frith, U., (2002) Autism, Asperger syndrome and brain mechanism for the attribution of mental states to animated shapes, in «Brain», 125

Courchesne, E., Karns, C.M., Davies, H.R. et al. (2001) Unusual brain growth patterns in early life in patients with autistic di-sorders: An MRI study, in «Neurology», 57

Courchesne, E., Carper, R., Akshoomoff, N. (2003). Evidence of brain overgrowth in the first year of life in autism. Journal of the American Medical Association

Dawson, G., Munson, J., Estes, A. et al. (2002) Neurocognitive function and joint attention ability in young children with autism spectrum disorders versus developmental delay, in «Child Development», 73

DiCicco-Bloom, E., Lord, C., Zwaigenbaum, L., Courchesne, E., Dager, S.R., Schmitz, C., Schultz, R.T., Crawley, J., Young, L.J. (2006). The developmental neurobiology of autism Spectrum disorder. Journal of Neuroscience. 26

Douglas, R.J., Koch, C., Mahowald, M., Martin, K.A.C., Suarez, H.H., (1995) Recurrent excitation in neocortical circuits, in «Science», 269

Emerson, R. W. *et al.* (2017) Functional neuroimaging of high-risk 6-month-old infants predicts a diagnosis of autism at 24 months of age. Sci. Transl. Med.9

Fletcher, P.C., Happé, F., Frith, U. et al. (1995) Other minds in the brain: A functional imaging study of «theory of mind» in story comprehension, in «Cognition», 57

Fombonne, E., Rogé, B., Clavierie, J., Courty, S., Fremolle, J. (1999) Microcephaly and macrocephaly in autism, in «Journal of Autism and Developmental Disorders», 29

Friston, K.J., Büchel, C. (2000) Attentional modulation of effective connectivity from V/2 to V/SMT in humans, in «Proceedings of the National Academy of Sciences of the United States of America», 97

Frith, U. (2001) Mind blindness and the brain in autism, in «Neuron», 32

Griffith, E.M., Pennington, B.F., Wehner, E.A., Rogers, S.J. (1999) Executive functions in young children with autism, in «Child Development», 70

Gusnard, D. A., Abudak, E., Shulman, G.L., Raichle, M.E. (2001) Medial prefrontal cortex and self-referential mental actity: Relation to a default mode of brain function, in «Proceeding of the National Academy of Sciences», 27

Happé, F., Ehlers, S., Fletcher, P.C., et al. (1996) «Theory of mind» in the brain. Evidence from a PET scan study of Asperger syndrome, in «Neuroreport», 8

Hardan, A.Y., Minshew, N.J., Mallikarjuhn, M., Keshavan, M. S. (2001) Brain volume in autism, in «Journal of Child Neurology», 16

Howard, M.A., Cowell, P.E., Boucher, J. et al. (2000) Convergent neuroanatomical and behavioural evidence of an amygdala hypothesis of autism, in «Neuroreport», 11

Huttenlocher, P. R. (1999) Dendritic synaptic development in human cerebral cor-tex: Time course and critical periods, in «Developmental Neuropsychology», 16

Huttenlocher, P. R., Dabholkar, A.S., (1997) Regional differences in synaptogenesis in human cerebral cortex, in «Journal of Comparative Neurology», 387

Keller, F., Persico, A.M. (2003). The neurobiological context of autism. Molecular Neurobiology. 28

Lainhart, J.E., Piven, J., Wzorek, M., et al. (1997) Macrocephaly in chilaren and adults with autism, in «Journal of the American Academy of Child and Adolescent Psychiatry», 36

Lamme, V.A.F., Roelfsema, P.R. (2000) The distinct mode of vision offered by feedforward and recurrent processing, in «Trends in Neurosciences», 23

Pierce, K., Müller, R.A., Ambrose, J., Allen, G., Courchesne, E., (2001) Face processing occurs outside the fusiform «face area» in autism: Evidence from functional MRI, in «Brain», 124

Salmond, C. H., Ashburner, J, Friston, K. J., Gadian, D. G., Vargha-Khadem, F. (2003) Investigating individual differences in brain abnormalities in autism, in «Philosophical Transactions of the Royal Society, London. B. Biological Sciences», 358

Zilbovicius, M., Boddaert, N., Belin, P., et al. (2000) Temporal lobe dysfunction in childhood autism: A PET study, in «American Journal of Psychiatry», 157

CHAPTER 7

From Criteria to Diagnostic Evaluation: Autism in DSM-5 and DSM-5 TER

In order to define a 'state of the person', it is necessary to start by clarifying the definition criteria. The effort is not only scientific, but above all linguistic, and as I said at the beginning, language, syntax and sentence construction, what we express is the reality in which we live.

In this chapter, I will first try to describe the criteria of 'medical diagnosis' currently adopted and the diagnostic instruments.

In the chapters immediately following, I will try to analyse the 'epidemic' phenomenon that has been perceived in recent years, when it seemed that autism had exploded, and the enormous progress in defining the subject by the movement of families of autistic persons and the definition of the concept of neurodiversity by Judy Singer.

It will be a tortuous path because we are talking about people, and not guinea pigs or patients. Because theirs are true stories, and not anonymous clinical reports or made-up stories.

It will be even more tortuous because we are used to the use of certain words, which have shaped our perception, and it is difficult to change habits.

I have personally known elderly people of great culture who had lived through the Second World War and colonialism still use the word 'nigger' out of habit, without any racist intention.

All this does not justify, but should make us reflect and take responsibility: when we create a 'way of saying' that expression becomes a 'way

M. Di Salvo, *Autism Research between Psychology and Neuroscience*,
https://doi.org/10.1007/978-3-031-68338-1_7

of being' in a society, and when these concepts are learnt as children, it is extremely difficult to eradicate them.

Perhaps this is also one of the gifts my friend Bruno gave me: he was only 'a little slower to write' in class, and therein lay the only difference between us.

From DSM-5 to DSM-5 TER

For the medical–clinical definition, I will refer to the definition of autism that emerges from the DSM-5 TER, the most recent version, and I will refer to the Italian reference edition.

DSM-5 is the acronym by which the edition is identified, and the changes approved by the APA on 1 December 2012 and therefore 'in force' from 2013.

Already with DSM-5 (APA, 2013), important changes have emerged compared to the previous diagnostic manuals. First of all, the multiaxial system is removed and by switching to a dimensional diagnosis, disorders begin to be considered as autonomous and independent 'dimensions'.

In this way, it is not only possible to assess whether a person has a particular disorder or not, but it is also possible to quantify its level of severity, thus enabling a more comprehensive individual diagnostic profile.

One begins to consider autism as a continuum of various levels of severity, giving more importance to individual relativity and the support needed by the person in various contexts. The various specifiers and the degree of support are described according to the three levels of severity, i.e. 'very significant support is needed', 'significant support is needed' and 'support is needed' (APA, 2013, p. 60). This is because the manifestations of the disorder vary widely depending on the severity of the autistic condition, the surrounding environment and the person's own development. The term spectrum, introduced in this manual, refers precisely to these variable characteristics, which depend on each person's particular circumstances.

It would also appear that the new criteria have provided a different and broader concept of autism than the DSM-III-R (APA, 1987), also signalling a concern for younger children. Various studies had noted the tightening of criteria for access to services as the nature of the diagnostic concept changed (Mcpartaland et al. 2012). This is because younger children may be particularly vulnerable to the limiting requirements of impairment of all three social communication symptoms and two restriction/repetition symptoms. The broadening of diagnostic boundaries and their lack of applicability also affected more cognitively able persons, including those diagnosed with Asperger's syndrome.

Following various observations, the APA commissioned two hundred experts to revise the manual, thoroughly analysing the literature of the last nine years and attempting to eliminate certain parts of the text that were considered obsolete (Bradley, Noble and Hendricks 2022). Four cross-cutting review teams (gender and sex, suicide, culture and forensics) were involved in revising all chapters to ensure good inclusion and avoid risk factors such as racism, discrimination and stigmatisation. The text was also revised by a working group focused on ethno-racial inclusion to achieve appropriate use of language.

In the DSM-5-TR (APA, 2022), we can see new diagnoses and changes to disorders with some coding updates. These renewals of diagnostic criteria and text in the DSM-5-TR are the result of two separate but concomitant processes, which allow for the addition or deletion of disorders and specifiers: an iterative revision process and the modification of diagnostic criteria on an ongoing basis.

In the 2013 manual, Autism Spectrum Disorder was defined by significant difficulties in the social use of verbal communication and interaction in multiple contexts (Criterion A). Criterion B, on the other hand, describes restricted and repetitive patterns of behaviour, interests or activities. While the minimum threshold for the Criterion B component was simple (at least two out of four), the minimum number required for deficits in social communication could be misinterpreted in DSM-5. In particular, the phrase 'as manifested by the following factors' (APA, 2013, p. 57) could be interpreted as 'any of those listed' or 'all of the criteria' (three out of three). As a result of this inaccuracy, the first change that can be noted in the new 2022 manual concerns 'Criterion A'; three areas of deficits are listed: deficits in social-emotional reciprocity, deficits in non-verbal communicative behaviour and deficits in the development, maintenance and understanding of relationships. In contrast to the DSM-5, all listed symptoms must be fulfilled to diagnose autism. The modification of this criterion allowed for greater clarity by changing the phrase to "as manifested by all of the following factors, present now or in the past" (APA, 2022, p. 66).

In both the DSM-5 (APA, 2013) and the DSM-5-TR (APA, 2022), autism spectrum disorder is placed in the category of 'Neurodevelopmental Disorders', i.e. those disorders that typically occur in the early period of a child's life. They are characterised by differences in brain processes or developmental deficits that cause significant impairment of tasks in the various contexts of life. In this category of disorders, one can go from more specific impairments to global limitations in social or intellectual abilities. Thanks to the recent dimensional approach, these disorders are no longer defined by static categories; this also makes it possible to determine a great deal of individual variability, dictated by the severity of the disorder and the presence of certain specifiers. Very often, there is comorbidity involving the overlapping of several neurodevelopmental disorders; for example, autism spectrum disorder can occur concurrently with intellectual developmental disorder, anxiety disorders, attention deficit/hyperactivity disorder (ADHD) and other disorders (APA, 2022).

The first information concerning autism to be found in both manuals concerns the diagnostic criteria, useful for diagnosing the disorder, which are described by five well-defined criteria. As can be read in the DSM-5-TR (APA, 2022), 'Criterion A' is characterised by persistent difficulty in communication and social interaction in several domains. This characteristic is described, in turn, by three points:

1. deficits in social-emotional reciprocity (*inability to respond to social interactions, lack of interest,* etc.);
2. deficits in non-verbal communicative behaviours useful for social interaction (*abnormalities in eye contact, body language, lack of facial expressiveness,* etc.);
3. deficits in the development, maintenance and understanding of relationships (*difficulties in making friends, lack of interest in peers* etc.).

Next, we find 'Criterion B', which is described by restricted, repetitive patterns of behaviour, interests or activities. Also in the DSM-5-TR, the presence of at least two of the following symptoms is required to fulfil this criterion:

1. movement, stereotyped or repetitive use of language and objects;
2. insistence on sameness of routine or certain verbal and non-verbal behaviour;
3. very limited interests that are abnormal in depth and intensity;
4. hyper- or hyporeactivity in response to sensory stimuli in the environment or unusual interests in certain aspects of the environment (APA, 2022).

Criterion C', on the other hand, specifies that these symptoms must occur in the early phase of development.

The last two criteria emphasise the great limitation that this disorder causes in the individual's various areas of life (Criterion D) and how "*these impairments are not best explained by intellectual disability (intellectual developmental disorder) or global developmental delay. Intellectual disability and autism spectrum disorder often occur concomitantly; to make a diagnosis of comorbidity of autism spectrum disorder and intellectual disability, the level of social communication must be lower than what is expected for the level of general development*" (APA, 2013, p. 58), made explicit by "Criterion E", also described in the new 2022 manual. In this specific point, the difference that can be noted between the two diagnostic manuals concerns the conditions for fulfilling 'Criterion A'. In the DSM-5 (APA, 2013), in fact, the phrase "as manifested by the following factors" (APA, 2013, p. 57) could be ambiguous, creating a misunderstanding as to whether the diagnosis was fulfilled. This description could be interpreted as 'all criteria' or 'any of those listed'. Consequently, in the DSM-5-TR, an improvement in the understanding of this point can be seen due to the phrase: "*as manifested by all of the following factors, present at present or in the past*" (APA, 2022, p. 66), which allows for immediate understanding. In both manuals, there is a specific note following the description of the diagnostic criteria, in which it is made explicit that "*the level of severity is based on impaired social communication and restricted, repetitive patterns of behaviour*" (APA, 2013, p. 58). This specification refers to a particular table that emphasises the level of support needed by the individual to fulfil everyday tasks based, precisely, on the severity of his or her condition. This support is described in three levels, which vary according to the person's communication and behavioural abilities, namely: 'very significant support is needed', 'significant support is needed' and 'support is needed' (APA, 2013, p. 60).

In both DSMs, there is a clarification emphasising that individuals with an established diagnosis of autistic disorder, Asperger's syndrome or developmental disorder not otherwise specified in DSM-IV (APA, 1994) should receive the diagnosis of autism spectrum disorder. Great care must be taken not to confuse social communication disorder with autistic disorder, placing great emphasis on meeting the criteria of the latter condition. In addition, specifics are added to the diagnosis; it is emphasised whether there is an intellectual or communicative impairment associated with the disorder, whether there is another neurodevelopmental disorder, another medical condition or the presence/absence of catatonia. These clarifications should not be underestimated in the overall description of the person, both to determine the right support but also to organise possible future treatment.

After the diagnostic criteria come the 'recording procedures', in which it is explained how to record the severity of the autistic condition and any concomitances with other disorders. In DSM-5, as in DSM-5-TR, it is made explicit that severity is recorded according to the level of support required for each of the two domains (Criteria A and B). It is specified whether the autism spectrum disorder has an associated intellectual impairment or not. Language impairment should then be recorded, specifying the current level of verbal functioning. The presence/absence of catatonia is recorded separately from the condition.

In the DSM-5-TR, clarifications are added regarding associations of the disorder with medical conditions, environmental factors or with a neurological, mental or behavioural developmental problem. In the new 2022 manual, it is made explicit that these particular specifiers apply when the listed conditions or problems are potentially relevant to clinical care. Clearly, if a particular associated problem (*neurodevelopmental, mental, behavioural,* etc.) meets the criteria for another disorder, it must be diagnosed in turn. For the registration of these specifiers, the autistic condition is named, followed by the associated disorder or factor (*e.g. Autism Spectrum Disorder associated with Rett syndrome*).

Severity specifiers are used to concisely describe the person's current symptoms, keeping in mind that the intensity of these criteria is context-dependent and not constant over time. This is because the autistic condition changes during the various stages of life and environment, being a variable disorder according to the various developmental stages. On this specific point, the two diagnostic manuals are rather similar; it is made explicit that the two main criteria (social communication and repetitive behaviour) are to be assessed separately and not as a whole, as these two elements are diagnosed independently of each other. The severity levels of these two criteria are not described and assessed for possible fulfilment of the use or provision of support services. It must be remembered that this kind of support can only be assessed on an individual level, based on the person's specific abilities, personal goals and current stage of development. This is to ensure adequate support and satisfaction in daily tasks by identifying specific personal services. The specification of the (often inhomogeneous) intellectual profile is also necessary to create a more comprehensive diagnostic picture in which strengths are determined on the basis of verbal and non-verbal abilities, both in children and adults. The same importance is given to language impairment in which receptive and expressive abilities, which may not coincide, are described separately. In some cases, the receptivity of language might be impaired but production might be good (*e.g. fluent speech*). The specifier 'associated with a known medical or genetic condition or environmental factor' may be used if the person has a known medical condition (*e.g. epilepsy*), a known genetic condition (*e.g. fragile X syndrome*) or an environmental factor (*e.g. foetal alcohol syndrome*). An important clarification not present in the previous manual emphasises that this specifier should not be regarded as synonymous with the causation of autism spectrum disorder.

A condition may be indicated as being associated with this disorder when it is considered to be potentially clinically relevant or to inform care and not because the physician is asserting a direct cause (APA, 2022). Also in the explanation of the specifier concerning the association with neurodevelopmental, mental or behavioural problems, additions appear that were not present in the DSM-5 (APA, 2013). It is made explicit that it "may be applied to indicate problems" (*e.g. irritability, sleep problems, self-injurious behaviour or developmental regression*) (APA, 2022, p. 70). If other neurodevelopmental disorders are diagnosed concurrently with autistic disorder, they will clearly be reported as separate diagnoses. A final addition in the new manual concerns a specification towards catatonia, which can also be a comorbid condition with autistic disorder. In addition to the typical symptoms of negativism, mutism and posturing, there could be a worsening of stereotypies, mutism or self-injury that may be part of the symptom set of catatonia in the context of autistic disorder.

In both DSM-5 (APA, 2013) and DSM-5-TR (APA, 2022), "*the essential features of autism spectrum disorder are a persistent impairment of reciprocal social communication and social interaction (Criterion A) and restricted and repetitive patterns of behaviour, interests or activities (Criterion B). These symptoms are present from early childhood and limit or impair daily functioning (Criteria C and D)*" (APA, 2022, p. 70). Thus, all the characteristics that affect the diagnosis of autistic disorder specifically are described. Not much difference can be seen between the two manuals in this description, with the exception of a final part added in the DSM-5-TR in which autism is described as a somewhat fluctuating disorder based on the individual characteristics of each person, which is precisely why the term 'autism spectrum disorder' began to be used to describe this condition. In fact, functional impairment only becomes known as a result of explicit demands from the environment that may be temporarily concealed by interventions and supports. This impairment varies according to the person's developmental stage, gender, autistic condition, etc. As a result, it is possible not to recognise the disorder in those who do not have an intellectual disability or an associated impairment, precisely because the symptoms are less explicit and more subtle. When referring to the deficits of 'Criterion A', one actually refers to less obvious symptoms if the person has good communication skills (*e.g. fluent speech*). On the other hand, the symptoms of this first criterion are pervasive and more long-lasting, which is why more sources of information are needed to obtain an accurate and reliable diagnosis because these specific features also vary according to age, intellectual level, current support, and previous treatment history. Autistic persons may go from total absence of language and total incomprehension to stunted or overly literal and repetitive language. Even when basic language skills are good (*e.g. the person has a good vocabulary*), there is still an impairment in the use of social language. The same applies to deficits in social-emotional reciprocity, i.e. the ability to engage with other people and share thoughts and feelings with them. This characteristic may be more evident in young children; in older persons, however, this difficulty may manifest itself in the elaboration and formulation of complex social responses appropriate to the situation (*e.g. knowing what not to say*). Adults can implement compensatory strategies to better adapt to these situations that are limiting for them. Since they do not have a spontaneous intuition for social situations like others, new situations can create great discomfort and loss of concentration due to thinking about one's actions and what to say. Compensation strategies acquired in a specific social context are not transferred to other similar situations. Non-verbal communication deficits, on the other hand, may include lack of eye contact, atypical use of gestures and facial expressions, unusual prosody, etc. An early feature of the disorder may occur in non-verbal attitudes such as the child's lack of joint attention (*not pointing at the object or not showing the object to someone to share their interest*). Some children, like adults, adopt functional strategies to integrate with their peers but their performance is always inferior to children with typical development. Deficits in maintaining and understanding relationships also vary by development, gender and culture and are most noticeable in younger children. This is because autistic children tend to avoid shared play and prefer solitary activities instead, very often focusing on certain interests. Social interest may be completely absent, reduced or atypical; an example of this is the one-sided relationship in which the child establishes a friendship for a sharing of particular interests without reciprocity, or the need to play with very rigid rules that cannot be ignored. This inflexibility in play also refers to the second criterion (Criterion B), in which autism spectrum disorder is defined as having restricted and repetitive patterns of behaviour, interests or activities that vary according to the person's characteristics. When talking about stereotyped or repetitive behaviour, reference is made to simple motor stereotypies (*clapping hands or feet*), repetitive use of objects (*putting objects in a certain way*), repetitive speech (*echolalia, repetition of previously heard words*), pronominal inversion (*use of 'you' when talking about themselves*), etc. There may also be an inflexibility in routines that may manifest itself in severe anxiety if a change in one's daily rituals occurs, even if it would seem to be a marginal variation. This inflexibility also affects eating, causing severe food restriction such as, for example, always eating the same thing or eating according to the colour of the food. Autistic persons may also have interests that are abnormal both in intensity and in the object of interest; these deficits may be less noticeable if they are more or less age-appropriate. Some behaviour may be the result of a reaction derived from hyper- or hypo-responsiveness to sensory stimuli, such as extreme and inappropriate responses to specific tactile stimuli or sounds and a consequent apparent indifference to other external stimuli (cold, heat, etc.). If no intellectual disability or language impairment is present, as has already been seen with 'Criterion A', strategies can be exercised and managed to suppress repetitive behaviour in public that is usually used by autistic persons to achieve a calming or anxiolytic effect. Furthermore, with the right support and focus, special interests can become a source of future employment as well as a source of pleasure in adults. Criterion D, on the other hand, requires that "*the characteristics cause clinically significant impairment of current functioning in social, occupational, or other important areas*". (APA, 2022, p. 72). This makes it explicit whether or not a particular person concerned by the diagnosis presents difficulties in several different everyday contexts, as a result of which he or she has difficulty integrating not only in a specific context but in several aspects of his or her everyday life. The last criterion ('Criterion E'), specifies whether social communication is impaired and not in line with the development of the persona, even though it is very often associated with an intellectual disability. Consequently, the limitations exceed the expected results according to the person's own level of development.

After describing all diagnostic features, the manuals refer to the standardised behavioural diagnostic instruments to obtain the diagnosis. With good psychometric properties, the reliability of the diagnosis is improved over time by means of caregiver interviews, questionnaires and clinical observation measures. In the DSM-5-TR, a small section is added at the end of the description of these diagnostic characteristics that are not present in the DSM-5. It is made explicit that the symptoms of autism spectrum disorder are presented as dimensions and not as categories; consequently, there are no universal 'cut-offs' to define this specific disorder. For this reason, the diagnosis remains clinical, always taking into account all the available information from the people in the person's life. Consequently, reference will not only be made to the score on a questionnaire or an observation but the complete picture concerning the person concerned by the diagnosis will be examined. In the DSM-5-TR, therefore, the relevance of using a dimensional rather than categorical approach is emphasised, making the individual diagnosis more precise and personal according to the identity and characteristics of the person.

When reference is made to the characteristics associated with supporting the diagnosis, those peculiarities concerning the condition that help in confirming and supporting it are taken into account. People with a diagnosis of autism may have language and/or intellectual limitations. Even those with medium/high intelligence, however, have an uneven set of abilities. In fact, the gap between intellectual abilities and adaptive functional abilities is frequently very wide. Very common is the theory of mind deficit, first described in the DSM-5-TR (APA, 2022), i.e. the difficulty in seeing reality from the point of view of a third person and understanding what the other person may be thinking. This obviously has repercussions in social interactions, which are often inadequate. Next, we also speak of central coherence deficits, i.e. the difficulty in understanding the context of reference by focusing on details and understanding concepts in a strictly literal and not general way. This specific difficulty was also not present in the 2013 manual but appears for the first time in the DSM-5-TR. Another recurring limitation for people diagnosed with autism concerns executive function, which does not have its own specificity but manifests itself at a general level. During childhood or adolescence, disruptive or defiant behaviour is more common than in other disorders, followed by catatonic motor behaviour, i.e. the presence of slowing down or static action. This risk of the onset of a catatonic episode is more frequent during adolescence and may include grimacing, muteness, unusual posture, etc. During the adolescent phase, there may also be a risk of the onset of some self-injurious behaviour.

The Diagnostic Assessment

The diagnostic assessment process in autism envisages the use of a multiplicity of instruments that make it possible not only to make a diagnosis but above all to identify a precise functioning profile of the person subjected to the assessment process.

As is clear at this point, there is a great deal of variability among autistic persons by virtue of their level of intellect and functioning and how the peculiarities typical of this condition are integrated with one another.

It is for this reason that the assessment tools are manifold.

Assessment is possible both through structured tests and observation protocols and through functional analyses of behaviour.

It is worth anticipating that since this is a condition that generally manifests itself at a very early age, especially when the autism spectrum disorder is mild or at any rate not accompanied by otherwise debilitating pathologies, through early intervention it is possible to greatly reduce the effects of some of the typical features and sometimes even significantly improve certain aspects of life.

The standardised tests, which have a clinical value, can be further subdivided into symptomatological tests (*capable of identifying the symptoms related to the autism spectrum condition*), tests that assess the adaptive and functioning level (*which allow a picture to be taken of the person based on how his or her abilities enable him or her to cope with the demands of everyday life and to be autonomous*) and tests that allow the intellectual level to be assessed.

Symptomatological Assessment

To diagnose autism, it is necessary to record the presence of specific patterns of behaviour. In order to increase the likelihood of the validity of the diagnosis, instruments have been developed that are able to systematically observe the presence of the characteristic behavioural patterns of autism. The choice of instruments varies depending on the age and characteristics of the person undergoing the diagnostic process: the specialist should be particularly careful to consider the right instrument for the person in front of him/her.

Some of the most commonly used tests are presented below.

Autism Diagnostic Observation Schedule—ADOS

The ADOS (Lord et al., 2000, 2001) is a semi-structured observation protocol capable of detecting the presence of characteristic behaviours typical of the autism condition by precisely investigating the area of communication and social interaction, play and the imaginative use of certain materials. Currently on the market is the second version of the instrument, the ADOS-2, which reflects the diagnostic changes that occurred with the publication of the DSM-5. It can be administered from 12 months onwards and consists of 4 different modules according to language ability and age: modules 1 and 2 are aimed at younger children who do not yet speak or use complete sentences but do not yet possess fluent language, modules 3 and 4 are aimed at children, adolescents or adults who possess fluent language. All modules consist of tests represented by interactive situations of play or social and communicative exchange that aim to observe the behavioural response. The results of the observations provide scores on language and communication, social reciprocity, use of imagination, repetitive behaviour, and the area of interests. The ADOS-2 not only allows a diagnosis to be made but also provides an individualised symptom picture from which, together with information derived from other instruments and observations by caregivers and professionals, an individualised treatment plan can be set up.

Autism Diagnostic Interview-Revised—ADI-R

The ADI-R (Lord et al., 1994) is an interview aimed at the parents, or caregivers, of a person with a suspected diagnosis of autism. The interview with the parents through the ADI-R allows information to be obtained on the child's behaviour, making it possible to record the possible presence of autistic peculiarities relating to language and communication, social reciprocity, repetitive behaviour and the area of interest. The interview takes into account the person's developmental process, so that not only the current behaviour is investigated but the interviewee is also asked to recall moments relating to the different stages of the developmental process. The person falls into the autism category if the scores obtained exceed the required threshold level. A diagnosis can only be made when scores exceeding the cut-off threshold are obtained in all areas. Although the ADI-R may allow information to be obtained such that a diagnosis of autism can be made, it is more frequently used as a tool to supplement the diagnostic assessment of a person already subjected to the ADOS test.

Autism Specrtum Quotient—AQ

The AQ (Baron-Cohen et al., 2001) is a self-administered questionnaire dedicated to the adult population. Although it cannot be considered an instrument that alone is capable of making a diagnosis, it has a good discriminative capacity. In order to take the test, it is necessary to have sufficient intellectual abilities to be able to adequately understand the statements to which it is possible to affix a degree of agreement according to a likert scale. There are 50 items in all and they investigate the sphere of communication and social skills, attention to detail and imaginative abilities, areas in which autistic persons are most impaired.

Cognitive Assessment

Wechsler Scales[1]

For the assessment of the intellectual level, the Wechsler scales, the most widely used in the clinical field, can be used. Currently, the most recent versions of these scales are the Wechsler Adult Intelligence Scale Fourth Edition—WAIS-IV (2013), which can be used with people aged 16–90 years, the Wechsler Intelligence Scale for Children Fourth Edition—WISC-IV, which can be used with children aged 6–16 years, and the Wechsler Preschool and Primary Scale of Intelligence—WPPSI-III (2008), which is for children aged between 2½ and 7 years. The first version of the Wechsler scales dates back to 1939 but many updated versions have followed since then. The initial theoretical approach involved calculating the intellectual level as a score to be identified through the administration of various types of tests that were then compared to a norm. The various subtests could be grouped into two areas, from which two different types of scores were then derived: verbal intelligence (verbal IQ) and non-verbal intelligence (performance IQ). The latest updates of the test have attempted to go beyond the dichotomy between verbal IQ and performance IQ by identifying different cognitive dimensions capable of representing the intelligence of the person in its complexity. The updated versions of the WAIS-IV and WISC-IV make it possible to identify four global areas: the verbal comprehension index, relating to general comprehension ability, the visual-perceptual reasoning index, which refers to the subject's ability to manipulate stimuli for non-verbal problem-solving, the working memory index, relating to the ability to manipulate information in short-term memory and the processing speed index, inherent to cognitive processing ability. In addition to a general score relating to total IQ, each cognitive dimension gets a score. The Wechsler scales provide a particularly nuanced picture of the subject's functioning. Indeed, it is possible to observe the level of ability taking into account the various dimensions as well as having the possibility of observing performance in each individual subtest. It is important to consider that the intellectual profile of autistic persons appears rather uneven. It is possible to identify falling peaks in some areas and above-average results in others. Sometimes, therefore, the overall result is indicative of an intellectual disability but cannot really be considered a representative score of that person's real abilities. In any case, the clinician's interest should be to understand what strengths and weaknesses need to be worked on to improve quality of life and to set up a dedicated intervention programme on the basis of this (and other valuable information).

Leiter International Performance Scale—Third Edition (LEITER—III)

The Leiter International Performance Scale—Third Edition (Roid, Miller et al., 1997) is an instrument that provides an index of a person's overall intelligence through the administration of non-verbal tests, thus measuring fluid intelligence, i.e. independent of cultural, educational and scholastic aspects. It is therefore suitable for all those with language difficulties or cognitive delays that would prevent them from taking a test in which the verbal component is essential for administration. The scale can be administered from 3 to 75 years of age and consists of 10 subtests: 5 of these make up the cognitive scale and provide a global index of non-verbal intelligence, inherent to reasoning and problem-solving skills; the other 5, on the other hand, make up the attention and memory scale.

Adaptive Assessment

Vineland Adaptive Behavior Scales—VABS

The VABS (Sparrow, Balla and Cicchetti, 1984) is among the most widely used instruments for assessing a person's adaptive behaviour from birth to adulthood. It consists of four scales: communication (*receptive, expressive and through writing*), socialisation (*interpersonal relationships, knowledge of and compliance with social rules, leisure time management*), motor skills (*fine and gross motor skills*) and daily living skills (*autonomy in personal, domestic and community activities*).

The assessment of these areas is carried out indirectly by means of a semi-structured interview that is administered to a parent or a person who is well acquainted with the person. The person's adaptive abilities are compared with those of a control group, thus yielding results that allow targets to be worked on in therapy.

Functional Assessment

Psychoeducational Profile—3rd Edition—PEP-3

The PEP-3 (Shopler et al. 2005) is among the most widely used instruments for the functional assessment of children aged 2–7 years with a diagnosis of autism. The test makes it possible to identify the child's abilities and strengths, allowing an individualised intervention programme to be constructed.

The test consists of observing the child's behaviour while performing a number of tests. Each of these represents a test item to which a score of successful, failed or emerging (*in the event that the required skill can only emerge with support*) is attributed. In addition to direct observation of the child, a questionnaire is administered to the parents with the aim of investigating the child's autonomy skills. Overall, the instrument's areas of investigation are those pertaining to the child's level of development ("*cognitive verbal-preverbal*", "*expressive language*", "*receptive language*", "*fine motor skills*", "*global motor skills*", "*visual-motor imitation*") and his behaviour with regard to emotional expression, social reciprocity and peculiar motor or verbal behaviour.

TEACCH Transition Assessment Profile—TTAP

The TTAP (Mesibov et al., 2010) is an instrument that allows the identification of a child's different abilities in various areas of functioning. The test allows for a formal and an informal assessment. In the former case, abilities are detected either through the direct administration of structured tests in the clinical setting, or through the administration to parents, caregivers and/or teachers of a questionnaire that allows for an understanding of the level of competence in the home, school and/or work environment. The areas of investigation of the tests and questionnaires relate to work attitudes, behaviour, independent functioning, leisure skills, functional communication and interpersonal behaviour. Each item can be scored in three different ways: passed (*in which case the skill investigated by the item is possessed by the subject*); not passed (*indicating that the child does not possess the skill investigated or is not yet ready to acquire it*); emerging (*indicating skills that have a way of emerging through support and can be improved*). The informal part of the test involves assessing behaviour and identifying which skills and abilities the person possesses that may be useful in work and residential environments.

Tools for Behaviour Observation

Additional information can be derived from behaviour observation. The preparation of observations, which should involve significant persons in the person's life, can be very useful in the case of adults and becomes even more necessary when it comes to children of developmental age. In this regard, it is possible to ask the people in the person's life to conduct an observational activity that can have various levels of structuring. On the one hand, one can ask for a narrative description of everything one observes at a given time, without referring to any particular coding systems; it is also possible to adopt a more systematic method and make use of more structured formats. In the latter case, a behaviour can be observed in consideration of its frequency of occurrence, the latency time between a possible stimulus and the observed behaviour, duration and intensity. It is clear that the adoption of observational instruments by third persons is essential, especially in cases where the person does not have a great capacity to refer to his or her own experience or where there are difficulties in telling in a clear and comprehensible way what happened at a given moment. In fact, it should be borne in mind that it is the person himself who is the best source of information about himself, and therefore, in the case of adolescents and adults, whose cognitive and functioning level allows it, it is important to be able to interview him directly. This does not mean bypassing relatives altogether, who may still be involved.

Note

1. See:

Wechsler, D. (1989). WPPSI-R Manual: Wechsler Preschool and Primary Scale of IntelligenceRevised. The Psychological Corporation, San Antonio, TX.

Wechsler, D. (1991). Manual for the Wechsler Intelligence Scale for Children-Third Edition (WISCIII). The Psychological Corporation, San Antonio, TX. (tr. it.: Manuale della Scala di Intelligenza Wechsler per Bambini—WISC-III. O.S. Organizzazioni Speciali, Firenze, 2006).

Wechsler, D. (1997). Manual for the Wechsler Adult Intelligence Scale-Third Edition (WAIS-III). The Psychological Corporation, San Antonio, TX.

Wechsler, D. (2003). WISC-IV. Wechsler Intelligence Scale for Children—Fourth UK Edition. Administration and Scoring Manual. Harcourt Assessment, London, UK. (tr. it.: WISC-IV. Wechsler Intelligence Scale for Children—Quarta Edizione. Manuale di somministrazione e scoring. Giunti O.S. Organizzazioni Spoeciali, Firenze, 2012).

Wechsler, D. (2008). WAIS-IV. Wechsler Adult Intelligence Scale—Fourth Edition. Technical and Interpretive Manual. Pearson Assessment.

Wechsler, D. (2013). WPPSI-IV. Wechsler Preschool & Primary Scale of Intelligence—Fourth UK Edition. Technical and Interpretive Manual. Pearson Assessment.

References

American Psychiatric Association (1987) Diagnostic and Statistical manual of mental disorder (3rd ed.—Text revision) Arlington, VA: American Psychiatric Publishing.

American Psychiatric Association (1994) Diagnostic and Statistical manual of mental disorder (4th ed.) Arlington, VA: American Psychiatric Publishing.

American Psychiatric Association (2000) Diagnostic and Statistical manual of mental disorder (4th ed.—Text revision) Arlington, VA: American Psychiatric Publishing.

American Psychiatric Association (2013) Diagnostic and Statistical manual of mental disorder (5th ed.) Arlington, VA: American Psychiatric Publishing.

Edizione italiana: Manuale diagnostico e statistico dei disturbi mentali (2014). Milano: Raffaello Cortina.

American Psychiatric Association (2022) Diagnostic and Statistical manual of mental disorder (5th ed.—Text revision) Arlington, VA: American Psychiatric Publishing.

Baron–Cohen, S., Wheelwright, S., Skinner, R., Martin, J., Clubley, E. (2001). The autism spectrum quotient (AQ): Evidence from Asperger syndrome/high functioning autism, males and females, scientist and mathematicians. *Journal of Autism and Developmental Disorders,*

Bradley, Noble & Hendricks (2022). DSM-5-TR: Salient changes. *The family journal: Counseling and Therapy for Couples and Families*

Lord, C., Rutter, M. e LeCouteur, A. (1994) Autism diagnostic interview-revised. A revised version of a diagnostic interview for caregivers of individuals with possible pervasive developmental disorders. *Journal of Autism and Developmental Disorders*

Lord C., Risi, S., Lambrecht, L., Cook, E.H., Leventhal, B.L., DiLa vore, PC., Pickles, A. e Rutter, M. (2000). The autism diagnostic observation schedule-generic: A standard measure of social and communication deficits associated with the spectrum of autism. *Journal of Autism and Developmental Disorders*

Lord C., Rutter M., DiLavore, P.C. e Risi, S. (2001). *Autism Diagnostic Observation Schedule*, Los Angeles, Calif. Western Psychological Services; trad. it. ADOS: Autism Diagnostic Observation Schedule, Firenze, Giunti O.S., 2005

McPartland, J., Reichow & Volkmar (2012). Sensitivity and Specificity of Proposed DSM-5 Diagnostic Criteria for Autism Spectrum Disorder. *Journal of American Academy of Child & Adolescent Psychiatriy*

Mesibov G.B., Thomas J. B., Chapman S.M. e Schopler E., (2010). TTAP-TEACCH *Transition Assessment Profile*, Firenze, Giunti OS (ed it. a cura di R. Faggioli, T. Sordi e M. Zacchini)

Roid, G. e Miller, L. (1997). Leiter International Performance Scale. Revised, Wood Dale, Ill., Stoelting; trad. it. Leiter R. Leiter International Performance Scale-Revised, Firenze, Giunti O.S., 2002.

Schopler, E., Lansing, M.D., Reichler, R.J., Marcus, L.M. (2005). *Psychoeducational Profile: TEACCH individualized assessment for children with autism spectrum disorders, Austin*, Tex., Pro Ed; trad. it. PEP-3: *Profilo psicoeducativo. Valutazione psicoeducativa individualizzata TEACCH per bambini con disturbi dello spettro autistico, Gussago, Vannini, 2006*

Sparrow, S.S., Balla, D.A. e Cicchetti, D.V. (1984). Vineland Adaptive Behavior Scales, Circle Pines, Minn. American Guidance Service; trad. it. Vineland: Adaptive Behavior Scales. Intervista, forma completa, Firenze, Giunti O.S., 2003

CHAPTER 8

The Infrastructure of Psychological Theories of the Second Autism and an Initial Epidemiology

We have left the aetiology of autism to the vaccine correlation thesis, but chronologically it is necessary to take a step back.

Up to DSM-III, we can speak of "first autism".

The distinction between "first autism" and "second autism" that I adopt in this book, in this chapter and in Chapter 2 was introduced by Bonnie Evans in 2013.

But following the definition of Bonnie Evans (2017), it is necessary to introduce a definition of "second autism".

In Chapter 7 we saw and examined the most recent diagnostic criteria considered by the medical clinic.

But we need to go back and try to gain a better understanding of what the new pathway was historically based on and the theoretical foundations on which the new paradigm was built.

In summary, this chapter will discuss: Lorna Wing's new definition, the new theory of mind, the theory of cognitive dysfunction, the theory of central coherence, the cognitive phenotype theory, the theory of the extreme male brain, the logical-mathematical model, all from the Anglo-Saxon school, and we will see how this new defining model has become epidemic and widespread globally.

M. Di Salvo, *Autism Research between Psychology and Neuroscience*, https://doi.org/10.1007/978-3-031-68338-1_8

The 'Second Autism'

The main feature of the change in the meaning of autism in the 1970s and 1980s was the externalisation of developmental psychology. Meaning was suddenly shifted from internal theories to external programmes and testing mechanisms.

Psychological knowledge about human relationships was no longer to be deduced from models of normal animal instincts but rather imposed by statistical knowledge of when everything went wrong. The concept of autism began to drive psychology rather than vice versa.

Historically, there could not have been a better time for such a change.

Children who would previously have been excluded from society because of their 'mental defect' were fully recognised as 'social beings' and fully integrated into the school system.

Between 1979 and 1981, Lorna Wing presented a critique of almost all previous classificatory systems and divisions drawn within child psychiatry and developmental psychopathology, including the work of Kanner, Heller, Mahler, de Sanctis and Van Krevelen.

Wing proposed to take descriptive psychopathology back to its foundations and reconstruct it around the concept of social disability.

Wing's work can be seen as a proxy critique of the entire field of child psychology and psychopathology. She presented a new argument that all psychological conditions of children should be conceptualised in relation to their social capabilities, as Susan Isaacs had also argued.[1]

Wing's rationale for focusing on social interaction problems was that 'all conditions in which the triad of language and social disorders occurs, regardless of the level of severity, are accompanied by similar problems affecting social and intellectual skills'.

The metamorphosis of autism meant that tests for sensory disorders, perception problems and language disorders were recast as social interaction problems. Different types of disabilities thus became 'social impairments'.

The innovative aspect of Wing and Gould's 1979 study is that it merged the new sociological category of 'social impairment' together with earlier classifications for autism and mental handicap. They thus generated a new model for merging mental handicap and autism into an epidemiological entity. Whereas Rutter, Tizard and Whitmore, in the 1960s and early 1970s, had used the concept of 'social impairment' as a sociological measure in relation to an ability to function in society, Wing turned it into a psychological entity.

Lorna Wing's (1981) 'rediscovery' of Asperger's cases was useful in this process because it helped overturn Kanner's hegemony on the concept of autism in childhood. It also allowed autism to be brought into the field of schizophrenia research.

It is not surprising that autism and schizophrenia have been confused in the past. Progress has been made in separating them and it is important to continue to improve the accuracy of diagnosis, despite the many difficulties encountered in clinical practice.

Whereas sensory impairment, as described in the theories of Hermelin and Uta Frith (1969, 1971) in the 1960s, was very specific in its definition relating to the child's inability to 'perceive' the external world through internal capacities, 'social impairment' expanded the limits of what could be conceived of as falling under that heading. Most importantly, it claimed the study of human relations as the science for the new psychology of autism, rather than a science of human relations of instinct theorists.

By linking cognitive tests for autism, whether for the detection of patterns or language abnormalities, under the umbrella of the concept of 'social impairment', Wing brought them into the field of human relations psychology and gave them a new legitimacy within a broader psychological theory.

Wing and Gould's work on 'social impairment' transformed Maudsley's psychology of autism into a theory of human relations.

Prior to the development of Wing and Gould's tests for social impairments and Newson's (1968) claims that children were biologically and innately social, psychologists trained by the Institute of Psychiatry in London had been limited in their ability to test for 'autism' or lack of social engagement.

Lotter (1966) had provided, as we shall see, epidemiological data that apparently could quantify the problem. Hermelin, O'Connor and Frith measured children's behaviour, the ability to encode and understand the world through language and images.

Rutter (1975) had developed tests to measure the exact impairment of language acquisition and unusual speech in autistics. But no one had yet developed a test for social impairment in all children.

Psychiatric epidemiology argued something different. Behaviourists argued that there was no way to measure human instincts, that it was simply impossible and that therefore psychologists should abandon their goal of exploring them in their theories of socialisation.

Anyone who presented a definition of autism that did not adhere to psychiatric epidemiology in the 1970s and 1980s was widely criticised.

For example, in the early 1970s, Nobel laureate Nikolas Tinbergen collaborated with his wife to develop a theory of autism based on Bowlby's work. He argued that autism was caused by environmental causes, stress within the family, hypothesising that stressful situations caused by 'modern life', such as long car journeys, skyscrapers, blockades and parental abandonment due to work pressures could be 'autismogenic'.

Although the Tinbergens put the word 'autism' in inverted commas, they clearly threatened the cognition and scientists who were working with a model of autism that focused on recognising the impairments and disabilities of autistic people, and thus contributing to the development of better educational and social services.

For the new cognitive autism researchers, Tinbergen syndrome is something else.

In a critical review of Tinbergen's book, Uta Frith (1984) argued that Tinbergen's claims that modern life was harming humanity "of a different class from that normally found in scientific discussions", that the authors failed to take into account the latest models of autism and that "this book presents nothing more than a veneer of science".

This criticism of the scientific credentials of a Nobel Prize winner must be seen in the context of the fact that the new model of autism, epidemiological autism, was advocating a new model of the rights of disabled children. Tinbergen's science had thus become politically problematic.

A New 'Theory of Mind'

The changes introduced by the 1981 Education Act were accompanied by new developments in the conceptualisation of autism within a general framework of theory of psychology. As might be expected, the intellectual environment that had been forged at the Institute of Psychiatry and University College London (UCL) was instrumental in shaping these new intellectual models or 'styles of reasoning'.

These theories would later become internationally dominant in defining a general theory of autism.

In July 1983, Uta Frith began writing notes on the basic definition of autism, stating that Rutter's three criteria for autism were only a working definition of autism and "there is still a need to cognitively relate linguistic/social/stereotypical phenomena in a meaningful way in order to explain the varieties of autistic behaviour".

This was what Wing was working on in her description of Asperger's syndrome.

The fact that there were many varieties of autistic behaviours encouraging a broader conceptual model for diagnosis. Frith's important work was to ask the question: "To what extent are social difficulties part of the cognitive deficit?".

Frith hypothesised that the 'cognitive dysfunction' of autism was due to a 'raw processing loop', or 'processing without representation', in which perceptual stimuli remain 'raw', 'unanalysed' and 'uncoded'. This short-circuit between input and output was also the reason why some autistic people showed excellent motor skills, memory skills and drawing abilities.

Based on Hermelin and O'Connor's (1964) method, Frith reasoned that autistic people perceived all stimuli as being of equal value: they lacked the ability to incorporate 'hierarchical salient features' leading to 'excessive segmentation'.

It was this reasoning that led her to hypothesise that autistic people have "no intentionality, no 'theory of mind', no person awareness, no social faculties".

This would then explain their "lack of pretence play", "inability to form and use symbols" and "inability to communicate linguistically in a sophisticated way".

It means that from the age of nine months to one year and beyond, autistic children show "an inability to consider another person's thoughts" and "an inability to know their own mental state".

When Simon Baron-Cohen, Alan Leslie and Uta Frith (1985) argued that autistic children lacked a 'theory of mind', they built their thesis on the post-1960s conception of autism, describing it as a 'profound disturbance' in understanding and coping with the social environment', in which the main symptom is 'impairment of verbal and non-verbal communication'.

Frith, Baron-Cohen and Leslie took the term 'theory of mind' from the work of Premack and Woodruff (1978).

Leslie conducted work on the abilities of neuronormal two-year-olds to understand make-believe play and argued that autistic children showed deficits in the ability to imagine. Baron-Cohen then posited that autistic children suffered from an impairment in the cognitive mechanism required to 'mentalise' or represent mental states.

This interpretation of autism as a deficit in 'theory of mind' is directly based on Lorna Wing's model of autism as a 'social impairment'.

This theory of autism was only made possible by the integration of all child populations into a theory of general psychology and the consequent removal of theories of adult psychopathology within developmental psychology.

Baron-Cohen, Frith and Leslie relied on Piaget's tests of egocentrism in their proposal that autistic children do not have a 'theory of mind', in particular, the 'three mountains test' (*in which a child had to describe what one might see from the point of view of another person seen on a mountain*). However, based on the work of Peter Hobson,[2] they argued that the three-mountain test only identified 'visual-spatial' whereas their test was able to identify whether or not a child was able to impute 'higher order' mental states to others. The children were then presented with tests using Sally and Anne's model.[3]

This reworking of Piaget's famous test for egocentrism was particularly significant as a reminder that it was Piaget who introduced the concept of autism into developmental psychology. For Piaget, autism existed before egocentrism and was not determined by it, but Frith, Baron-Cohen and Leslie argued that the Sally-Anne test was able to delve deep into a child's mental capacity to understand the thoughts of others and that this was the explanation for all their problems.

In effect, the 'theory of mind' tests were a way of describing an individual's relationship with other minds that completely bypassed the thought processes that could mediate that relationship. This model or 'style' of reasoning about autism represented exactly what was needed at that moment in history.

Frith's work, with Baron-Cohen and Leslie, was important in forging new concepts.

They thought about the problem in relation to broader theories of child development, they reasoned that even in early childhood human beings show species-specific sociality, behaviour in terms of 'interest in people, faces, voices', 'imitation', 'sociability', 'early pointing' and that therefore deficits in primary social responses may be indicative of autism.

However, what they needed to explain was the primary deficit and the correlates of second-order impairments. In their search for the basic features of the perceptual problem in autism, Alan Leslie argued that this should be seen not as a failure of selective attention, but rather as an "unusual awareness" of lower-level perceptual products".

Frith, Baron-Cohen and Leslie were aware that these concepts posed a challenge to psychoanalytic models of development because they argued that social and imaginative thinking skills were not necessarily related to psychological faculties.

While Isaacs, Klein and Anna Freud had built their theories on an edifice of human fantasies, such as drives towards human relationships, Frith and colleagues argued that human relationships depended on an individual being able to adopt another's perspective. Instead of sexual desire, it was the ability to see another's point of view that acted as the glue and created the social dimension and world.

It was no longer just a question of how autism was diagnosed, but also how society perceived people with autism, 'the non-functional subjects'.

Instead of a social model built on the reproduction and exclusion of non-functional elements, Frith and colleagues espoused a social model built on 'social skills' and the willingness to support and encourage such skills.

By the mid-1980s, work on autism had stimulated the study of normal child perception in an attempt to understand the subtle differences that might cause the development of an atypical way. Leslie (1988) became interested in the testing of cause and effect perceptions, employing increasingly sensitive tests to identify perceptions from an early age. He used these studies to demonstrate that even six-month-old infants were capable of perceiving cause and effect, which were then incorporated into the innate visual apparatus.

As with the 'theory of mind' tests, these studies also attempted to overturn Piagetian models of child perception by presenting a model of children's thinking that could be tested and measured by proto-behavioural measures.

The First Epidemiological Studies

It was only after the publication of Wing and Gould's 1979 epidemiological study that international epidemiology and autism studies began to develop, and only after this study did the reported rates of autism began to increase. These increases were associated with a widespread change in the description of the condition.

In 1984, Victor Sanua of the Department of Psychology at St John's University in New York wrote to child neuropsychiatrists around the world asking for an answer to the question, "Is childhood autism a universal phenomenon?".

The Chairman of the Department of Psychology at the University of Sao Paulo, Lourencao Van Kolck, had replied: 'Actually, autism is a rare disease in our country'.

Dr Chakraborty, a professor of psychiatry in Calcutta agreed that even in India cases were rare, and Dr Mohammed Fakr el Islam, who worked in the Middle East, argued that 'cases of autism are very few in this part of the world'.

There was no way to validate these claims, but the point is that there is also no way to confirm that everyone was actually talking about the same thing.

However, Sanua concluded that "*childhood autism seems to be a disease of occidental civilisation, and seems relevant in high-tech countries where the nuclear family dominates*".

The idea that autism was a disease of occidental civilisation appeared increasingly shaky as parent groups began to form and demand that autism be properly investigated in their countries and that the international rights of children be respected.

In the early 1970s, Peter Venables, dean of psychophysiology research, conducted a major longitudinal study on child health and development.

The WHO Scientific Group recommended that any longitudinal study method should be conducted in a developing country and that it should investigate three-year-old children, establish nursery schools for these children and use psychophysiological research to identify risk factors for later mental illness.

Mauritius was chosen as the study country because it was a small island with low emigration. This study by Venables has become a model for health epidemiology in developing countries and a symbol of international collaboration.[4]

In 1989, Autism Europe presented a charter of rights for autistic people.

These included 'the right of persons with autism to live independently' and to 'live a full life to the limit of their potential', 'the right to sexual and other relationships, including marriage, without exploitation or coercion' and 'the right to freedom from fear or threat of unjustified incarceration in psychiatric facilities, hospitals or other restrictive institutions'. The declaration also stated that 'at least one million EU citizens are "affected" by autism, which is a mental disability'.

This was an important document as it recognised that autistic children grow up to become autistic adults and thus extended rights to a wider age group.

In this respect and point of view, autism, as a globalised category, is important because it provides unique rights to those diagnosed with it and also allows parents of children diagnosed with autism to become part of a wider network of global institutions interested in treating developmental problems and conditions of children in general, as well as in providing care and support.

Towards the end of the 1980s, it became evident that researchers in different countries, and even different regions, were giving very different results on the prevalence of autism and different interpretations of what autism was.

When the DSM-IV was published in 1994, it retained the 'triad' for 'autistic disorder' but also included 'Asperger's disorder'.

The criteria for the diagnosis of Asperger's were similar to those for 'autistic disorder', but for Asperger's, they did not contain the 'qualitative communication difficulties' element of the triad. In fact, the criteria specifically state that in Asperger's cases, "there is no clinically significant general delay in language" and "no clinically significant delay in cognitive development", thus in effect encouraging the diagnosis in children who presented only a "qualitative impairment of social interaction" and "repetitive and stereotyped patterns of behaviour, interests or activities".

Theoretically, if a child meets only three behavioural criteria, such as 'failure' to develop developmentally appropriate peer relationships', 'lack of social or emotional reciprocity' and 'persistent preoccupation with certain objects', then he or she meets the criteria for an Asperger's diagnosis.

In 1991, Gillian Baird, Uta Frith, Lorna Wing, Simon Baron-Cohen, Christopher Gillberg, Patricia Howlin and others wrote an open letter arguing that while it was a good thing that autism was recognised as a developmental disorder, and therefore associated with mental retardation and other developmental problems rather than emotional problems, this was overshadowed by the fact that it was classified as a 'pervasive' developmental disorder.

'Pervasive Developmental Disorder' (Waterhouse et al. 1992) was the final term for the definition of 'autism' but was a step towards creating a broader definition of autism.

The DSM-IV resolved many of these concerns by adding Asperger's to the list, thus making it very clear that the pervasive syndrome is a health problem.

The term 'neurodevelopmental disorder' has often been used as a synonym for 'autism spectrum disorder'. At the same time, the term 'neurodevelopment' was increasingly applied to describe all psychological aspects of childhood that were thought to have some correlation with chemicals involved in brain development.

It was only with the increase in diagnoses of autism and related conditions that 'neurodevelopmental' psychology, along with a number of neuroscientific studies, focused on broader problems of social interaction and 'social disability'.

This burgeoning scientific field of genetic and neurochemical research on 'social impairment' championed by Frith, Baron-Cohen and Leslie's 'new Theory of Mind' hypothesis involved attempts to identify 'social phenotypes' within neurodevelopment.

Autism served as the central concept through which such studies were approached.

International studies on autism grew dramatically in the 1990s, and some pointed out that the change in diagnosis seemed to lead to an increase in reported cases, which in turn stimulated further epidemiological studies.

In the late 1990s, Simon Baron-Cohen with Gillian Baird (Baird et al. 2000) used new methods of analysis to argue that, in fact, there were 57.9 cases of children with 'autism spectrum disorders' per 10,000, and the term autism was increasingly being used to refer to almost all 'pervasive developmental disorders'.

Until then, the figures ranged from around 2 to a maximum of 8 cases per 10,000.

After the publication of this study and its conclusions, a flood of international studies confirmed that the rate for all 'autistic spectrum disorders' was very high. A study from the same year in New Jersey recorded a figure of 67.4 per 10,000, a Danish study in 2002 recorded 30 per 10,000, and in the same year a Cambridge study recorded 58.3 cases per 10,000 and an Atlanta study in 2003 recorded 34 per 10,000.

After the UK Department of Health admitted to being puzzled and concerned by these incidence levels Gillian Baird et al. examined the South Thames population in the United Kingdom and came up with an even higher figure, i.e. 116.1 per 10,000.

In 2006, two US studies conducted in different states showed a rate of 66 per 10,000 and 67 per 10,000 respectively, in Denmark 53.3 per 10,000 and in Canada 64.9. In 2008, a Japanese study went as high as 181.1 per 10,000.[5]

Lorna Wing's autism, which Bonnie Evans calls second autism, is largely dependent on the science of epidemiology and has become an international phenomenon thanks to the DSM-III-R and ICD 10 that have followed her definitions. Wing's definition of 'autism', what can be called the 'population-wide definition of autism', was brought forward in the interest of establishing a comprehensive and inclusive epidemiology of child development studies. However, it then triggered a series of international interpretations, analyses, qualifications, explanations and rationalisations.

The first decade of the new millennium witnessed more global reflection on whether or not this epidemic was 'real'.

As Gil Eyal (2010, 2013) and Eric Fombonne (2009) argued, these higher rates were largely a reflection of the fact that children with higher intelligence and lower intelligence, or retardation, were included in the studies.

The development of Simon Baron-Cohen's Checklist for Autism for Younger Children (CHAT) also encouraged earlier diagnosis, serving as a pre-positive purpose. ADOS has improved the techniques used to accurately measure autism, in many ways self-reporting: the more ADOS is used in studies and confirms results, the more it is recognised as an accurate measure of autism, the more it is used in future studies.

Changes in the institutionalisation, schooling and treatment of children with developmental atypicalities have contributed to the increased observation of autism.

Wing, Gould and other psychiatrists developed a way to calculate the 'social impairment' of entire societies. They devised a method to measure the limits of this entity called 'society' through the study of some of its most vulnerable members.

The 'triad of impairments' was a powerful concept of organisation that came to replace the concept of intelligence as a way to measure the limits and potential of intelligence in advanced social organisation. A 'society' is only as good as its ability to identify and assist all its members who are 'socially disabled'.

The Epidemiology of Autism

According to Bonnie Evans, "*low rates of autism have become synonymous with a lack of education about disability rights. This is why many non-Western researchers have tried to emulate the epidemiological rates of autism in industrialised countries*".

The use of diagnostic instruments designed and functional to occidental culture only makes sense in the cultural context of occidental society. Pretending to extend cognitive and behavioural tests as well as relational tests to different cultures risks creating methodological nonsense.

And this risks becoming the legacy of autism spectrum psychology.

Researchers no longer seem interested in intelligences and individual intelligences (Di Salvo, 2024a) but want to know how much 'autism' there is in their nations because it has become a symbol of advanced nations, of altruistic ethics towards children.

It is also a symbol of a new version of humanitarian commitment. Developing countries declare that their autistic population can no longer be hidden, and medical research companies support this approach because it promises to bring new discoveries. According to Evans, "This represents a new version of psychology in a post-colonial context".

This new psychology of autism allows atypical children to have their own autonomy but only through this particular set of psychological schemas, systems and concepts around the idea of the autism spectrum.

China is a case in point where growing media attention and international consensus have encouraged and enabled the adoption of epidemiological methods developed in the United Kingdom.

In 2012, the Chinese versions of the ADI-R and ADOS were produced and approved by the publisher, Western Psychological Services, thus providing the ideal tool for such analysis. This model of psychology is often presented as the basis for humanitarian approaches to psychological conditions in children.

Similar developments have taken place in Brazil. In April 2010, the first national meeting for autism research was held, bringing together representatives of government departments and research centres. Here, the then President of the Department of Psychology at the University of São Paulo, Lourencao Van Kolck, had stated in 1984 that 'autism is actually a rare disease in our country'. Many of the participants at the 2010 meeting concluded instead that 'autism can no longer be considered a rare disorder and more than one million Brazilians are likely to be affected'. They argue that it is therefore a major public health problem.

The extent to which the 'autism spectrum' model overcomes the psychology of human relationships depends on how many children are considered autistic, or how much the category is applied to high-functioning cases.

Lorna Wing's model of the 'autism spectrum' has led to new visions of society, social organisation and social care in developing countries.

The fact that epidemiological studies have detected an increase in autism rates, while genetic studies detect an increasing complexity of the disorder and those on pathogenic elements are still poorly understood, only aggravates the situation and supports the idea of the autism spectrum and encourages further anxieties around 'what is it all about'.

Health care in developing countries had focused on infant mortality and basic childhood illnesses until the early 2000s, when increasing attention to autism as a global health crisis encouraged international organisations to shift their focus to mental health as well.

Some Explanations for the Increase in Autism Cases

To state that the 'autism epidemic' is a historical product is not to say that there are no environmental factors that may influence the development of autism spectrum disorders as they are currently defined.

Gil Eyal's work has been fundamental in exposing the ways in which diagnostic change has influenced the 'autism epidemic' phenomenon.

The main focus of this work, mainly from a historical and sociological point of view, is the investigation of the increase in autism diagnoses, showing in many ways that this was not an epidemic, but rather a change in the conditions, including institutional as well as medical, under which developmental disorders in childhood are perceived and treated.

In the work of Gil Eyal, Brendan Hart, Emine Onculer, Neta Oren and Natasha Rossi, the very category of autism "emerges in the interstice between retardation and mental illness, as an element intended to stitch them together as part of a comprehensive childhood control programme".

In their research in an attempt to "piece together the puzzle of contemporary autism, the best place to start was to ask the question of why autism was rare in the past and with the existence of a population of individuals housed in custodial institutions for the mentally retarded".

If, as Lorna Wing suggested, we focus not on a global diagnosis but on specific handicaps, then there is little reason to distinguish so sharply between mental retardation and autism.

Rather, it was more useful to see them as overlapping conditions, different combinations of symptoms drawn from the same pool of deficits.

Understanding autism as a spectrum meant defining it not as a specific constellation of symptoms, but increasingly as the deficit pool itself. Therefore, multiple combinations of symptoms could occur and be represented as regions of the spectrum.

Gil Eyal also highlights how 'the massive diagnostic substitution in the 1990s gave rise to the perception of an "epidemic" autism. A diagnostic substitution had parental demands as a precondition. The line between autism and mental retardation was blurred, making substitution possible, but it was consequential for an ethical economy of treatment.

"*The treatment of mental retardation is still perceived and institutionally defined as uneconomical, especially after a certain age, whereas the 'critical window' involved in autism, and the new ethical model of intervention it triggers, makes its treatment economical*". Unlike mental retardation, the diagnosis of autism delays the medical diagnosis of social destiny.

However, it is not only the diagnostic change over time but also broader historical changes that have influenced the very thing the researchers claim to be looking for.

As Paul Shattuck (2006) pointed out, in the United States, diagnoses of 'mental retardation' and 'learning disabilities' have decreased while those of 'autism' and 'developmental delay' have increased. However, no one was interested in why 'developmental delay' occurred, while there was almost mass hysteria over the 'autism epidemic'.

This is because autism has become emblematic of children's right to a 'typicality', or rather a statistical 'typicality' of development.

Knowledge about autism is part of a new global project to understand child development. This new project also has the potential to reformulate broader claims about cause and effect in the development of social skills in children. Indeed, it could be argued that there has been so much controversy over the autism 'epidemic' precisely because it was a category that was built on the idea that child psychology could be fully understood through epidemiological research.

The new criteria for diagnosing autism are intimately linked to epidemiological methods.

If Susan Isaacs, Melanie Klein (1961) and other psychologists who espoused the first definition of autism could be criticised for their focus on the impact of instincts on thought processes in infants, one could similarly criticise researchers who focused entirely on epidemiological data without considering the multiple factors involved in their production, on the assumption that numbers speak for themselves and that there is no need to challenge their validity in the search for causal hypotheses.

This is largely due to the fact that the main components of autism have shifted from the study of internal hallucinations to the study of epidemiology itself.

There has been a radical change in the 'reasoning styles' used to understand autism, and this change has affected the types of arguments about causation.

Epidemiological sciences have replaced 'case reasoning', and reflection on individual lives and this has had a great impact on the way autism, and child development in general, has been understood across multiple fields of enquiry.

In this sense—and I agree with Evans—the Wakefield method (the one on the investigation of the correlation between vaccines and autism) with statements explaining an epidemic without any epidemiological support should not be forgotten, as it represents an example of the way in which epidemiological data and 'styles of reasoning' can be distorted far beyond the limits of an investigation.

Brief Note on the Use of AI in Scientific Research

As I have argued elsewhere (Di Salvo, 2024b), it makes no sense to talk about Artificial Intelligence. It would be more appropriate to call it Augmented Computational Capability (*but the name does not lend itself to cyber-utopian Silicon-Valley-style marketing*).

However, it is undeniable that never have we had such a powerful and sophisticated integrated analysis and processing system at our disposal to which we can delegate data analysis.

The additional value of using these technologies in the study and research applied to complex structures such as autism is inherent in the need to create ex ante and in plain text the logical analysis criteria that the machine must adhere to.

This implies that we have to write the criteria for searching inputs, and also the weight to be assigned to the values, the exclusion and control criteria, the relevance parameters, the control parameters... all commands and settings that follow a logic of the type "if...then..."/"if...do..."/ "if... do not do..." etc.

Making these logical decisions also means motivating them, and what is most relevant is precisely the emergence of the motivation of the decision behind the choice of computational command.

In this way, any subjective biases, mental frames and human logical errors would immediately emerge, allowing the analysis to be rethought in a more correct manner.

Mathematical, Logical and Scientific Autism

At the end of the 1990s, researchers working in the field of autism quickly became authorities in the field of all-round child development, when, in fact, their field of study was very specific. An important growth in authority and expansion of the field occurred when new autism researchers quickly had to better clarify the meaning of 'social disability'.

What has happened is that the second autism, formulated through the logic of epidemiological studies in the 1960s on characteristics such as the absence of imagination, imagination and hallucination, has been broadened and globalised to such an extent that its representation as a simple absence or 'social impairment' has become increasingly difficult to justify.

In this social absence, researchers have begun to explain the new abilities and skills that autistic individuals may have.

For instance, it is no coincidence that the idea that autism is characterised by marked mathematical and scientific skills arose in the 1980s in line with the expansion of the Asperger's category and the growth of a new theory of human relationships. It is also no coincidence that this model arose in the context of a mass collection of global epidemiological data on an aspect of child development that would otherwise have been meaningless.

The problem of autism has always been associated with its role as a central organisational concept in broader theories of child development. Gaps in knowledge about this central concept have increasingly been filled with the knowledge produced by the instruments used to measure it.

A perfect example of how some scientists have attempted to mimic and surpass the intelligence test is undoubtedly the creation of Simon Baron-Cohen's (et al. 2001) 'Autism Spectrum Quotient', also known as the AQ, developed as a 'challenge' to the supremacy of the Intelligence Quotient (IQ) as a means of measuring 'the sum of human psychology and its genetic basis'.[6]

Central Coherence

Based on his studies of sense-motor testing and pattern detection in autistic children, Frith had argued that 'central coherence' was a normal cognitive tendency to focus on extracting general meaning and substance or gestalt from information.

Frith and Happé (2001) went on to argue that the autistic inability to extract general information could be termed 'weak central coherence', which refers to a focus on parts, details and particulars rather than on the whole.

Is it possible that the weakness in central coherence in autism is another aspect of executive function disorder? It is important to clarify at the outset that the brain basis of the ability to extract meaning from context is still unknown.

According to Uta Frith, it does not seem forced to link this ability with a component of the supervisory system. For example, in order to make good decisions, the supervisory system needs to integrate information from various sources. Therefore, a component of executive functions may certainly have something to do with multimodal integration.

Perhaps stimulus overload is common due to the accentuation of elementary stimulus processing. Small differences in the quality and quantity of sensory information may remain relevant and impossible to ignore. The failure of mentalisation may cause a lack of general categories in social situations, and this may again contribute to overload.

As we have already seen, it appears that two mechanisms of attention work well in autism: that of focusing on a stimulus and that of maintaining that focus.[7]

Ideally, attention should be a flexible tool that serves as a guide for meaning. For example, after focusing on certain details, this tool should automatically return to the service of the overall view.

The links with the ability to mentalise are less clear. Brain lesions of the frontal lobes can cause patients to fail in executive function tests and theory of mind tests, but performance in these tests is not correlated. This suggests that these tasks are based on independent brain systems. A patient with a lesion in the left amygdala present from birth failed spectacularly in theory of mind tests but did well in a large battery of tests that tested his executive functions. The links between mentalisation and executive functions could, however, exist in the case of normal development.[8]

Josef Perner and Birgit Lang (1999) reviewed various hypotheses regarding possible connections and highlighted the fascinating possibility that self-awareness of mental states is a prerequisite for the emergence of self-control in normal 4-year-olds. They plausibly hypothesise that the more one understands one's own mind, the better one can exercise self-control.

Stem Universe

The autistic inability to extract general information called 'weak central coherence' by Frith, as we have seen, refers to a focus on parts, details and particulars rather than on the whole.

A form of thinking that in itself could be useful, allowing people to focus on details without worrying about the overall gestalt, Baron-Cohen identified engineers, mathematicians and physicists as having autistic traits.

He thus constructed a model of 'autism' that beautifully reflects the tools that have been used to measure it. The upper end of the 'autism spectrum' is now considered to be a condition characterised by mathematical abilities, scientific skills and logical thinking.

This was exactly what the new autism psychologists were always looking for in order to measure quantities and detect reasons for abilities such as logic, competence, rigour, numbers and certainty. Because autism was such an elusive concept, it gave psychologists more room to build a model supported by the tools used to measure it.

Genetic studies supported this epidemiological statistic.

Attempts to delineate the broader 'phenotype' of autism increased, such as in Baron-Cohen's (1997) work on the 'cognitive phenotype' of Asperger's syndrome.

The more children were diagnosed, the more difficult it was to challenge the diagnosis.

An increasing number of studies described different psychological profiles in family members as part of a broader 'autistic phenotype'.

Another convincing argument for some sort of genetic cause came from the fact that all epidemiological studies found a higher percentage of male cases in their samples, and when it came to the broader category of autism spectrum disorders including Asperger's syndrome, the figures were often even higher.

This led Baron-Cohen and Hammer (1997) to develop their theory of autism as being caused by the 'extreme male brain'.

They studied the effects of testosterone on brain development in foetuses and infants to explore why male and female brains develop differently.

One of Baron-Cohen's most formidable experiments involved accessing a bank of frozen amniocentesis samples to measure foetal testosterone levels and then analysing infants in their psychological performance. They first measured the frequency of eye contact with the mother versus mechanical contact with moving objects, then later language skills, finding a correlation between high testosterone levels and low eye contact with the mother and limited language skills.

This finding supports various theories proposed since the late 1950s, which relate language and cognitive abilities to sex differences. It also adds to a growing number of studies examining the effects of testosterone on foetal brain development in both humans and rats following the work of Geschwind and Galaburda (1987).

Baron-Cohen's work on autism and the extreme male brain uses the concept of the 'male brain' in autism to understand gender differences in infants and children in ways very similar to the use of the concept by earlier authors, such as Anthony and Winnicott, to understand unconscious childhood drives for human relationships; however, instead of attempting to theorise sexual instincts in the context of individual child cases, Baron-Cohen theorises the broader concept of child gender in infant populations. Baron-Cohen's work on autism attempts to explain an issue concerning sex differences and the impact of hormones on the functioning of psychological drives towards particular objects.

For the new autism psychologists, what was hypothesised in the biological were not desires, drives and motivations, but cognitive desires, tracking devices that drove individuals to be social. This generation of autism psychologists began to bind themselves to a theory of sociality as an innate ability to read minds, to detect patterns, to observe cues, to understand configurations and to map the thinking of others.

This was the new, epidemiologically defined autism that would become a dominant model for thinking about child thinking in the 1990s, the dominant model for reasoning about child thinking style and child development.

Statistics are now supposed to show the movements and compulsions of unconscious thought driven by instinctual and hormonal forces, to reveal the depths of man and explain them without having to resort to description and conscious thought. They should explain instinct as a sum of statistical measures.

* * *

Before proceeding to the further discussion of possible causes or concomitant causes related to autism and addressing the field of genetic research, it is necessary to return to the neuroscientific basis underlying these and other aetiological theories that have been put forward on autism.

Only with an overall picture of the causal correlations and the many hypotheses that have been put forward will it be possible to have a clearer idea of the epidemiological findings and to enter into the data on the prevalence of the autistic condition.

Notes

1. See in particular:

   Isaacs, S. 'An acute psychotic anxiety occurring in a boy of 4 years'. *International Journal of Psychoanalysis* 24 (1943)

   Isaacs, S. 'Anxiety in the first year of life'. Talk given at the British Psychoanalytical Society, London, 1934.

   Isaacs, S. *Intellectual Growth in Young Children*. London: Routledge & Kegan Paul, 1930.

   Isaacs, S. 'The nature and function of phantasy: 1943'. In *The Freud–Klein Controversies, 1941–45*, edited by P. King and R. Steiner. London: Tavistock/ Routledge, 1991.

   Isaacs, S. *Social Development in Young Children: A Study of Beginnings*. London: George Routledge & Sons Ltd, 1933.

2. For a comparison with Hobson's views see:

   Hobson, R. P. (1990). On acquiring knowledge about people and the capacity to pretend: Response to Leslie (1987). Psychological Review, 97

   Hobson, R. P. (1990) On psychoanalytic approaches to autism. American Journal of Orthopsychiatry, 60

3. The false belief test was developed in 1983 by Joseph Perner and Heinz Wimmer: its most famous formulation is also known as the Sally-Anne test. It serves to test the development of metarepresentational capacity in humans, i.e. the development of a theory of mind.

   It is constructed to test children's ability to attribute a belief to the minds of other people. The crux of the matter lies in the role played by the attribution of a false belief: in fact, predicting the behaviour of another individual, based on a belief that the child knows to be false, constitutes proof that he or she is not trivially projecting his or her own opinion of reality onto the other subject.

   This means that the test subject identifies that particular behaviour and considers it to be 'causally determined' by an 'intentional' mental state of the other person. This mental state, which coincides with the false belief, exists only in the mind of the other person and not in that of the child, who is therefore able to distinguish it and knowingly attribute it to the other person.

   The test takes the form of a game in which the subjects are presented with two dolls: one, Sally, carries a basket and the other,

Anne, has a box. A pretend game is then staged, in which Sally goes out for a walk after putting a marble in her basket and covering it with a cloth.

Meanwhile, Anne takes the marble from the basket and hides it in her own box. At this point, Sally comes back with the intention of playing with the marble and the examiner asks the child where he would look for Sally to take the marble.

In order to pass the test, the child must understand that Sally will look for the marble where she believes the marble is placed, and that she believes the marble is still where she left it, i.e. in the basket. She must understand that, in that particular situation reproduced, Sally has no way of seeing what has happened, which is why Sally has formed an incorrect representation of reality, a false belief.

In responding correctly, the child has to assume the other's position and thus momentarily suspend his own perception of things, in order to represent the content of his mind, i.e. a false belief with respect to reality, so that he can predict what the other will do precisely on the basis of his false belief.

Children up to the age of three systematically give the wrong answer, so they will say that Sally will look for the marble where it really is, in the box. At the age of three and a half, some children start giving the correct answer, from the age of four onwards the correct answer becomes prevalent. Developmental psychologists believe that after the age of 4, they begin to develop tacit knowledge of the behaviour of others as determined by internal mental states, this ability has been named theory of mind. One of the reasons why the ability to pass the false belief test has been deemed to identify a specific cognitive capacity, namely theory of mind, is that children with autism have been found to have a great difficulty or lag behind.

For a review, even in a partially critical sense, see most recently.

Carlsson E, Miniscalco C, Gillberg C, Åsberg Johnels J. Assessing False-Belief Understanding in Children with Autism Using a Computer Application: A Pilot Study. J Psycholinguist Res. 2018.

4. In the context of this field of research and this protocol also in Mauritius was the research on crime and the biological correlates of violence by Adrian Raine summarised in a beautiful book (Raine, Adrian (2013). The Anatomy of Violence: The Biological Roots of Crime (Pantheon Books). Raine took over from Peter Venables in the direction of the project.

5. Among others see:

   Wignyosumarto et al., 'Epidemiological and clinical study of autistic children in Yogyakarta, Indonesia'; Honda et al., 'Cumulative incidence and prevalence of childhood autism in children in Japan', 234.—Magnusson and S.mundsen, 'Prevalence of autism in Iceland—NAS archives: Box 4. File: Autism Europe Publications/ Charters/Leaflets. 'Building a Brighter Future' pamphlet, 10/9/97, p. 17.—Baird et al., 'A screening instrument for autism at 18 months of age'.—Centers for Disease Control and Prevention (CDC), *Prevalence of Autism in Brick Township, New Jersey, 1998*; Department of Health, 'Medical research council to re-examine autism'; Baird et al., 'A screening instrument for autism at 18 months of age'; Madsen et al., 'A population-based study of measles, mumps, and rubella vaccination and autism'; Scott et al., 'Brief report: Prevalence of autism spectrum conditions'; Yeargin-Allsopp et al., 'Prevalence of autism in a US metropolitan area'; Baird et al., 'Prevalence of disorders of the autism spectrum'; Kawamura et al., 'Reevaluating the incidence of pervasive developmental disorders'; Fombonne, 'Epidemiology of pervasive developmental disorders'.
6. Good reasons lead one to believe that intelligence is associated with better judgements. Intelligence is correlated with good performance in virtually all fields; other things being equal, it is associated not only with higher academic achievement but also with higher professional performance (Kuncel, 2010).

   There is much debate, not always lucid, about how intelligence or general mental ability (also called GMA, a term that is preferred today to intelligence quotient or IQ) should be measured. There are deep-rooted clichés about the innate nature of intelligence, but in reality, tests measure abilities that have developed, which are partly derived from character traits and partly influenced by the environment, including educational opportunities.

   There are also various forms of intelligence that the GMA tests do not measure, such as practical intelligence and creativity. Psychologists and neuroscientists make a distinction between crystallised intelligence, i.e. the ability to solve problems from a background of knowledge about the world (including arithmetic operations), and fluid intelligence, i.e. the ability to solve entirely new problems (Nisbet, 2012; Schmidt, 2004).

Apart from mental abilities, people differ in their cognitive style, i.e. their approach to judgement tasks. Many instruments have been developed to identify cognitive styles, many of which are related to AD but measure different elements.

One of these measures is the cognitive reflection test (or cRT).

The cRT is considered by many to be an instrument to measure a broader concept: the propensity to adopt reflective rather than impulsive thought processes. The cRT can be considered a measure of the propensity to rely on slow system 2 thoughts rather than fast system 1 thoughts. (Patel et al. 2019).

Other tests attempt to measure abilities instead of resorting to self-descriptions.

One example is the ADMC (Adult Decision Making Competence) scale, which measures a person's propensity to make typical errors of judgement such as overconfidence or inconsistency in risk perception. Another is the Halpern Critical Thinking Assessment scale, which focuses on critical thinking skills, including the propensity for rational thinking and a whole range of skills that can be learned. (Bruine et al. 2007; Butler, 2012).

Those who score well on either of these scales seem more likely to make better judgements in real life: they experience fewer negative events due to poor choices, such as having to pay a surcharge for returning a rented film late or dealing with an unwanted pregnancy.

The only measure of cognitive style or personality that they found reliable for predicting performance was another scale, developed by Jonathan Baron to measure so-called 'active open-mindedness', i.e. the aptitude to actively seek out information that contradicts one's pre-existing assumptions, such as the contrary opinions of others and the careful weighing of new data that contradicts old beliefs.

While cognitive reflection and cognitive need scores measure the propensity to engage in slow and careful thinking, active open-mindedness goes further: it is the humility of one who is always aware that one's thinking is in the process of being defined and wishes to correct oneself.

See:

N.R. Kuncel, S.A. Hezlett, Fact and Fiction in Cognitive Ability Testing for Admissions and Hiring Decisions, in "Current Directions in Psychological Science", 19(2010), n. 6, pp. 339–345.)

R.E. Nisbett et al., Intelligence: New Findings and beoretical Developments, in "American Psychologist", 67 (2012), n. 2, pp. 130–159.

F.L. Schmidt, J. Hunter, General Mental Ability in the World of Work: Occupational Attainment and Job Performance, in "Journal of Personality and Social Psychology», 86 (2004), n. 1,

N. Patel, S. Glenn Baker, L.D. Scherer, Evaluating the Cognitive Reflection Test as a Measure of Intuition/Reflection, Numeracy, and Insight Problem Solving, and the Implications for Understanding Real-World Judgments and Beliefs, in "Journal of Experimental Psychology: General", 148 (2019), n. 12, pp. 2129–2153

W. Bruine De Bruin et al., Individual Differences in Adult Decision-Making Competence, in "Journal of Personality and Social Psychol-ogy", 92(2007), n. 5, pp. 938–956.

H.A. Butler, Halpern Critical Thinking Assessment Predicts Real-World Outcomes of Critical Thinking, in "Applied Cognitive Psychol-ogy", 26(2012), n. 5, pp. 721–729.)

7. On the circuit of attention from a neuro-psychological point of view, see as references Broadbent's (1958) filter theory and Treisman's (1960) attenuated filter theory, both largely based on the study of the so-called cocktail party effect, first described in the early 1950s by Colin Cherry.

Currently, in the psychological framework, attention is not a unitary concept but concerns a variety of psychological phenomena that are also very different from each other. We distinguish between: 1. selective attention, 2. Control processes (executive functions), 3. Sustained attention and 4. Vigilance.

The study of the attention circuit is linked to at least three particularly active research strands that received new impetus at the end of the 1980s: on the one hand, it is central to research on consciousness and conscious states, on the other hand, it has become central again to the theory of mind, and finally in the field of psychological research in the economic sphere.

See:

Cherry, C. (1957). On human communication; a review, a survey, and a criticism. The Technology Press of MIT.

Broadbent, D. E. (1958). Perception and communication. Pergamon Press.

Treisman, A. M. (1960). Contextual cues in selective listening. The Quarterly Journal of Experimental Psychology, 12

Tononi, G. (1998). "Consciousness and Complexity". Science. 282

Bargh, J., (2018) Before You Know It: The Unconscious Reasons We Do What We Do Windmill Books

Seth, A., (2021) Being You: A New Science of Consciousness. Faber & Faber

Kahneman, D., Sibony, O., Sunstein, C. R., (2021) Noise. William Collins

8. See also:

A.D. Rowe, P.R. Bullock, C.E. Polkey, R.G. Morris, «Theory of mind» impairments and their relationship to executive functioning following frontal lobe excisions, in «Brain», 124, 2001

C. Fine, J. Lumsden, R.J.R. Blair, Dissociation between «Theory of mind» and executive functions in a patient with early left amygdala damage, in «Brain», 124, 2001

M. Turner, Toward an executive disfunction account of repetetive behaviour in autism, in J. Russell (a cura di), Autism as an Executive Disorder, 1998

References

Baird, G., T. Charman, S. Baron-Cohen, A. Cox, J. Swettenham, S. Wheelwright and A. Drew. (2000) A screening instrument for autism at 18 months of age: A 6-year follow-up study. *J Am Acad Child Adolesc Psychiatry* 39

Baron-Cohen, S. (1997) Parents of children with Asperger syndrome: What is the cognitive phenotype?'*J Cogn Neurosci* 9

Baron-Cohen, S. and J. Hammer. (1997) Is autism an extreme form of the male brain? *Advances in Infancy Research* 11

Baron-Cohen, S., A. M. Leslie and U. Frith. (1985) Does the autistic child have a "theory of mind"? *Cognition* 21

Baron-Cohen, S., S. Wheelwright, R. Skinner, J. Martin and E. Clubley. (2001) The autism-spectrum quotient (AQ): Evidence from Asperger syndrome/ high-functioning autism, males and females, scientists and mathematicians. *J Autism Dev Disord* 31

Di Salvo, M. (2024a). Artificial Intelligence and the cyber utopianism of justice. Why AI is not intelligence and man's struggle to survive himself. Russian

Journal of Economics and Law, 18(1), 264–279. https://doi.org/10.21202/2782-2923.2024.1.264-279
Di Salvo, M. (2024b). Multiple intelligences and specialization. A return to Gardner and the wisdom of rethinking psychology in Hillmanian transparency. - Revista De Estudios Socioeducativos. ReSed, 1(12). https://doi.org/10.25267/Rev_estud_socioeducativos.2024.i12.18
Evans, B. (2013) How autism became autism: The radical transformation of a central concept of child development in Britain. Hist Human Sci. Jul;26(3)
Evans, B. (2017). The transformation of social life and the transformation of autism in the 1960s. In The metamorphosis of autism in Britain (pp. 189–247). University of Manchester Press.
Eyal, G. (2013) For a sociology of expertise: The social origins of the autism epidemic'. *American Journal of Sociology* 118
Eyal, G., B. Hart, E. Onculer, N. Oren and N. Rossi (2010) *The Autism Matrix*. Cambridge: Polity
Fombonne, E. (2009) Epidemiology of pervasive developmental disorders. *PediatrRes* 65
Frith, U. and B. Hermelin (1969) The role of visual and motor cues for normal, subnormal and autistic children'. *J Child Psychol Psychiatry* 10
Frith, U. and B. Hermelin (1971) Psychological studies of childhood autism: Can autistic children make sense of what they see and hear. *Journal of Special Education* 5,
Frith, U. (1984) 'Review of Tinbergen, Autistic-children: New hope for cure', 461, 463.
Frith, U. Archive (1983) Uta Frith, *Notebook: History of Ideas Autism*, 1968, 1970, 1983–1984. Basic Diagnosis: 7.7.83
Geschwind, N. and A. M. Galaburda (1987) *Cerebral Lateralization: Biological Mechanisms, Associations and Pathology*. Cambridge, MA: MIT Press
Happe, F., J. Briskman and U. Frith (2001) Exploring the cognitive phenotype of autism: Weak "central coherence" in parents and siblings of children with autism: I—Experimental tests. *J Child Psychol Psychiatry* 42
Hermelin, B. and N. O'Connor. (1964) Crossmodal transfer in normal, subnormal and autistic children. *Neuropsychologica*
Klein, M. (1961) *Narrative of a child analysis: The conduct of psycho-analysis of children as seen in the treatment of a ten-year-old boy*. Basic Books.
Leslie, A. M. and U. Frith.(1988) Autistic children's understanding of seeing, knowing and believing. *British Journal of Developmental Psychology* 6
Lotter, V. (1966) Epidemiology of autistic conditions in young children'. *Social Psychiatry* 1
Newson, J. and E. Newson (1968) *Four Years Old in an Urban Community*. London: Allen & Unwin

Perner, J., Lang, B. (1999) Development of theory of mind and executive control, in «Trends in Cognitive Science»

Premack, D., & Woodruff, G. (1978) Does the chimpanzee have a theory of mind? *Behavioral and Brain Sciences, 1*

Rutter, M., D. Shaffer and M. Shepherd. (1975) *A Multi-Axial Classification of Child Psychiatric Disorders: An Evaluation of a Proposal.* Geneva: World Health Organization

Sanua, V. D. (1984) Is infantile autism a universal phenomenon? An open question. *Int J Soc Psychiatry* 30

Shattuck, P. T. (2006) Diagnostic substitution and changing autism prevalence. *Pediatrics* 117

Tinbergen, E. A. and N. Tinbergen. (1972) *Early Childhood Autism: An Ethological Approach.* Berlin: Parey

Waterhouse, L., L. Wing, R. Spitzer and B. Siegel. (1992) Pervasive developmental disorders: From DSM-III to DSM-III-R. *J Autism Dev Disord* 22

Wing, L. and J. Gould. (1979) Severe impairments of social interaction and associated abnormalities in children: Epidemiology and classification. *Journal of Autism and Developmental Disorders* 9

Wing, L. (1981) Asperger's syndrome: A clinical account. *Psychological Medicine* 11

CHAPTER 9

The Neuroscientific Infrastructure of Second Autism Theories

After the cognitive revolution (Gardner, 1987), the neuroscientific revolution is probably the area in which the integration of transversal specialisations has been most witnessed.

Virtually all disciplines participate in this area and all may have 'something to say'.

The term neuroscience historically refers to the set of disciplines that study the various morpho-functional aspects of the nervous system through the contribution of numerous branches of biomedical research, from neurophysiology to pharmacology, from biochemistry to molecular biology, from cell biology to neuroradiology techniques.

Historically, they arose with the identification of the neuron as an autonomous and functionally independent cellular unit of the nervous system. The studies carried out to define the properties of the neuron have benefited from advances in various disciplines, in particular, using methodologies to measure ionic and molecular displacements at the subcellular level and, thanks to original approaches in psychopharmacology, from advances in our knowledge of the integrated systems underlying behavioural variations in the individual.

In practice, neuroscience has contributed, as far as the biomolecular aspect is concerned, to defining the functioning of the neuron, the role of neurotransmitters, neuromodulating molecules and those with trophic action. In parallel, the biomedical approach has made it possible

M. Di Salvo, *Autism Research between Psychology and Neuroscience*,
https://doi.org/10.1007/978-3-031-68338-1_9

to employ various investigative techniques to explore the anatomo-functional structure of the nervous system as an integrated unit, both under normal and pathological conditions. In this sense, progress has been made in neuroradiology and in those particularly refined techniques that are widely used in clinical diagnostics (computed tomography, magnetic and positron emission nuclear resonance).

The approach was then extended to the description of molecules capable of controlling the genesis of certain brain proteins (genetic engineering). In particular, molecular biology has made it possible to study amino acid sequences of peptides that appear to play a physiological or pathological role, in relation to different conditions of isolation and characterisation.

One of the main topics of discussion for philosophers and scientists in the twentieth century has been whether 'mental' activities such as thought, emotions, self-awareness and will are different functions from 'brain' activities such as the movement of a limb, the perception of a colour, etc., or whether they too represent functional expressions of the neurons that make up the brain. The distinction between mental and cerebral activities, in the light of current knowledge, appears artificial to those practising one of the many disciplines that constitute neuroscience.

Mental and brain activities are, in fact, simply the unique and indivisible expression of the activities of the neuronal and glial elements that make up the brain organ. Although the expression is different in quality and in the ways in which it manifests itself, both activities are due to a single mechanism by which neurons communicate with each other and with the rest of the organism.

According to this conception, so-called mental activities must be considered as emergent properties, the result of such a complex sum of simpler neuronal activities that they constitute a quantitative leap that is essentially still undecipherable. After all, life itself has the same characteristics. None of the chemical activities that take place within a cell, taken individually, has the connotations of life but, taken together, all the molecular reactions that are coordinated and organised within the confines of a cell membrane allow vital functions to emerge and assert themselves.

With multidisciplinary and interdisciplinary approaches, the possibility of elucidating the mechanisms by which neurons, organised in three-dimensional structures of varying nature and magnitude, process incoming information, store it, if necessary, and emit a behavioural response is thus envisaged in the future.

These studies are also beginning to provide fundamentally important information on the nature of mental processes such as consciousness, will and memory, enormously complex issues that form the core of the third level of brain functions. It is probable that once the first two levels, that of the functions of individual neurons and that of the activities of neuronal networks, have been fully understood, it will be possible to arrive at the elucidation of the type of circuits or nervous activities by which we are able to decide on a given motor act or recall of the past and the mechanisms by which our encephalon, at the same time as it is processing sensory input, makes us conscious of all these operations.

It is not surprising, in this context, that more and more neuroscientists and cross-disciplinary research groups engaged in neuroscience have become interested in the great issues of the human mind, from consciousness to free will, from the redefinition of apex concepts such as life, death, intelligence, the unconscious, memory and not least autism, with the search for neuronal correlates and the neurophysiological basis of psychological theories.

In this chapter, I will attempt to describe the neuroscientifically based theories on autism, for which the neurophysiological and structural foundations have been explored in Chapters 5 and 6.

Social Understanding and the Theory of Mind

The golden rule of moral conduct is common to most philosophical and religious traditions and may serve as a typical example of the way in which the understanding of society intervenes in the lives of humans who live in society. David Premack and Guy Woodruff wondered in 1978 '*whether the chimpanzee possesses a theory of mind*'.

That is, they questioned whether or not the ability to interpret one's own and others' behaviour by drawing inferences about the mental states of others—desire, intentions, beliefs and knowledge—is a peculiarity of human beings.

According to them, the expression theory of mind is justifiable as long as the mental states of others cannot be directly observed by the subject and must therefore be represented in hypothetical or theoretical form in order to make predictions about their behaviour.

In 1996, John Barresi and Chris Moore posed the problem of the theory of mind in an evolutionary and hierarchical framework of intentional relations directed at real or imaginary objects capable of intervening in social understanding. They identified four levels of representation of intentional relations based on the distinction between perspective representation of self and other, whether real or imaginary.

Level 1 includes organisms capable of anticipation and of distinguishing between self and others, although without understanding the possible similarities in action, but incapable of imitation or sharing. These are, for example, rats subjected to classical or instrumental Pavlovian conditioning, which brings anticipation and reinforcement into play but without leading to the development of any coordinated associated life.

Level 2 is that of numerous organisms capable of sharing a representation of themselves and others, but only in the present and without mutual understanding. The sharing of common activities and the reciprocal passing of information can be found ranging from the formation of schools of fish and herds of mammals to the collective croaking of frogs and toads, the polyphonic howling of wolves and the communication systems with cries of vervet monkeys. Dorothy Cheney and Robert Seyfarth distinguished in 1990 different alarm cries in vervets (*Chlorocebus pygerythrus*), which express different types of fear when faced with leopards, eagles or snakes, respectively, and which stimulate different synchronised responses: climbing up a tree if it is out of fear of leopards, looking skyward if it is out of fear of an eagle, looking down if it is out of fear of a snake. These cries of alarm are produced independently of any collective control or teaching of the cubs. Vervets communicate elaborately within their group through grunts directed according to a hierarchy from subordinate to dominant, but also in reverse and across different groups. They may elicit responses such as a return grunt or an exchange of glances as if engaging in conversation. At this rudimentary level, however, they are not signs of imitation or mirror recognition, empathy or sympathy. One is unable to compare oneself with others and understand such relationships.

Level 3 involves self-recognition. When an animal is capable of intelligent imitation, at the opposite extreme of parrot-like repetition, a device for interpreting one's own and others' intentional relations in a shared conceptual system and the imagined representation of self and others as intentional agents emerge. Premack and Frans de Waal (1996) noted, for example, that captive chimpanzees are capable of understanding another chimpanzee's purpose, imitating and counteracting it. They are also capable, like whales and dolphins, of providing mutual aid if one of them is injured. They show solicitude towards their young and maimed fellows and generally show sympathy. Bonobos, in particular, the so-called pygmy chimpanzees, are famous for their emotional life and intense sexual activity, as well as their ability to reconcile after scenes of violence, usually by resorting to sex. Moreover, chimpanzees pass the mirror test: they look at their own image reflected in water and can spend hours contemplating a speck of paint that has been made on their forehead or ear without their noticing it. They can even recognise themselves in a distorting mirror and in a comparison with multiple images.

Level 4 is perhaps reached by chimpanzees to the extent that they vividly represent, or conceptualise, the intentional relations of others and their own towards others. Like the human being, the chimpanzee is capable of introspection and is able to attribute its own mental states to other congeners and thus to the human being. For de Waal, it is capable of deception and develops a genuine theory of mind that Barresi and Moore cast doubt on, believing it to be an exclusively human characteristic.

Incidentally, if these data are correct, as also appears from numerous control studies, if the rats reach level 1 as indicated, the thesis of the existence of autistic rats seems highly questionable—on neuroscientific grounds—at least according to the criterion of the theory of mind, which we shall see in the chapter on genetic studies.[1]

In human beings, social understanding proceeds in stages. From the age of 2 months, and perhaps even earlier, reciprocal communication is established between mother and child, and by the end of the first year an exchange of looks is established between the child and those around it. The little one communicates with gestures and fixes objects or situations in a proto-diagnostic manner. He can use visual and auditory information. He is also able to represent intentional relationships between the first and third person: the human being starts at level 2 from the very beginning.

In the second year, the child reaches level 3 and in the fourth year develops a true theory of mind. During the second year, he begins to search for hidden objects, uses imitative mechanisms and plays "do as if", develops language and refers to stored representations to interpret perceptual events and react to them. Uses imagination to compare objects stored in the past with the present situation. Recognises itself in the mirror, like adult chimpanzees. Children of less than 18 months perceive suffering in another child and start crying out of sympathy. A little later their behaviour changes and they manifestly try to comfort another child in distress.

What Piaget (1945) and Kohlberg (1973) called 'decentralisation' is thus realised. The child understands that other people's feelings can be different from his own and that his behaviour can change them. He imagines other people's mental states and acts in accordance with these (level 3). From the age of two onwards, the child is able to attribute the faculty of seeing, wanting and believing to intentional objects. According to Heinz Wimmer and Joseph Posner in 1983, the diriment test is that of false belief.

A child may imagine that another child does not have adequate knowledge of a new situation that is known to him. In his own imagination, he compares a representation of his own knowledge with that of the other (level 4). As James Baldwin recognised in 1894, the understanding of the self develops in parallel with the simultaneous understanding of others on an imaginative level. A clear relationship is thus established between self-knowledge and empathy, or sympathy.

In 1995, David and Ann Premack used an elegant video animation to demonstrate that infants of 10 months or older attribute 'human' intentions to very simple moving objects, such as images of balls 'caressing' or 'bumping' into each other. The child encodes caresses positively and bumps negatively. In another scenario, a ball that helps another to escape from a box is evaluated positively, while hindering its escape is evaluated negatively. The child attributes an internal cause to intentional objects and understands reciprocity mechanisms. For example, if A caresses B, then B should act positively towards A. He also appreciates the difference between a ball bouncing well or badly. The young child thus possesses a spontaneous moral value system that favours cooperation and sympathy, and even aesthetic concerns. In contrast, autistic children develop severe cognitive dysfunctions that affect social communication and affective contacts, empathy and sympathy.

According to Baron-Cohen, Alan Leslie and Uta Frith (1985), autistics lack theory of mind. They cannot infer information to the first and third person and are therefore reduced to level 1. Several attempts have been made to identify neuronal correlates of the theory of mind by means of brain visualisation techniques.

Based on psychological tests built on the recognition of terms specifying particular mental states, in 1994, Baron-Cohen and his co-workers recorded increased activity in the right orbitoprefrontal cortex.

In 1995, Vinod Goel and his co-workers tested knowledge of twentieth century man and his motives and showed selective activation of the left prefrontal cortex. Despite the differences, both experimental conditions see the prefrontal cortex involved in the theory of mind. This result is not unexpected, as it is the most recently evolved part of the human brain.

In 1966, Lorenz described an inhibitor of violence in animals, part of the cognitive mechanism of non-verbal communication, which Irenäus Eibl-Eibesfeldt returned to in 1970.

The victim produces signals, e.g. of submission—the dog shows its throat—which inhibit attack behaviour in the aggressor. Blair extended this conception to the child in the form of a model for the development of the moral sense. His theories favoured a model of spontaneous activation of the violence inhibitor during development. Between the ages of four and seven, the child reacts to an expression of sadness, cries and tears by desisting from violent behaviour. This is followed by what can be called moral emotion: empathy, sympathy, guilt and remorse. The violent act is inhibited. As much as autism seems to result from a selective alteration of the theory of mind, psychopathic children present according to Blair a selective deficit of the inhibitor of violence. Consistently, the psychopathic child shows no emotional reaction to the distress of others. He is violent and aggressive without remorse or guilt, despite knowing he is causing suffering, which shows that his mind is intact. Several authors, such as Hans Eysenck in *Crime and Personality* of 1964 and Gordon *Trasler in The Exploration of Criminality of* 1962, have argued that fear of punishment for transgressing moral principles will encourage the child to develop moral behaviour.

Turiel (1966, 1978) conducted a seminal study on moral and social reasoning in children and adolescents, showing that they distinguish between judgements of moral necessity that are obligatory and unrelated to contingency and judgements that are related to social conventions, voluntary and subordinate to contingencies. He studied children from different fundamentalist religious communities, Amish Mennonites, Dutch Reformation Calvinists, Conservative and Orthodox Jews. He asked them precise questions about norms of conduct regarding the day of prayer, baptism, the way women, or men, cover their heads, and food. Turiel discovered that non-moral religious norms are attributed to a specific religious group and subordinated to the word of God. Thus, these children think that religious norms are not applicable to people of other religions and that members of their own religion are not obliged to follow norms for which there is no specific reference in the Bible. Judgements on moral standards imply a different relationship with religion. They think that people outside a religion are equally obliged to follow these norms, and moral judgements are not considered dependent on the word of God.

These children therefore distinguish between compulsory moral norms and non-moral, conventional norms.

The Neuronal Basis of Theory of Mind and Attribution

Human beings are social animals that interact differently than individuals of other species. The necessary conditions for a direct reciprocity theory of cooperation were proposed by Axelrod in 1984.

A cooperative strategy such that the 'prisoner's dilemma' game can be solved requires several conditions that include mutual recognition of the participants, recollection of previous encounters, reward or punishment for cooperating or not and a place to meet. Unlike other species, human beings also possess the combination of rationality and sociability that we might call social intelligence. According to Chris and Uta Frith (2010), this involves the recognition of one's place in society, the ability to acquire knowledge from others and to teach them new tasks and knowledge: in other words, pedagogy. Human beings have the unique ability to understand and manipulate the mental states of others and thus modify their behaviour; they are aware that others have knowledge, beliefs and desires—similar to or different from their own—on the basis of which they organise their behaviour. As we have already seen, these characteristics can be understood within the theory of mind or attribution, i.e. under the banner of what we can call intentional attitude. This calculation of others' intentions can now be studied in children.

Sally and Anne's test illustrates the notion of false belief: Sally puts a marble in a basket and leaves the room. Anne takes the marble and moves it into a box. The children involved in the experiment are asked where Sally should look for the marble. The children have to understand that Sally will probably look in the basket because she has a false belief that they have to detect.

The neuronal correlates of the theory of mind are numerous.

Patients with prefrontal lesions, particularly in the orbital and medial areas, have serious problems with social behaviour. Furthermore, brain visualisation reveals medial prefrontal, temporoparietal and inferolateral frontal cortical activation during tests involving theory of mind. Recordings in primates demonstrate primitive systems underlying theory of mind. In the superior temporal cortex, there are distinct responses to hand and face movements, but not to inanimate objects, and responses to the direction of another's gaze directing one's own. There are mirror neurons for intentional performance in the inferior prefrontal cortex without the subject distinguishing between self and other. Responses that take this distinction into account are found in the superior temporal cortex; according to Perrett, such responses distinguish sounds or images that come from others from those produced by the subject himself. One can also find activation of neurons in the medial prefrontal cortex and anterior cingulate cortex that anticipate the production of self-generated movements and thus suggest explicit self-representation (Changeux 2008).

As Rizzolatti and Sinigaglia (2019) state in their articulate recapitulation of some thirty years of mirror neuron studies, *'some have postulated a "division of labour", with mirror responses and corresponding transformations primarily involved in the attribution of purpose, and a network of areas, including the posterior portion of the superior temporal sulcus and the temporo-parietal junction, primarily involved in the attribution of mental states such as beliefs, desires and intentions as motivating reasons for performing observed actions' (Kilner and Frith,* 2007; *Keysers et al.* 2014).

'Without entering into the vexed quaestio as to whether or not there exists a network of areas specifically deputed to mentalisation (Saxe et al. 2004), *we cannot, however, fail to note that the aforementioned studies never involved tasks that explicitly (and in a controlled manner) required participants to represent other people's mental states as reasons motivating the execution of a given action. It is true that the observation of unusual or, in any case, implausible actions may lead the observer to infer the reasons that would have induced such behaviour. But inferring the reasons for a given behaviour does not mean eo ipso inferring the mental states that would have motivated the agent to perform the action he performed. If the car in front of you does not start at the green light, you are likely to attribute this to a mechanical*

reason or to the driver being distracted. Of course, you might also believe that the driver in front of you thinks it is better to wait a while before starting, because that intersection is particularly dangerous. But the first kind of reason does not necessarily imply the other'.

Mirror responses and corresponding sensorimotor transformations may relate to the purpose of the observed action, regardless of how that purpose is realised. More than one study has shown activation of the most posterior portion of the temporo-parietal junction in tasks that, while having some social significance, did not involve any mentalisation (Cook et al. 2014; Sowden and Catmur, 2015; Bardi et al. 2017). If we then consider the posterior portion of the superior temporal sulcus in the right hemisphere, it is known to be involved in the representation of movements made with the hand or other parts of the body, and although this representation may contribute in some way to the attribution of mental states such as beliefs or intentions (Frith and Frith, 2003), it is nevertheless not sufficient on its own to give us anything like a full understanding of action.

In this context, Rizzolatti and Sinigaglia explicitly propose that '*instead of focusing on the "division of labour", it might be much more interesting to examine whether and to what extent variations in the ability to motorically represent specific action goals might have a content-specific effect on the attribution of mental states motivating actions directed towards those goals. [...] The reverse is also true. Indeed, it would be of extreme interest to measure the effect that the attribution of mental states might have on mirror responses. In other words, it would be a matter of measuring such a response in tasks that do not consist in the mere observation of the actions of others, but presuppose the exercise of mentalisation skills, in particular the attribution of (false) beliefs. Our hypothesis is that, although the processes and representations involved in mirror responses and the corresponding sensorimotor transformations are different in format from the representations and processes that characterise mental states such as beliefs, desires or intentions that have (or may have) propositional content, there may nevertheless be instances in which what the observer believes about the beliefs, desires or intentions of others depends, at least in part, on his or her ability to motorically represent the purpose or purposes of the observed actions*'.

The 'Fragmented World' of Uta Frith

In autism, the extraordinary ability with puzzles is a common endowment. The way autistic children construct a puzzle is markedly different from the way a neuronormal child does it. To investigate this, Beate Hermelin and Uta Frith (1969) conducted an experiment in which they compared performance in two types of puzzles.

In one type, the children were provided with rectangular pieces with straight edges of a jigsaw puzzle representing a picture, in the other with typically jagged pieces but without a picture. In the second test, the autistic children were better at it. They enjoyed fitting one piece with another, even when no picture came out in the end.

In most jigsaw puzzles there is an image that is broken up regardless of its natural shape or the boundaries of its content. Only the reassembled pieces will take on a legible character and will acquire meaning: isolated, a puzzle piece means nothing.

According to Uta Frith, it is likely that the autistic child can see the individual pieces within the completed jigsaw puzzle. Perhaps completing the jigsaw is really a piecemeal exercise that starts with small sections and eventually results, almost incidentally, in a large picture. If this were the case, then building jigsaw puzzles without pictures could be just as easy.

As a metaphor, the jigsaw puzzle that remains a collection of fragments, even when they are strung together, symbolises, according to Frith, the effect of autistic detachment. In contrast, for the non-autistic person, once the fragments have been put together and form a single image, they lose their meaning as fragments and gain meaning through the meaning of the larger unit to which they belong: the image as a whole.

This is a context effect on meaning, or what he called 'strong central coherence'. The lack of context effect and, consequently, the lack of guidance for meaning can be defined as a weakness of central coherence.

The difficulty of finding puzzle pieces is reminiscent of the old pastime of 'find the hidden figure'. Amitta Shah (1983), at the time he was doing his PhD at the MRC Cognitive Development Unit in London, had the idea of investigating how well autistic children were able to locate hidden figures.

He used a standardised test, the children's hidden figure test, and found that autistic children had an average score for their mental age, indeed, they were faster and more accurate than normal children of the same mental age.

Witkin (1971, 1981) and co-workers claimed, on the basis of unjustly forgotten earlier studies, that people capable of finding hidden figures also perform well in other tests designed to demonstrate 'field independence', i.e. not being influenced by the context in both visual perception and social interaction. These studies suggest that field-dependent people are easily influenced by the opinions of others and tend to make the prevailing views in their group their own; field-independent individuals, on the other hand, are not influenced by current fashions and do not care much for the opinions of others. Individuals with a high degree of social detachment are generally able to spot hidden figures well. The terms strong and weak central coherence are very similar in meaning to the terms field dependency and field independence.

Central coherence can thus be seen as a force that binds large amounts of information together. If it is powerful, this force may be responsible for the difficulty of seeing the figure stand out from the whole. On the other hand, a weak central coherence may explain the ease of this test. Judging by the wide individual differences in performance on the hidden figure test, some people have a bias towards strength and others towards weak central coherence.

At the two extremes, weak coherence means not seeing the forest for the trees; strong coherence, not seeing the trees for the forest.

One of the best-documented strengths of autism is proficiency in the cube drawing test. The crucial feature that links drawing with cubes to the hidden figure test is that a large geometric figure (*gestalt*) must be broken up into small shapes. The large shape must be copied with small building blocks: in fact, the drawing elements that correspond to the blocks are analogous to the hidden figures.

They must be 'identified' before the design can be reconstructed. So the first problem is to separate the given drawing into appropriate segments, which has little to do with what is generally understood as 'spatial ability', but may have much to do with resistance against the force that drives coherence in higher-level central thought processes. If autistic children show weak coherence, then they must show less tendency to perceive the image as a whole, and segmentation will be easy for them.

Amitta Shah (1993) experimentally introduced such a manipulation with new drawings to be copied by autistic and non-autistic children.

The superior automatic memory, which is based on small elements, and the ordinary memory, which is based on large coherent units, give us an idea of the advantages and disadvantages of both strategies, the fragment-based and the global one.

When one has to search for a key in a bunch of other similar keys, the fragment strategy is useful. On the other hand, when one has to extract the general sense from a long text, the global strategy is helpful.

Consequently, the problem of categorisation and generalisation arises. Human beings continuously categorise things and events and generalise from one event to another. In categories useful in real life, kitchen and bathroom are considered similar.

Clearly, in order to see similarities on a global level, many differences in details must be ignored. Tasks that are set in context (*such as performing a subtraction in mind while giving a remainder*) are very different from doing, say, a subtraction during a written exam. The same person can perform extremely differently in these contrasting situations, even when the arithmetic problem is identical. It appears that the unschooled individual is much more affected by the influence of context. The cube drawing test and the hidden figure test are all designed to test abilities that are uncoupled from the context of everyday life. Performance in these tests is higher in people who have enjoyed regular education. School promotes the acquisition of abstract knowledge, which is not dependent on a specific context. In this sense, school promotes weak central coherence.

According to Uta Frith's hypothesis, in the normal cognitive system, there is an inherent tendency to form coherence along as wide a range of stimuli as possible and to generalise over as wide a range of contexts as possible. It is this force that gives rise to the great systems of thought. It is precisely this capacity for coherence that is impaired in autistic children. The result is that their information processing systems, and their very existence, are characterised by detachment, understood technically, in the sense of the quality of thinking, which could be due to the lack of global coherence or resistance to it.

In common parlance, detachment means the absence of cognitive distortions and non-impressionability, as well as a certain social isolation. This connection is no mere coincidence. Nevertheless, autistic detachment has different causes: it is not based on reflection and results from a lack of coherence.

The central coherence deficit theory may well explain the excellent performance on the cube drawing and hidden figure tests, as well as the excellent performance in the automatic memory of autistic children.

The normal operation of central coherence forces human beings to prioritise the understanding of meaning. This is why we are able to easily extract what is meaningful from meaningless material.

Despite the processing effort involved, we remember the sense of the message and not the message word for word. Moreover, the sense is best remembered if it can be placed in a broader context. The need to place information in a wider context is another way of considering the effects of higher-level central cohesion.

As proof of our ability to make sense of what we are dealing with, Uta Frith devised experiments in which meaning and sequence structure are manipulated. Very different materials are used in these experiments, but they all have a common denominator: the contrast between stimuli that are separated from each other and stimuli that are held together tightly. In one type of experiment, we compare unconnected and connected words (*based on the underlying meaning*), in another type there are nonsense sounds in an unpredictable order and a predictable one (*based on the underlying structure of the sequence*). The third type of experiment employs isolated and hidden shapes (*global design*). All results show high-performance of autistic children in tasks requiring isolation of stimuli and favouring detachment, compared to their poor performance in tasks requiring connection of stimuli and favouring coherence. In contrast, in the case of neuronormal young children, those with retardation as well as unschooled older children with a different cultural background, performance is opposite. Therefore Frith assumed that a central cohesive force is a natural (and useful) feature of the cognitive system, but is significantly disturbed in autism.

This 1989 hypothesis was enriched and partly modified by new work, in particular, that of Francesca Happé (1996, 1999) who showed that the effects of central coherence can be observed not only in higher-level tasks, such as extracting meaning from a sentence, but also in low-level visual tasks. She showed that there is a type of central coherence that acts on the immediate visual context by creating optical illusions.

Sample is the Ebbinghaus illusion.

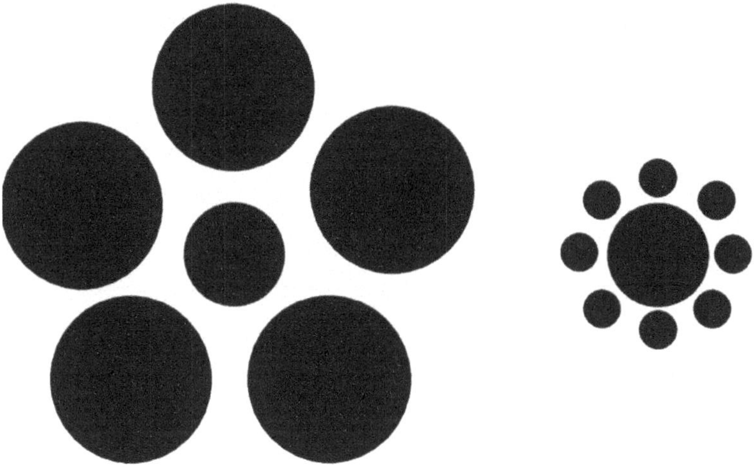

Here, the central circles of the two drawings are actually the same size; however, one is seen as wider than the other. This is because the context of the circles surrounding the central ones affects the perception of the latter.

If the circles around are small, the central one appears relatively larger, and vice versa. Autistics, compared to control subjects, more easily resist this type of illusion, which is determined by the context.

These kinds of optical illusions are no mere theoretical speculation, since in the natural environment such illusions are adaptive and widely used. For example, the Ebbinghaus illusion is used by fiddler crabs (*Uca mjoebergi*) for both defence and courtship purposes.

It is important—here and elsewhere—to give due consideration to when a psychological experiment has a completely abstract basis in relation to the ecological environmental reality or adheres to and tends to reproduce and study behavioural models existing in nature, also in order to validate the results and adaptive purposes of behaviour with cross-experience.

Happé et al. (2001a, b) collected evidence on detail-focused processing in autism in three rather different areas: visual perception, auditory perception and verbal semantics. He formulated the hypothesis that central coherence is a cognitive style that varies from weak to strong in the normal population and is reflected in a normal bell-shaped distribution of scores on relevant tests. In the population of autistic persons, one would also expect to see a similar bell-shaped distribution, but with the mean shifted towards the weak extreme. This interpretation makes it possible to explain inconsistencies in the results, as there is a clear overlap between the two populations, which can lead to negative results when comparing small groups. (*However, as we shall see in Chapter 11, all human skills have—given a statistically adequately large population—a Gaussian distribution.*)

The 'Extreme Male Brain' Theory

Simon Baron-Cohen (2002) with his Cambridge group, starting from slightly different premises, reported similar results on relatives of autistic persons. He identified a style of information processing that he called 'systemising', based on an intuitive understanding of how mechanical objects work and a preference for information concerning the physical rather than the psychological world.

In this way, it takes into account a particular style of information processing, combined with a particular content. Style and content not only suit autistic individuals but also many normal people and certainly modern 'geeks'.

Baron-Cohen contrasts 'systemising' with 'empathising', which concerns the attribution of mental states and is the same as mentalising. Using experimental tests investigating these two capacities, he found, as he expected, that autistic individuals exhibit an excellent ability to systemise, but poor empathy. The two cognitive styles are among the differences traditionally attributed to males and females and between the natural sciences and the humanities. Baron-Cohen found that there is a preponderance of fathers of autistic people in the engineering, computer and science professions. He is currently looking for markers to identify systemisers.

The Theory of Levels of Empathy

Baron-Cohen (2012) contrasts 'systemising' with 'empathising', and in his book 'The Science of Evil' introduces a scale of values of empathy, a reconstruction of related behavioural patterns, places a number of pathologies in the categories of empathy (zero-negative) and in the extreme zero-positive the autistic spectrum.

Let's see the thesis in detail.

Through the investigation of the EQ (emotional quotient) one can determine the person's level of empathy, there are 7 different conditions.

Level 0—The person has no empathy, the emotions, feelings and thoughts of others are worthless and unfortunately they are capable of murder, torture and aggression. Fortunately, not everyone with an EQ level of 0 carries out these actions, in fact, some isolate themselves from the world and social relationships because they have no desire to harm them.
Level 1—The person is capable of hurting others but is also capable of feeling regret. The subject when he hurts the other does so because he cannot stop himself, his empathic circuit is not capable of inhibiting the tendency to hurt others. The subject when he has these urges does not consider the other, he has the one and only purpose of attacking and destroying him.
Level 2—The person has enough empathy to control themselves and not attack other human beings. However, the person has difficulties in social relationships, not realising that he/she is exaggerating in hurting those around him/her with words. He says and does inappropriate and inappropriate things without even perceiving that he has gone too far. This kind of problem could lead him to social isolation and loss of job.
Level 3—The person is aware that he/she has problems with empathy, in fact, masks it and avoids its use by attending jobs that do not require its use. He avoids social relationships a lot and does not understand jokes. He feels much relief in being alone because then he does not have to pretend to be 'normal'.
Level 4—The person has a low to medium empathy value. Empathy does not influence his or her daily behaviour but the person feels uncomfortable talking about emotions, much preferring to discuss sports or other things. A large number of men fall into this bracket compared to a few women. Social relationships are based on technical–practical activities.
Level 5—The person's empathy is above average, there are many women here and fewer men. Those with this level prefer confrontation and discussion on issues concerning emotions, feelings and also friendships are based on sharing and support. People with this level are not always attuned to others, but they are able to put everything aside to listen to those in difficulty.

Level 6—The person with this level of empathy is constantly focused on the emotions and feelings of others, experiencing a perpetual over-excitement of the empathic circuit.

Grade zero has no empathy, this lack causes an inability to relate to other people, selfishness and makes it inaccessible to reach the emotional world of others. Those with compromised empathy are unaware that different points of view may exist and judge those who think differently from their own as stupid.

Degree zero constitutes a kind of solitary existence, condemned at best to misunderstanding with others, at worst to selfishness. Those who do not have a functioning empathic circuit are unable to restrain their own behaviour, and therefore say and do whatever comes into their minds without caring about the reactions of others. The worst cases of this lack of empathy can lead to murder, torture and abuse.

This condition is not attained by chance, but it is necessary for certain things to happen during childhood, adolescence and later life when real psychic pathologies are formed in order to develop it.

In turn, degree zero is divided into two extremes.

The zero-negative includes people with a borderline diagnosis, psychopathy, narcissists.

The zero-positive includes people diagnosed with Asperger's and autism spectrum disorder.

According to Baron-Cohen's theory, the empathic circuit is under-activated in the person diagnosed with autism, and mirror neurons would not be structured enough to allow the reproduction of expressions observed in other people.

People diagnosed with autism find it difficult to understand the emotional state of others as well as their own because the components of the brain deputed to perform these tasks are the same whether they are talking about others or themselves, resulting in both skills being under-activated.

Systematisation on the other hand means analysing changes in structure to understand how things work: if these changes are not random, patterns are formed that are repetitions of previously verified information. People with Asperger's syndrome are very good at spotting patterns.

Why do poor empathy and strong systemisation occur together in people with this syndrome?

Model building allows for prediction. If the model undergoes a change in a single variable, this change causes a change in the system; thus enabling change and access to the truth (by truth is meant a controllable and repeatable model).

Truth can reflect natural and social patterns; it is the repeatability and control of a given pattern that gives it the character of truth.

The systematisation mechanism is made up of parts aimed at perceiving patterns and understanding how things work. As with the empathic circuit, it does not have the same level in all people and 7 levels have also been identified.

Level 0—They do not perceive any kind of pattern, are unable to analyse changes and have no difficulty in coping with changes.
Level 1—They perceive simple and very repetitive patterns, struggle to understand new systems and do not like mathematics and tend to avoid it in educational settings.
Level 2—They can notice new patterns but only if someone leads them to notice them. They are not able to make new patterns work on their own.
Level 3—They can cope with simple systems but if a complicated one arises, they struggle much more. In this level, there are many more females than males.
Level 4—Has high functioning in juggling systems. There is a majority of males compared to females within this category.
Level 5—Their life is very much based on research and the study of patterns. They do not, however, spend their whole lives systematising; they are, however, capable of handling the unexpected.
Level 6—These are people with autism or Asperger's syndrome, they spend their entire lives systematising their surroundings.

The Neuroscientific Basis of the Empathy Circuit

In the empathic process, it has been discovered that not the whole brain is involved but about ten areas, interconnected with each other, and it is likely that others are also involved, sometimes in an additional way.

Medial Prefrontal Cortex—Processes social information and enables comparison of one's own viewpoint with that of other humans. The medial prefrontal cortex is divided into a dorsal and a ventral part.

Dorsal: is involved in the process of representing the thoughts and emotions of others as well as when we think about our own thoughts and feelings.

Ventral: plays an important role in self-awareness and is used when thinking of one's own mind rather than that of someone else.

Orbitofrontal Cortex—Has the function of making social judgements and assessing whether a certain event causes pain or not.

The Frontal Operculum—Has the function of encoding the intentions of other people, in fact, if we take two subjects and ask one to do an action and the other to simply observe it in both we will have the activation of the same brain cells. Another function it performs is the production of fluent language.

The Inferior Frontal Gyrus—This is the area used for emotion recognition. The processing of emotions observed in others is located in different areas depending on the emotion itself; disgust is processed in the anterior insula, happiness in the ventral striatum, anger in the supplementary motor cortex and sadness in the hypothalamus and other areas.

The Anterior Cingulate Cortex and Anterior Insula—This region is activated not only when the subject experiences pain but also when he or she observes someone else experiencing it. The activation of this area is modulated according to how much one imagines oneself in someone else's shoes. The more sympathetic the person is to the subject, the more the subject will get an activation of this area. Men find it less difficult not to be empathic with those they do not consider to be correct. These two areas are fundamental within the empathy circuit.

The Temporoparietal Junction—Aims to judge the intentions and beliefs of others. It is the seat of the theory of mind process: ability to understand and represent emotional states, beliefs, desires and predict the behaviour of others. It is also concerned with directing the subject's attention. If stimulated, it may cause dissipation in the subject.

The Superior Temporal Sulcus—Performs the function of understanding where the other is looking, it is also involved when observing the autonomous movements of living beings (biological movements).

The somatosensory cortex—It is involved in encoding one's own sensory experiences and those of the person being observed, in fact, when one observes discomfort in the other there is identification at the sensory level. There does not necessarily have to be a conscious and voluntary activation, because the empathic subject winces when he sees someone else get hurt.

The Inferior Parietal Lobule—This is an area that is activated both when actions are performed and when they are performed by others.

The Amygdala—This is one of the most important regions when it comes to empathy because it is involved in people's emotional learning and regulation. It enables eye contact between individuals and thus recognition of the emotions of others; if there is damage to the amygdala, the person is unable to make eye contact and thus will not be able to understand the emotions of others.

Mirror neurons are located in some of these areas. The regions in which they have been found are the frontal operculum, the inferior frontal gyrus and the inferior parietal lobule. Mirror neurons are not comparable to empathy, they are simply components of a system.

It has been hypothesised that emotional contagion may be a form of empathy. Feeling sad when another person is sad, yawning when another person does it even if we are not sleepy, smiling when we see someone else do it, are all examples of empathic contagion somehow driven by the mirror neuron system.

Many regions that are activated when we encode our experiences are also activated when we perceive that others have similar experiences. The ten regions we have discussed have a different level of activity from person to person, which is why not everyone would have the same degree of empathy. One might expect those with low empathy to have lower neural activity in all or part of the empathy circuit.

According to Milton (2012, 2014), the idea that autistic people lack sociality is problematised, suggesting that autistic people are not well described by concepts such as 'social brain' or 'zero degrees of cognitive empathy'. He argues that there is a qualitative difference in autistic sociality, and questions the extent to which these differences are biological or cultural in nature, and the extent to which interactional competence can be acquired on both sides in interactions between autistic and non-autistic people. Milton argues that autistic people have often become distrustful of researchers and their goals and are often excluded from knowledge production processes. Such a context results in a negative feedback loop with further damage to the growth of interactional competence between researchers and autistic persons, and a breakdown in trust and communication leading to increased tension between the stakeholder groups. The involvement of autistic scholars in research and the improvement of participatory methods can therefore be seen as a requirement if social research in the field of autism is to claim ethical and epistemological integrity. These will be addressed in detail in Chapter 12.

Gazzaniga's Theory of the Interpreter

Michael Gazzaniga is a neuroscientist who has extensively studied split brains and constructed the 'interpreter theory' (Di Salvo, 2024).

In simple terms and in extreme synthesis, it was by studying split brains (*a surgical technique adopted for extreme cases of epilepsy*) that he formulated the hypothesis that what we know as mind is the result of the interaction between two distinct ways of proceeding of the two hemispheres that would follow two different ways of perceiving and interpreting reality.[2]

In Gazzaniga's words, "*…we must now clarify where exactly this 'entity' is located within the left cerebral hemisphere. We are in the presence of an area that is not yet fully delineated, but which certainly involves the basal ganglia, the amygdala and other sites that are still being identified experimentally. Finding the exact site of the interpreter means playing a chess game on an almost limitless chessboard, although certain clinical data may help. For example: the most devastating neurological diseases, leading to disturbances in the production of thought and inability to solve problems, arise from lesions of the middle cerebral artery. Therefore, it is logical to assume that the interpreter must be somewhere along the path of this artery. However, autistic children—who never get to construct a mental picture of themselves and others—have not confirmed this damage. Thus, some have suggested that the interpreter in their left hemisphere is simply 'sick'*".

Experiments conducted on autistic children by means of positron emission tomography (PET) have, however, confirmed the existence of hypometabolism in that same region. Medical research, which has developed in this direction, therefore does not rule out the possibility that certain clinical pathologies are, at least in part, diseases of the interpreting function of the left cerebral hemisphere.

The Relationship Between Weak Central Coherence and Systemisation

Systemising is highly compatible with weak central coherence: systemisers typically collect information within autonomous taxonomies, which do not correlate with one another. An example of an autonomous and independent taxonomy is the structure that 'calendar genes' build from details, starting with single days, and then moving on to weeks, then months and finally years.

The fascinating example of Elly Park (Park and Youderian 1974) illustrates the coexistence of a weak central coherence and a strong local one.

Elly paid close attention to shadows because they were important and significant to her and very relevant to her moods. When Elly travelled to a different time zone, she was alarmed that her shadow at six o'clock in the afternoon was not where she would be at home. She was unable to relax until her mother explained that six o'clock on her watch meant only five o'clock in the new place. This example suggests that Elly's pattern of linking shadow size with times of day was governed by consistent rules. When unexpected discrepancies occurred, it was a real problem for her. Nevertheless, her scheme was narrow and autonomous.

Elly was unable to see his scheme for what it was: a small fragment in a larger model of reality, susceptible to expansion and modification. This awareness would depend on the possession of a strong central coherence, together with a large amount of information.

Heightened Perceptual Processing

Autistic people have often been attributed with the ability to distinguish very fine details with their eyes and ears. Is this a consequence of weak central coherence, or is it a phenomenon in its own right? It is the result of super-efficient low-level perceptual processing, as Mottron and Burack (2001) suggest.

Autistic children may not be ready to grasp similarities at a more abstract level, and thus to place stimuli within the same category, even when they differ only in very small details.

Kate Plaisted (2000) and Michelle O'Riordan et al. (2001) conducted groundbreaking studies showing that higher discrimination hinders category learning and generalisation. Children with autism learned a specific dot pattern very well but then failed to generalise it to another set of dots, arranged in a slightly different way.

On the other hand, the autistic children succeeded brilliantly in difficult visual search tasks, where the target stimulus (*the one to be detected*) differed from other distracting stimuli by only a small detail.

This work promises to capture fundamental features of perceptual processing.

One such characteristic may be a visual system that functions inefficiently: the magnocellular system, named after the large size of its constituent neurons. An inefficiency of this system leads to a reduced ability to perceive the direction of rapidly moving stimuli.

Indeed, this deficit has been found in numerous studies of different groups of children with developmental disorders, including autism.

Elizabeth Milne with John Swettenham and colleagues (2002) proposed that the magnocellular system processes patterns in terms of a general gestalt, while a different vi sive system, the parvocellular (so-called because of the small size of its cells) processes stimuli according to their individual characteristics.

If there were only a 'magno-system' the world would appear soft and blurred, without contour lines. If there were only a 'parvo-system' the contours and details would stand out sharply. It is possible that these two systems in autism are out of balance, which would, in fact, lead to a different way of seeing things.

Difficulties in integration and generalisation have also been documented by Laura Klinger and Geraldine Dawson (2000) at the University of Washington's Centre for Autism. These researchers found that autistic children have difficulty applying a previously learned concept to a new situation in which that concept would be appropriate.

Instead of assuming that this problem arises from superior low-level processing, they propose that the problem is the result of a defect in high-level processing. This defect would prevent the formation of abstract mental categories. Using two types of tests, they showed that autistic children were not able to abstract an implicit prototype, but were capable of forming categories if it was possible to do so by memorising rules.

John Bargh (2017) has argued that our brains are wired to absorb two different streams of information from our eyes: one is for understanding and knowledge, the other for appropriate behaviour. The first flows into a conscious estuary and the second into a more automatic, unconscious estuary.

The two visual streams were discovered in the 1990s by two neuropsychologists, David Milner and Melvyn Goodale (1991).

Both start in the retina and then travel to the primary visual cortex, where the information is analysed in greater depth. At this point, one stream channels to the brain regions responsible for knowledge—for example, identifying an object—and provides the type of information that is used to answer questions about it; the other stream flows directly to a region responsible for action. The action channel operates mainly outside consciousness, while the understanding channel is usually accessible to consciousness.

According to Bargh, this circuit in autistic people does not modulate the flow of information.

While many studies have focused on vision, Laurent Mottron and his colleagues (2000), in Montreal, focused on hearing, and again found evidence of accentuated low-level processing. In a musical experiment, they set up an ingenious situation that produced a conflict between global and local processing. The transposition of a melody into a different key can be conceived of as a gestalt, which involves global processing, whereas the detection of a single out-of-tune note involves local processing.

The autistic persons were perfectly able to transpose a melody, showing good global processing. They were also superior to the control group in recognising a single wrong note, even when the error was very slight, showing good local processing. Given that many individuals with autistic disorder can remember the absolute pitch of individual notes, Mottron and his colleagues proposed that autistics exhibit heightened processing of the elemental physical properties of sounds.

Early investigations using the technique of 'neural network modelling' (Gustafsson, 2000; McClelland, 2000) demonstrate the accentuation of visual and auditory processing, an argument in favour of a modification of the original formulation of the central coherence theory. Other experiments, however, argue against such a modification.

This unresolved issue reminds us that information processing occurs in a vast dynamic system. It is conducted both 'top-down', i.e. top-down (*e.g. controlled by prior knowledge*) and 'bottom-up', i.e. bottom-up (*e.g. controlled by incoming data*). The key problem in the autistic brain may be to unify these two channels of information. If this is the case, the old hypothesis and the new can be reconciled by assuming that bottom-up processing is redundant precisely because there are problems with top-down processing.

Relationship Between Mind Reading and Central Coherence

Happé (2001) hypothesises that the ability to develop sophisticated mentalisation strategies is facilitated by the ability to make use of context. According to Baron-Cohen, there is a trade-off between empathising and systemising; she argues that the systemising strategy is not suitable for understanding social relations. Theories about a heightened capacity for discrimination tell us little about mentalisation.

This opens up two hypotheses: on the one hand, the activity of mentalising may not require large amounts of information integration, and on the other hand—contrary to Uta Frith's 1989 theory—some autistics may, in fact, have a preference for strong, rather than weak, central coherence. If this is so, then Happé's hypothesis implies that people with stronger coherence are more capable of acquiring social understanding and can more readily compensate for their mentalisation deficits.

Happé points out that lesions of the right hemisphere often lead to a decrease in global perception. A fragmented style has also been identified in children with Williams syndrome, and this finding may be a clue to its brain basis. (I will discuss this in more detail in Chapter 14.

The 'Mirror Neuron Dysfunction' Hypothesis and the 'Distorted Relevance Landscape' Hypothesis

The ability to have a theory of mind is based not on our general intelligence but on a specific set of brain mechanisms that have evolved to endow us with a no less important degree of social intelligence.

The idea that there are specialised circuits for social cognition was first suggested by psychologist Nick Humphrey (1983) and primatologist David Premack, and now enjoys much empirical support. Thus Uta Frith's intuition about autism and theory of mind seems to be confirmed: perhaps the severe deficits of autistic children in social interactions stem from an impairment of theory of mind circuits. This is undoubtedly a sensible idea, but, come to think of it, to say that autistic children are unable to interact socially because they have a defective theory of mind in practice describing their symptoms in other words. This is a good starting point, but what is really needed is to identify the brain systems whose known functions correspond to those that are disturbed in autism.

Many neuroimaging studies have been conducted on autistic children, some of which neuroscientist Eric Courchesne (see Chapter 6 pioneered). It has, for example, been observed that these children have larger brains, with larger ventricles (*brain cavities*). The same group of researchers also noted considerable differences in the cerebellum. These are interesting observations that will certainly have to be taken into account when we begin to understand autism better, but they do not explain the typical symptoms of the disorder.

In children with lesions of the cerebellum caused by other organic diseases, very characteristic symptoms are observed, such as intentional tremor, nystagmus (*rhythmic movements of the eyeballs*) and ataxia (*lack of coordination of movements when walking*), but none of these symptoms are typical of autism. Conversely, typical symptoms of autism are never observed in dysfunctions of the cerebellum.

One reason could be that the abnormalities observed in the cerebellum of autistic children are perhaps the unrelated side effects of abnormal genes whose other effects are the real causes of the disorder. If so, what could these other effects be? If we want to explain autism, we must look for neural structures in the brain whose specific functions correspond exactly to the particular symptoms that are characteristic of this condition.

The clue comes to us from mirror neurons. In the late 1990s, Ramachandran's group thought that such neurons were ideal candidates for research. The discovery of mirror neurons was crucial, as they provide the physiological basis for certain higher abilities that neuroscience had long been trying to explain. They were struck by the fact that precisely the supposed functions of mirror neurons, such as empathy, interpretation of intentions, imitation, pretence play and language learning, were dysfunctional in autism.

According to Ramachandran "*If we wrote two lists next to each other, the first with the known characteristics of mirror neurons and the second with the clinical symptoms of autism, we would find an almost exact match*".[3]

It therefore seemed reasonable to hypothesise that the main cause of autism was a dysfunction of the mirror neuron system. This hypothesis has the advantage of explaining many apparently unrelated symptoms with a single cause. His hypothesis is that, in essence, the main cause of autism is a dysfunction of the mirror neuron system.

To intercept the activity of this class of neurons, Ramachandran and co-workers used the electroencephalogram (EEG), which picks up brain waves through a grid of electrodes applied to the scalp and was the very first scanning technique invented, well ahead of CT and MRI scans.

The brain generates revealing patterns of electrical waves of different frequencies. It has been known for over half a century that a particular brain wave, the 'mu', is suppressed every time a voluntary movement is made, even one as simple as opening and closing one's fingers.

It was later discovered that mu-wave suppression also occurs when watching someone else perform the same movement. Ramachandran hypothesised that mu-wave suppression could be a simple, inexpensive and non-invasive 'probe system' to monitor mirror neuron activity.

In the famous pilot experiment with Justin, an autistic child, as in neuronormal children, the mu-wave appeared conspicuous when he sat idly by doing nothing, while it was suppressed whenever he was asked to perform simple voluntary movements. But when he watched someone else perform the action, there was no suppression. This observation significantly supported the hypothesis: the child's motor command system was intact (*in fact, Justin was perfectly capable of opening doors, eating French fries, drawing pictures, climbing stairs and so on*), whereas the mirror neuron system showed a deficit.

Ramachandran presented a single pilot case study at the annual congress of the Society for Neuroscience in 2000 and that was followed by ten more in 2004. The results were identical. In the intervening years, these observations have been widely confirmed by many different groups using a variety of techniques.

A team of Finnish researchers led by Riitta Hari at the Technical University of Helsinki, for example, confirmed our conjecture using magnetoencephalography (MEG). Michele Villalobos and her colleagues at San Diego State University used functional magnetic resonance imaging to show that in autistic patients there is a reduction in functional connectivity between the visual cortex and the prefrontal region of mirror neurons.

Other researchers have tested the hypothesis using transcranial magnetic stimulation (SMT). In a way, SMT is the opposite of the EEG: instead of passively intercepting electrical impulses from the brain, it creates electrical currents using a powerful magnet above the head: one can artificially induce neural activity in whatever brain region is near the scalp; many areas are hidden in the deep circumvolutions of the brain, but several others, including the motor cortex, are located right at the edge of the skull, where SMT can easily 'hit' them.

In this case, SMT was used to stimulate the motor cortex, recording electromuscular activation as subjects watched other people perform actions. When a neuronormal subject watches another person perform an action, such as clutching a tennis ball with his right hand, the muscles of his right hand register a small increase in the usual electrical 'noise': the mere act of watching the action produces a small but quantifiable increase in the readiness to contract of the muscles that would contract if he were performing the act himself. The motor system automatically simulates the perceived action, but at the same time automatically suppresses the spinal motor signal to prevent it from being fully activated; however, a small percentage of the suppressed motor command manages to filter out and reach the muscles. The autistics involved in the test, on the other hand, showed no increase in muscle potentials while they watched others perform the actions: their mirror neuron system did not become active.

The mirror neuron hypothesis can explain several of the strangest manifestations of autism. It has long been known, for example, that autistic children often have difficulty understanding proverbs and metaphors. When they are told to 'get a move on', they may literally start moving around the room.

This difficulty with metaphors requires an explanation.

A branch of cognitive science called 'embodied cognition' argues that human thought is profoundly shaped by its interconnectedness with the body and the intrinsic nature of sensory and motor processes. This theory is in contrast to what we might call the 'classical view', which dominated the cognitive sciences from the mid- to late twentieth century and argued that the brain resembled very much, in practice, a multi-purpose 'universal computer' casually connected to a body.

People who, as a result of lesions to the left supramarginal gyrus, suffer from apraxia find it difficult to interpret action-based metaphors such as 'looking for the moon in the well'. Since the supramarginal gyrus also contains mirror neurons, experiments on circuits equipped with mirror neurons have led to the suggestion that, in humans, this system is not only responsible for performing coordinated actions but also for understanding action-based metaphors and other aspects of embodied cognition.

The mirror neuron hypothesis also allows us to better understand autistic language difficulties. Mirror neurons almost certainly come into action when a young child first repeats a sound or a word that it has heard. If we observe a child between 3 and 6 months of age, i.e. when he has just discovered the capacity for vocal modulation but before he begins to experience intentional articulated speech, we notice that he is engaged in an 'imitation game' of lip movement with the adult who is talking to him. In this context, the role of mirror neurons is essential to enable the acquisition of the modulatory motor skills suitable for the production of a shared language. This activity may require an internal translation: the mapping of acoustic patterns onto corresponding motor patterns and vice versa.

Research conducted on persons diagnosed with autism spectrum disorders has shown how, not only at a cerebral level but also at a cognitive level, the understanding of the vital form (in Daniel Stern's sense, 2010) characteristic of action is different from and in part independent of the understanding of the purpose to which that action is directed, to the extent that one can occur without the other [by 'vital form' here we mean the 'quality understood as manner' of an action; for example, a grip can be strong, firm, light...].

Magali Rochat (Rochat et al. 2013) compared the performance of persons diagnosed with autism with that of neuronormal persons of the same chronological and verbal age in purpose recognition and vital form tasks of observed actions similar to those used in the fMRI studies cited above. The results showed that individuals diagnosed with autism were not significantly less accurate when it came to identifying the purpose of observed actions and judging whether they were the same or different purposes. Things changed, however, and not a little when it was the life forms that were the subject of identification and judgement. People diagnosed with autism were far less accurate; in particular, they made far more errors when the actions differed in purpose or life form, i.e. when it was a question of recognising the same life form in different actions or different life forms in actions that were identical in terms of the type of purpose.

Rochat and co-workers' finding that persons diagnosed with autism did not differ in the recognition of the purpose of the observed actions is in line with the results obtained in a study conducted by Sonia Boria and co-workers, in which children diagnosed with autism were required to observe images depicting manual actions directed at objects of different types (e.g. a hand touching or picking up a glass or a pen) and judge what the observed hand was doing (in the case of the glass, was it touching or picking it up?) or why it was doing what it was doing (if it was picking it up, was it picking it up to drink or to move it?) (Boria et al. 2009).

Children with autism spectrum disorder did not have great difficulty identifying the immediate purpose of the observed action, whereas they had greater problems identifying why the observed action was performed, tending to be guided by the type of object rather than the type of action. Similar results regarding the ability of individuals diagnosed with autism to recognise the immediate purpose of observed actions were obtained by Antonia Hamilton and collaborators, who showed that these children have no difficulty imitating the purpose of the observed action, so much so that they do so even when the movements performed are not the same as those exhibited by the demonstrator (Hamilton et al. 2006).

Peter Hobson and Anthony Lee have shown that children diagnosed with autism, while imitating the purpose of observed actions without major problems, tend not to imitate the vital form that characterises those actions, especially when that form does not directly relate to the purpose (Hobson and Lee, 1999).

Similar results were obtained by Peter and Jessica Hobson in a subsequent study aimed at systematically investigating the ability of autistic children to imitate the vital forms characteristic of observed actions (Hobson and Hobson, 2008). To this end, the authors made use of a rather simple set of actions, separately testing the ability to imitate the presented action with respect to the purpose to which it was directed or with respect to the vital form that characterised it, and distinguishing between vital forms that were not incidental to the purpose, vital forms that served as a means to the attainment of the purpose and vital forms that themselves constituted the purpose or were part of it. As in the previous study, it was found that the autistic children did not differ in imitating the purpose of a given action; when it came to imitating the vital form of that action, they were significantly less able, particularly when the style to be imitated was incidental to the purpose of the action imitated.

Taken together, the above data suggest that the understanding of the vital forms of others is different from and, to a large extent, independent of the understanding of the purposes of others' actions, both of which refer to sensorimotor transformations of a different kind.

This indicated that the children diagnosed with autism were not able to motorically represent the action of reaching for and taking the object as part of an overall action that had the purpose of prehension of that object to bring it to the mouth instead of putting it in a container. It is not surprising, therefore, that they did not activate the mylohyoid muscle in any way while watching others perform prehension actions aimed at bringing it to their mouth instead of putting it in a container. Since they could not have those kinds of motor representations in first-person execution, they could not recruit them when the action was observed instead of performed. This would also explain why, in the study by Boria collaborators (Boria et al. 2009), they had problems identifying the distal purpose of the observed action (to do what the experimenter takes a given object), especially when this purpose was not suggested by the type of object taken.

There are two ways in which such a system could have been formed. The first is that, as soon as the word is heard, a mesthetic trace of the phonemes is formed in the auditory cortex. The child then tries to articulate various sounds and, using the error feedback from the mesthetic trace, gradually refines the output to match the memory (*this is something we all do, when we whistle to ourselves a tune we have just heard and then sing it out loud, refining the output to match the soft whistle*). The second is that the networks for translating heard sounds into spoken words are innate and specified by natural selection. In both cases, the net result would be a system of neurons with properties like those we attribute to mirror neurons.

If the child could, without trial and error, repeat a group of phonemes that it has just heard for the first time, we would have proof of the existence of a translation mechanism inscribed in the genes (*which somewhat disproves, from a neuroscience perspective, the early Chomsky's theory of language innatism*).

Regardless of how this extraordinary mechanism might have arisen, the results of this class of experiments suggest that a defect in its initial configuration causes the severe deficit of autism. Empirical results with mu-wave suppression corroborate the hypothesis and also provide a unified explanation for a series of seemingly unrelated symptoms.

Although it originally developed to elaborate an internal model of the actions and intentions of others, perhaps in humans the mirror neuron system evolved further, turning inwards to represent the mind to itself. The theory of mind not only serves to intuit what is going on in the minds of friends, strangers and enemies, but, in the uniqμe case of man, greatly enhances the understanding of how his own mind works. This development most likely occurred during the mental phase transition that we underwent some two hundred thousand years ago and marked the dawn of accomplished self-awareness.

If the mirror neuron system presupposes a theory of mind, and if the theory of mind in humans also has the additional feature of being able to apply to the inner, i.e. the self, it is understandable why autistic people struggle to have social interactions and strong self-identification, and why so many autistic children struggle to use the pronouns 'I' and 'you' correctly in conversation. In all likelihood, they do not have adequate mental self-representation to grasp the distinction. If this hypothesis were true, even otherwise high-functioning autistics, who are able to speak normally, would struggle to make conceptual distinctions between words such as "self-esteem", "pity", "compassion", "forgiveness" and "embarrassment", not to mention "self-pity", which is a meaningless term in the absence of a full sense of self.

It should first be noted that small groups of cells with mirror neuron-like properties are found in many parts of the brain and should be considered parts of a vast interconnected circuit, a kind of 'mirror network'. We must also be careful not to attribute all puzzling aspects of the brain to mirror neurons. Finally, ascribing certain cognitive capacities to certain neurons (in this case, mirror neurons) or certain brain regions is only a start; we still need to understand how those neurons perform their calculations and how they communicate them, as well as to understand in detail with which and how many neurons they are connected and from where they acquire the inputs to which they respond.

Understanding anatomy is a fundamental guide and helps to reduce the complexity of the problem. In particular, anatomical data limit theoretical speculation and help eliminate many hypotheses that initially seemed interesting. To say 'mental capacities emerge in a homogeneous network' not only adds nothing but contrasts with the empirical evidence of the admirable anatomical specialisation of the brain, which is not to be confused with locationalism: the brain operates through complex integrated circuits integrating highly specialised components. Just think in absolute terms of the proportionally small size of Broca's and Wernike's areas in relation to the extreme complexity of the task, and despite this high degree of specialisation, the system of comprehension and modulation of language and expression and articulation of thought would not be possible without a vast interconnected network.

The mirror neuron hypothesis explains well the main features of autism: the lack of empathy, pretence play, imitation and theory of mind. However, the spectrum is not complete, because there are other common (*though not decisive*) symptoms of the disorder on which mirror neurons have no obvious influence. Some autistics, for example, rock back and forth, avoid making eye contact, show hypersensitivity and aversion to certain sounds, and often indulge in tactile self-stimulation, sometimes beating themselves up, perhaps to alleviate auditory hypersensitivity. Can we put this idea in the context of the other considerations we have made so far on this topic?

In the early 1990s, Ramachandran's research group in collaboration with Bill Hirstein and Portia Iversen, co-founder of Cure Autism Now, reasoned about how these other symptoms could be explained and came up with what was called the 'relevance landscape theory'. When a person looks at the world, he or she faces a potentially disconcerting sensory overload. Information about the world is first sifted by the sensory areas of the brain and then relayed to the amygdala which, as the gateway to the emotional core of the brain, performs emotional surveillance of the outside world, assesses the emotional importance of everything we see and decides whether it is something mundane and routine or something relevant. In the latter case, it tells the hypothalamus to activate the autonomic nervous system in proportion to the excitement that the triggering vision is capable of arousing (*in a range from mildly interesting to absolutely terrifying*). Thus the amygdala creates a 'relevance landscape' of our world, with hills and valleys corresponding to the large or small importance of what is perceived.

Sometimes it is possible for this circuit to go haywire. The autonomic response to something exciting manifests itself in increased sweating, heart rate, muscle readiness, and so on, all of which prepare the body for action. In extreme cases, the wave of physiological arousal reverberates in the brain and prompts the amygdala to 'sound the alarm'. The result is a blitzkrieg of the autonomic nervous system. Many adults are subject to such panic attacks, but most people do not risk being overwhelmed by such a maelstrom (on the details of how this circuit works in PTSD, in particular, see Di Salvo, 2024).

With all this in mind, Ramachandran's team analysed the possibility that the relevance landscape of autistic children was distorted. The problem could be partly due to indiscriminately strengthened (*or reduced*) connections between the sensory cortices and the amygdala and perhaps between the limbic structures and the frontal lobes. Because of such abnormal connections, every slightest event or object would have triggered an uncontrollable storm of the autonomic nervous system, which would have explained why autistics prefer trance and routine. If, on the other hand, the emotional arousal had been less pronounced, autistic people would perhaps have attached absolutely excessive significance to some unusual stimuli, which would have explained their strange interests, including occasional savants. Conversely, if some of the connections from the sensory cortex to the amygdala had been partially obliterated by relevance landscape distortions, autistic people would have ignored things, such as eyes, that most normal children find very interesting to look at.

To test the relevance landscape hypothesis, the psychogalvanic reflex was measured in a group of thirty-seven autistic and twenty-five neuronormal children. As expected, the children in the control group showed arousal for certain categories of stimuli, but not for others (*they had the psychogalvanic reflex in front of photos of their parents, but not in front of photos of pencils*). The autistic children, on the other hand, showed generally higher autonomic arousal, which was further amplified by the most mundane objects and events, while some very relevant stimuli, such as eyes, had no effect on them.

If the relevance landscape theory were to go in the right direction, it would be logical to expect abnormalities of visual pathway 3 in the brains of autistic persons. Pathway 3 not only projects to the amygdala, but also crosses the superior temporal fissure, which, together with the neighbouring insula, is rich in mirror neurons. Mirror neurons in the insula have been shown to be responsible for the empathic perception and expression of certain emotions, such as disgust (also social and moral). Thus a lesion in these areas, or perhaps a deficiency or dysfunction of the mirror neuron system within them, may not only distort the landscape of relevance but also reduce empathy, social interaction, imitation and pretence play.

The relevance landscape theory could also explain two other bizarre aspects of autism that have always baffled scientists. The first is that, as some parents have reported, a child's autistic symptoms temporarily disappear during an attack of high fever. As a rule, fever is caused by bacterial toxins that act on thermoregulatory mechanisms in the hypothalamus, at the base of the brain, an area that is part of pathway 3 (*certain dysfunctional behaviours, such as hysterical seizures, originate from networks bordering the hypothalamus*). So the fever could have had an 'overflow' effect that dampened activity in one of the bottlenecks in the feedback loop that generates the autonomic arousal storms and associated hysterical crises. This is a hypothetical explanation, but if it turns out to be true, it would provide another basis for therapeutic intervention. The second bizarre aspect is that autistic children often hit and slap themselves. This behaviour is called 'somatic self-stimulation'. In light of this theory, it can be hypothesised that self-stimulation mitigates the autonomic arousal storms these children suffer from. Ramachandran's research team found that it not only has a calming effect but also leads to a measurable reduction in the psychogalvanic reflex. In this theoretical context, rocking back and forth could serve a similar purpose. We know that it most likely stimulates the vestibular system (*sense of balance*) and we know that balance-related information at some point splits off to travel along route 3 to the insula. So repetitive rocking perhaps dampens arousal in the same way that beating yourself up dampens it. It might also help in terms of perception to anchor the self to the body and provide coherence to an otherwise chaotic world.

Almost one-third of autistic children suffered from temporal lobe epilepsy (*TLE—the percentage could be much higher if we included complex partial seizures that were not clinically identified*) in childhood. In adults, TLE manifests with overt emotional disturbances, but since the brain is fully mature, the disease does not lead to profound cognitive distortions. Less well known, however, are the effects of temporal lobe epilepsy on the developing brain. Seizures are triggered by repeated random bursts of nerve impulses passing through the limbic system. If they occur frequently in the infant brain, they can lead, through a process of synaptic reinforcement called kindling, or 'recurrent stimulation', to a selective, but diffuse and indiscriminate strengthening (*or sometimes cancellation*) of the connections between the amygdala and the visual, auditory and higher somatosensory cortices. This would explain both the frequent false alarms aroused by the sight of trivial or insignificant things and by sounds that are for all intents and purposes neutral and, conversely, the lack of reaction to socially salient information, two typical features of autism.

In more general terms, the feeling of being an integrated and embodied self, which each of us has, seems to depend above all on a reciprocal, echo-like 'reverberation' between the brain and the rest of the body, and also, through empathy, between the self and others. An indiscriminate disturbance of the connections between the higher sensory areas and the amygdala, and the resulting distortions of one's own landscape of relevance, could, as part of the process itself, result in a disturbing loss of the sense of bodily identity that allows human beings to feel a distinct and autonomous self, anchored to a body and embedded in a society.

Perhaps somatic self-stimulation represents an attempt by some autistic children to regain bodily identity by restoring and reinforcing body-mind interactions while attenuating amplified autonomic signals. A fine balance of such interactions is perhaps crucial to the normal development of an integrated self, which we usually take for granted is, axiomatically, the foundation of being a person. It is therefore not surprising that the very sense of being a person is profoundly disturbed in autism.

Both theories—the mirror neuron dysfunction hypothesis and the distorted relevance landscape hypothesis—attempt to find a unified rational reason to explain the bewildering array of seemingly unrelated symptoms that characterise the autistic condition. Of course, the two hypotheses are not necessarily mutually exclusive; indeed, it is known that there are connections between the mirror neuron system and the limbic system. It may be that it is distortions in the limbic-sensory connections that ultimately produce a deficit in the mirror neuron system. Whatever the mechanisms behind the disorder, the results clearly suggest that autistic children have a dysfunctional mirror neuron system that can explain many features of the syndrome.

Autism reminds us that the uniquely human sense of self, despite its pronounced tendency to assert its privacy and independence, actually emerges from reciprocal interactions with others and with its host body. When it distances itself from society and withdraws from its own body, it almost ceases to exist, or at least the self that defines our lives as human beings no longer exists. In this sense, autism could ultimately be considered a disorder of self-consciousness and, if it were, research in this field would help us understand the nature of consciousness itself.

Notes

1. In 2013, Marcus Wohr of the University of Marburg, and Maria Luisa Scattoni of the Istituto Superiore di Sanità in Italy, argued that "when tested in an open field or an observation box, rodents with "autism" they would demonstrate that they have a triad social behaviour deficit through the limitation of "social interaction" and "sociality" play". According to them, they demonstrate the communication deficits of the triad through the lack of scent marking, the absence of interaction and induction by the female through ultrasonic "vocalisations". Finally, "motor stereotypes", "repetitions" and "self-cleaning", the repetitive burying of marbles and "limited exploration" would demonstrate the stereotyped and repetitive behaviours of the triad of impairments. A Franco-German project has also developed an autism severity scoring system to measure the degree of autism in mice, measuring the degree of skill in nest building and other activities and the degree of atypical behaviours such as excessive burying. These standardized approaches have helped consolidate the idea that rodents can suffer from autism. Quite different, of course, are the mouse models in which—having identified a genetic correlation—one intervenes to replicate and study autism-related gene interaction and expression. Considerable studies have demonstrated the validity and usefulness of genetic experimentation in this regard. Much less certain is the behavioural analysis, especially in the case of complex states such as autism that involve a broad and heterogeneous spectrum.
2. See:

   Gazzaniga, M. (1970) The Bisected Brain, Appleton-Century-Crofts, New York.

   Gazzaniga, M. LeDoux, J., (1978) The Integrated Mind, Springer US.

   Gazzaniga, M. (1985) The Social Brain, Basic Books, New York.

   Gazzaniga, M., Senior, C., Russell, T., (2006) Methods in Mind-Cognitive Neuroscience-MIT Press.

   Gazzaniga, M. (2007) L'interprete. Come il cervello decodifica il mondo, Di Renzo Editore, Roma.
3. For an overall summary of the topic see.

   Ramachandran, Vilayanur (2012) The Tell-Tale Brain: A Neuroscientist's Quest for What Makes Us Human—W W Norton & Co Inc.

In specific, see:

Ramachandran, Vilayanur (2000). "MIRROR NEURONS and imitation learning as the driving force behind 'the great leap forward' in human evolution.

Ramachandran, VS; Hirstein, W (1998). "The perception of phantom limbs. The D. O. Hebb lecture". Brain. 121 (9): 1603–30.

Ramachandran, V. S; Yang, T. T; Gallen, C. C; Cobb, S; Schwartz, B. J; Bloom, F. E (1994). "Noninvasive detection of cerebral plasticity in adult human somatosensory cortex". NeuroReport.

Ramachandran, V; Rogers-Ramachandran, D; Stewart, M; Pons, Tim P (1992). "Perceptual correlates of massive cortical reorganization". Science. 258.

Ramachandran, Vilayanur S; Oberman, Lindsay M (2006). "Broken Mirrors: A Theory of Autism". Scientific American. 295.

Oberman, Lindsay M; Hubbard, Edward M; McCleery, Joseph P; Altschuler, Eric L; Ramachandran, Vilayanur S; Pineda, Jaime A (2005). "EEG evidence for mirror neuron dysfunction in autism spectrum disorders". Cognitive Brain Research. 24.

Oberman, Lindsay M; Ramachandran, Vilayanur S (2007). "The simulating social mind: The role of the mirror neuron system and simulation in the social and communicative deficits of autism spectrum disorders". Psychological Bulletin. 133.

Oberman, Lindsay M.; McCleery, Joseph P.; Ramachandran, Vilayanur S.; Pineda, Jaime A. (2006). "EEG evidence for mirror neuron activity during the observation of human and robot actions: Toward an analysis of the human qualities of interactive robots". Neurocomputing. 70.

Oberman, Lindsay M.; McCleery, Joseph P.; Ramachandran, Vilayanur S.; Pineda, Jaime A. (2006). "EEG evidence for mirror neuron activity during the observation of human and robot actions: Toward an analysis of the human qualities of interactive robots". Neurocomputing. 70.

On research on mirror neurons related to the autism spectrum, see in particular:

Whiten, A. Imitation, Pretence and Mindreading: Secondary Representation in Comparative Primatology and Developmental Psychology. In: A. E. Russon, K.A. Bard and S.T. Parker (Eds.) Reaching Into Thought: The Minds of the Great Apes, pp 300–324. Cambridge: Cambridge University Press; 1996.

Whiten A. The Machiavellian mindreader. In: A.Whiten R.W.Byrne (Eds.) Machiavellian Intelligence II.: Evaluations and Extensions. Cambridge University Press; 1997.

Smith I.M. & Bryson S.E. Imitation and action in autism: a critical review. Psychol Bull. 1994.

Rogers S. An examination of the imitation deficit in autism. In: J. Nadel & G. Butterworth (Eds.) Imitation in infancy. Cambridge: Cambridge University Press. 1999.

Rogers S.J., Bennetto L., McEvoy R. & Pennington B.F. Imitation and pantomime in high-functioning adolescents with autistic spectrum disorders. Child Dev. 1996.

Perrett D.I., Smith P.A.J., Potter D.D., Mistlin A.J., Head A.S., Milner A.D. and Jeeves M.A. Neurons responsive to faces in the temporal cortex: Studies of functional organization, sensitivity to identity and relation to perception. Human Neurobiol. 1984.

Perrett D.I., Smith P.A.J., Mistlin A.J. Chitty A.J., Head A.S. Potter D.D. Broennimann R., Milner A.D. & Jeeves M.A. Visual analysis of body movements by neurons in the temporal cortex of the macaque monkey: A preliminary report. Behav Brain Res. 1985.

Perrett, D.I., Harries, M.H., Benson, P.J., Chitty, A.J. & Mistlin, A.J. Retrieval of structure from rigid and biological motion; an analysis of the visual response of neurons in the macaque temporal cortex. In: Troscianko, T. & Blake, A. (Eds.) AI and The Eye. Chichester: J. Wiley & Sons. 1990.

Gallese V., Faddiga L., Fogassi L. & Rizzolatti G. Action recognition in the premotor cortex. Brain. 1996.

Rizzolatti G., Fadiga L., Matelli M., Bettinardi V., Paulesu E., Perani D. & Fazio F. Localisation of grasp representations in humans by PET: 1. Observation vs execution. Exp Brain Res. 1996.

Rizzolatti G., Fadiga L., Matelli M., Bettinardi V., Paulesu E., Perani D. & Fazio F. Localisation of grasp representations in humans by PET: 1. Observation vs execution. Exp Brain Res. 1996.

Carey D.P. 'Monkey see, monkey do' cells. Current Biol. 1996.

Hari R., Forss N., Avikainen S., Kirveskari E., Salenius S. & Rizzolatti G. Activation of human primary motor cortex during action observation: a neuromagnetic study. Proc Natl Acad Sci. 1998.

Fadiga L., Fogassi G., Pavesi G. & Rizzolatti G. Motor Facilitation during action observation: a magnetic stimulation study. J Neurophysiol. 1995.

Grafton, S. T., Arbib, M. A., Fadiga, L., & Rizzolatti, G. Localization of grasp representations in humans by positron emission tomography. 2. Observation compared with imagination. Exp Brain Res. 1996.

Rizzolatti, G., Fadiga, L., Gallese, V., & Fogassi, L. Premotor cortex and the recognition of motor actions. Cog Brain Res. 1996.

Iacoboni M., Woods R.P, Brass M., Bekkering H., Mazziota J.C. & Rizzolatti G. Cortical mechanisms of human imitation. Science. 1999; 286.

Fogassi L., Gallese V., Fadiga L., Rizzolatti G. Neurons responding to the sight of goal-directed hand/arm actions in the parietal area PF (7b) of the macaque monkey. Soc Neurosci Abstracts. 1998; 257.

Gallese V., and Goldman A. Mirror neurons and the simulation theory of mind-reading. Trends Cog Sci. 1998; 2

Hatfield E., Cacioppo J.T., & Rapson R.L. Emotional Contagion. New York: Cambridge University Press, 1994.

Surakka V. Hietanen JK. Facial and emotional reactions to Duchenne and non-Duchenne smiles. Int J Psychophysiol. 1998; 29.

Whiten A. & Ham R. On the nature and evolution of imitation in the animal kingdom: Reappraisal of a century of research. Advance Study Behav. 1992; 21.

Hughes C., Russell J. & Robbins TW. Evidence for executive dysfunction in Autism. Neuropsychologia. 1994; 32.

Hughes C. Planning problems in autism at the level of motor control. J Autism Dev Disord. 1996; 26.

Ozonoff S., Pennington B. & Rogers S.J. Executive function deficits in high-functioning autistic individuals: relationship to theory of mind. J Child Psychol Psychiatry. 1991; 32.

Ozonoff S. & Jensen J. Specific executive function profiles in three neurodevelopmental disorders. J Autism Dev Disord. 1999; 29.

Perner J., Lang B. Development of theory of mind and executive control. Trends Cog Sci. 1999; 3

Griffith E.M., Pennington B., Wehner E.A. & Rogers S.J. (1999). Executive functions in young children with autism. Child Development 70.

Ozanoff S. & Strayer D.L. (1997). Inhibitory function in non-retarded children with autism. J Autism Dev Disord; 27.

Hughes C., Leboyer M., Bouvard M. (1997). Executive function in parents of children with autism. Psychological Medicine 27.

Avikainen S., Kulomaki T. & Hari R. Normal movement reading in Asperger subjects. Neuroreport. 1999; 10.

References

Axelrod, R., (1984) The Evolution of Cooperation. New York: Basic Books

Bardi, L., Six, P., & Brass, M. (2017) "Repetitive Ms of the temporo-parietal junction disrupts participant's expectations in a spontaneous Theory of Mind task" - Social Cognitive and Affective Neuroscience, 12, 11, pp. 1775–1782

Bargh, J., (2017) Before You Know It: The Unconscious Reasons We Do What We Do—Atria Books

Baron-Cohen, S., (2002) The extreme male brain theory of autism. «Trends in Cognitive Science» 6

Baron-Cohen, S. (2012) Zero Degrees of Empathy: A new theory of human cruelty and kindness—Penguin

Baron-Cohen, S., & Ring, H. (1994) A model of the mindreading system: Neuropsychological and neurobiological perspectives. In C. Lewis & P. Mitchell (Eds.), Children's early understanding of mind: Origins and development. Lawrence Erlbaum Associates, Inc.

Baron-Cohen, S., A. M. Leslie and U. Frith. (1985) Does the autistic child have a "theory of mind"?' Cognition 21

Barresi, J., Moore, C. (1996) Understanding self and other. Behavioral and Brain Sciences 19

Boria, S., Fabbri-Destro, M., Cattaneo, L., Sparaci, L., Sinigaglia, C., San-Telli, E., Cossu, G., Rizzolatti, G. (2009), "Intention understanding in autism". In PLOS ONE, 4, 5, e5596

Changeux, J.P., (2008) Du vrai, du beau, du bien. Une nouvelle approche neuronale. Odile Jacob

Cheney, D. L., & Seyfarth, R. M. (1990). How monkeys see the world: Inside the mind of another species. University of Chicago Press

Cook, R., Bird, G., Catmur, C., Press, C., & Heyes, C. (2014), "Mirror neurons: From origin to function". In Behavioral and Brain Sciences, 37, 2, pp. 177–192

de Waal, F. (1996) Good Natured. Harvard University Press. Cambridge, MA
Di Salvo, M. (2024) Psicoterapia e plasticità cognitiva e neuronale. Cervello sociale e disturbo post traumatico da stress.—Ricerca Psicoanalitica Journal of Relations in Psychoanalysis
Eibl-Eibesfeldt, I., (1970) Ethology: Biology of Behaviou. Holt, Rinehart and Winston
Frith, U., Frith, C.D. (2003) "Development and neurophysiology of mentalizing". Philosophical Transactions of the Royal Society B: Biological Sciences, 358, 1431, pp. 459–473
Frith U, Frith C. (2010) The social brain: allowing humans to boldly go where no other species has been. Philos Trans R Soc Lond B Biol Sci.
Gardner, H. (1987) The Mind's New Science. New York: Basic Books
Goel, V. (1995) Sketches of Thought. MIT Press
Goodale, M.A. et al. (1991) Percewing the World and Grasping It. A Neurological Dissociation, in «Nature», CCCIL
Gustafsson, L. (2000) Inadequate cortical feature maps: A neural circuit theory of autism, in «Biological Psychiatry», 42
Hamilton, A.F.C., Wolpert, D.M., Frith, U., Grafton, S.T. (2006) "Whe-re does your own action influence your perception of another person's action in the brain?". In Neurolmage, 29, 2, pp. 524–535
Happé, F. (1996) Studying weak central coherence at low levels: Children with autism do not succumb to visual illusions, in «Journal of Child Psychology and Psychiatry», 37
Happé, F. (1999) Autism: Cognitive deficit or cognitive style, in «Trends in Cognitive Science», 3
Happé, F. (2001) Social and non-social development in autism: Where are the links?, in J.A. Burack, T. Charman, N. Yirmiya, P.R. Zelazo (a cura di), The development of Autism: Perspectives from Theory and Research, Erlbaum, Mahwah (NJ)
Happé, F., Briskman, J., Frith, U. (2001a) Exploring the cognitive phenotype of au-tism: Weak «central coherence» in parents and siblings of children with autism. Experimental tests, in «Journal of Child Psychology and Psychiatry», 42
Happé, F., Briskman, J., Frith, U. (2001b) Exploring the cognitive phenotype of au-tism: Weak «central coherence» in parents and siblings of children with autism 2. Real-life skills and preferences, in «Journal of Child Psychology and Psychiatry», 42
Hermelin, B., Frith, U. (1969) The role of visual and motor cues for normal, subnormal and autistic children, in «Journal of Child Psychology and Psychiatry», 10
Hobson, R.P., Hobson, J.A. (2008) "Dissociable aspects of imitation: A study in autism". In Journal of Experimental Child Psychology, 101, 3, pp. 170–185

Hobson, R.P., Lee, A. (1999). "Imitation and identification in autism". In Journal of Child Psychology and Psychiatry and Allied Disciplines, 40, 4, pp. 649–659

Hunphrey, N., (1983) Consciousness Regained: Chapters in the Development of Mind, Oxford University Press, (in via ricapitolativa)

Keysers, C., Perrett, D.I., Gazzola, V. (2014) "Hebbian learning is about contingency, not contiguity, and explains the emergence of predictive mirror neurons". In Behavioral and Brain Sciences, 37, 2, pp. 205–206

Kilner, J.M., Frith, C.D. (2007) "Action observation: Inferring intentions without mirror neurons". In Current Biology, 18, 1, R32–33

Klinger, L.G., Dawson, G. (2000) Prototype formation in autism, in «Development and Psychopathology», 13

Kohlberg, L. (1973) The Claim to Moral Adequacy of a Highest Stage of Moral Judgment. The Journal of Philosophy, Vol. 70

Lorenz, K. (1966) On aggression. London: Methuen.

McClelland, J.L. (2000) The basis of hyperspecificity in autism: A preliminary suggestion based on proper-ties of neural nets, in «Journal of Child Psychology and Psychiatry», 30

Milne, E., Swettenham, J., Hansen, P., Campbell, R., Jeffries, H., Plaisted, K. (2002) High motion coherence thresholds in children with autism, in «Journal of Child Psychology and Psychiatry», 43

Milton, D. (2012). On the ontological status of autism: The 'double empathy problem. Disability & Society, 27(6), 883–887.

Milton, D. (2014). Autistic expertise: A critical reflection on the production of knowledge in autism studies. Autism, 18(7).

Mottron, L., Burack, J. (2001) Enbanced perceptual functioning in the development of autism, in J.A. Burack, T. Charman, N. Yirmiya, P.R. Zelazo (a cura di), The Development of Autism: Perspectives from Theory and Research, Erlbaum, Mahwah (NJ)

Mottron, L., Perez, I., Ménard, E., (2000) Local and global processing of music in high-functioning persons with autism. Journal of Child Psychology and Psychiatry 41

O'Riordan, M.A., Plaisted, K.C., Driver, J., S. Baron-Cohen, S. (2001) Superior visual search in autism, in «Journal of Experimental Psychology—Human Perception and Performance», 27

Park, D., Youderian, P., (1974) Light and number: Ordering principles in the world of an autistic child, in «Journal of Autism and Childhood Schizophrenia», 4

Piaget, J. (1945). La formation du symbole chez l'enfant. Tr. It. La formazione del simbolo nel bambino. La Nuova Italia, Firenze, 1972.

Plaisted, K.C., (2000) Aspects of autism that theory of mind cannot explain, in S. Baron-Cohen, H. Tager-Flusberg, D.J. Cohen (a cura di), Understanding OtherMinds: Perspectives from Developmental Cognitive Neuroscience, Oxford University Press, Oxford

Premack, D., & Premack, A. J. (1995). Origins of human social competence. In M. S. Gazzaniga (Ed.), The cognitive neurosciences The MIT Press.

Premack, D., & Woodruff, G. (1978). Does the chimpanzee have a theory of mind? Behavioral and Brain Sciences, 1

Rizzolatti G., & Sinigaglia C. (2019). Specchi nel cervello. Come comprendiamo gli altri dall'interno. Raffaello Cortina Editore Milano

Rochat, M.J., Veroni, V., Bruschweiler-Stern, N., Pieraccini, C., Bonnet-Brilhault, F., Barthélemy, C., Malvy, J., Sinigaglia, C., Stern, D.N., Rizzolatti, G. (2013), "Impaired vitality form recognition in autism". In Neuropsychologia, 51, 10, pp. 1918–1924

Saxe, R., Xiao, D.K., Kovacs, G., Perrett, D.I., Kanwisher, N. (2004) "A region of right posterior superior temporal sulcus responds to observed intentional actions". Neuropsychologia, 42, 11, pp. 1435–1446

Shah A., Frith U., (1983) An islet of ability in autistic children: A research note, in «Journal of Child Psychology and Psychiatry», 24

Shah A., Frith U., (1993) Why do autistic individuals show superior performance on the Block Design task?, in «Journal of Child Psychology and Psychiatry», 34

Sowden, S., Catmur, C. (2015) "The role of the right temporoparietal jun-ction in the control of imitation". In Cerebral Cortex, 25, 4, pp. 1107–1113

Stern, D. (2010) Le forme vitali. L'esperienza dinamica in psicologia, nell'arte, in psicoterapia e nello sviluppo. Tr. it. Raffaello Cortina, Milano 2011

Turiel, E., (1966) An experimental test of the sequentiality of developmental stages in the child's moral judgments

Turiel, E., (1978) Social interactions and the development of social concepts in preschool children published in Child Development

Wimmer, H., Perner, J. (1983) Beliefs about beliefs: representation and constraining function of wrong beliefs in young children's understanding of deception, Cognition, 13

Witkin, H.A., Goodenough, D.R., (1981) Cognitive Styles: Essence and Origins, International University Press, New York

Witkin, H.A., Oltman, P.K., Raskin, E., Carp, S. (1971) Children's Embedded Figures Test, Consulting Psychologists Press, Palo Alto

CHAPTER 10

Genetic Research on Autism: From the 'Autism Gene' to the 'Genetic Susceptibility Base'

According to Joseph LeDoux (2002), "*because psychiatric disorders involve complex circuits within the brain, any genetic contribution is likely to involve interactions between multiple genes. Since the expression of each gene also involves interactions with environmental factors, polygenetic inheritance greatly expands the way in which the environment can influence genetic predispositions. Moreover, in some cases, individuals with the gene for a disease show little or no evidence of having symptoms of the disease itself. This phenomenon is called non-penetrance. In other cases, genetic diseases are variably expressed, where some members of a family may manifest the full-blown disease, while others only show mild symptoms. Finally, mental disorders probably have multiple independent causes*".

This author fully subscribes to this view.

In particular, with regard to autism, genetics as we shall see was understood as that branch of studies that would find 'the cause' and propose 'the cure'. In reality, it was very soon realised that this would not be the case.

Genetics is to medicine and biology as quantum physics is to physics and astronomy. From nuclear physics, we should have learnt that infinitely small can be infinitely powerful and also uncontrollable. While it is true that with the power of the atom, we can power entire cities, it is equally true that we can destroy the entire planet.

M. Di Salvo, *Autism Research between Psychology and Neuroscience*,
https://doi.org/10.1007/978-3-031-68338-1_10

The Higgs boson (*which we have already met in the introduction*) has also become known to the general public and the media as the 'God particle', derived from the title of Leon Lederman's popularisation book 'The God particle: If the universe is the question, what is the answer?' published in 1993. This title was derived from a change by the publisher of the nickname 'Goddamn particle', originally chosen by the author in reference to the difficulty of its identification (*Higgs stated that he disagreed with this expression, finding it potentially offensive to people of religious faith*). These and much greater misunderstandings often give rise to those delusions of omnipotence I have already mentioned. Yet star physics is beginning to confirm that for every 'black hole' there can be a 'white hole' (and vice versa).[1]

We have played God with the power of the atom, we keep imagining we can make God with life.[2]

The limit should not be set by the word 'God' nor by a religious belief. Scientists often say they are atheists or agnostics, when, in fact, their choices and actions display a religiosity and dogmatism typical of a religious cult, that of the science of man of which they are the priests.

Genetics was born as the study of genes. By stroking the lamp of the DNA helix (*a discovery acknowledged to Watson and Crick but previously photographed by Rosalind Franklin with the famous 'photo 51'*), first the chromosomes and then the 'genes' emerged.

To these genes, man has expressed virtually all his desires for solving any problem.

And when the genes—apparently—stopped giving the answers, at least the desired ones, genetic manipulation was born, sometimes without paying too much attention to what could happen outside the laboratories.

Genetics, genetic analysis and genetic medicine, have very high costs: they require highly specialised operators, particularly expensive laboratories, continuous investment in technological innovation. And especially in the past, in order to secure these resources with a few too many simplifications, researchers have promised private associations, public institutions and pharmaceutical companies rapid progress and fabulous cures.

Molecular Genetics and the Emergence of the Autism Spectrum

At the beginning of the twenty-first century, many researchers thought that the genes for autism could be discovered quickly; thanks to the major advances following the establishment of the Human Genome Project (Rutter, 1999).

It quickly became apparent that no single-gene mutation could be linked to autism and the focus shifted to full genome scans and candidate gene studies. Genetic research helped reinforce the idea of a broad autism spectrum, with complex genetic correlates.

As several autism sociologists have pointed out, autism research has attracted significant funding due to the popularity of the diagnosis.

Much of this funding went to genetic research, particularly that focusing on 'susceptibility genes', as well as brain imaging studies, because many companies had a direct interest in identifying autism susceptibility genes and demonstrating that the brain has atypicalities that could potentially be corrected with the use of pharmaceuticals.

In 1997, the first molecular genetic studies on autism looked to the serotonin transporter 5-HTT to find some biological basis, serotonin having become central in the description and treatment of depressive disorders.

Interest in 5-HTT had arisen in relation to autism as part of a broader theory of psychological atypicality, particularly reports on its association with anxiety-related traits (Cook et al. 1997; Lesh et al. 1996).

In 1999, Zhong and colleagues dismissed the association between 5-HTT and autism, particularly with a US study of an ethnically heterogeneous population.

The triad of impairments introduced by Lorna Wing (*Chapter 7*), in particular, as measured in the ADI, have been linked to discrete fragments of the human genome.

Other early molecular genetic studies on autism examined the region of the dopamine β-hydroxylase promoter gene and found that mothers of autistic children appeared to exhibit abnormalities that may create a 'suboptimal uterine environment'. The dopamine hypothesis (Robinson et al. 2001) emerged from other biochemical studies on depression and other mental illnesses in adults, just as had happened with the serotonin hypothesis.

Many other genes have been proposed as candidates for autism. For example 15q11-13 because it was found to be a common site for chromosomal abnormalities in autism, although early studies were inconclusive.

In 2006, Sabhrangshu Guhathakurta published the results of a major study on autism in which he argued that the disagreement between reports on 5-HTT could be explained by the 'complexity of serotonergic neurotransmitters', the system in which the synthesis, uptake, neurotransmission, metabolism and regulation of 5-HTT involve a number of proteins and the effect of each protein may contribute to pathology.

These studies highlighted the complexity of genetic data and dispelled the idea that a single gene would be discovered for autism. Yet, as with all genetic studies, the belief remained that genetic research on the autism spectrum would yield results over time.

The more research that has been conducted, the more complex it has become.

In a 2007 study, Guhathakurta and other Indian researchers found no association between autism and glutamate 6 receptors, or GluR6 or GRIK2 polymorphisms, despite the results of a study by the Paris Autism Research International Sibpair that had found an association to exist, as found in a 2004 study in China (Shuang et al. 2004).

Neuroscientific research on humans has been accompanied by new explorations of 'autism' and 'social distress' in other animals, such as mice, in order to facilitate large-scale testing of genetic atypicalities that allow genetic research.

The conceptual problems involved in identifying autism in a mouse are obvious but have usually been overshadowed by the broader implications of research and the discovery of a possible genetic correlate.

Therefore, a version of the ADI or ADOS was devised for mouse populations and researchers attempted to categorise the behaviours associated with the triad of impairments as present in rodents.

In 2013, Marcus Wohr of the University of Marburg, and Maria Luisa Scattoni of the Istituto Superiore di Sanità in Italy, argued that "*when tested in an open field or observation box, rodents with 'autism' would demonstrate triad social behaviour deficits through the limitation of 'social interaction' and 'social play'*".

According to them, they would demonstrate triad communication deficits through lack of scent marking, absence of interaction and induction by the female through ultrasonic 'vocalisations'. Finally, 'motor stereotypies', 'repetitions' and 'self-cleaning', repetitive burying of marbles and 'limited exploration' would demonstrate the stereotyped and repetitive behaviours of the triad. A Franco-German project has also developed an autism severity scoring system to measure the degree of autism in mice, measuring the degree of skill in nest building and other activities and the degree of atypical behaviours such as excessive burying. These standardised approaches helped consolidate the idea that rodents may suffer from autism.

The experimentation, to an open mind and a trained eye, decidedly similar to Skinner's pigeon experiment, amply highlights a perhaps somewhat excessive tendency towards anthropomorphism, which borders on ignoring human typicality (*also in terms of neurocircuit complexity*) especially on the levels of the imagination competence plane, of the construction of narrative in the relationship with the other and even of a theory of the mind, all things that are glossed over (*to come back to Skinner's pigeons*) in these attempts to bring the autistic spectrum into the animal world and to rethink early animal development accordingly.[3]

Very large datasets containing WGS data of persons with autism and their families are now being assembled: MSSNG (10,000 'autistic genomes'), SPARK (genomic data of 50,000 families) and Spectrum10K (with 10,000 potential participants). In 1977, an anonymised DNA sample could not yet be used to identify an individual. Today, it is possible, thanks to the proliferation of genealogical databases, DNA sampling in the criminal justice system and the pervasive use and storage of patient-related genetic data in medicine. Genetic research can never again be truly anonymous, and consent forms that promise anonymity or minimise risk do not provide participants or their proxies with accurate information to support fully informed consent.

Moreover, researchers themselves cannot accurately predict what uses genetic data might be put to in the future. Thus, although Simon Baron-Cohen has stated that he opposes the use of genetic data to create prenatal diagnostic tests for autism, if data from his study are available for re-use, this risk remains (Evans, 2017).

The crux of the problem is the mismatch of research priorities between autistic people and biomedical researchers. In modern biomedical research, there has been a movement to involve patients in clinical trials for many years; 'participatory' methods, such as co-creating interventions with end-users, are an established practice in social science research. Part of the reason is cost: looking at drug development alone, pharmaceutical companies spend millions to develop new products, few of which reach the market. Social science research is labour-intensive and also requires significant resources. Pharmaceutical companies have resources that far exceed those of researchers developing new techniques in areas such as special education, speech therapy or community support, but they all want their work to be cost-effective.

When patients or end-users are involved in research from the outset, setting priorities and ensuring that continuing research projects are geared towards meeting real needs, results are much more likely to be achieved (*and perhaps even marketable, although the profit from improved services or support provision is usually indirect, e.g. observed years later in higher employment levels or reduced need for social benefits*).

With regard to autism research, in the words of Milton (2014) "*autistic people have often become distrustful of researchers and their goals, and are often excluded from knowledge production processes. Such a context results in a negative feedback loop with further damage to the growth of interactional competence between researchers and autistic people, and a breakdown in trust and communication leading to increased tension between stakeholder groups. The involvement of autistic scholars in research and in the improvement of participatory methods can therefore be seen as a requirement if social research in the field of autism is to claim ethical and epistemological integrity*".

In the United States, genetic research on autism in human and animal forms is particularly flourishing, mainly due to the efforts of parent organisations, as Chloe Silverman pointed out, in particular, the National Alliance for Autism Research (NAAR), founded in 1994, and Cure Autism Now (CAN), founded in 1995. CAN established the AGRE, a gene archive of families with two or more members diagnosed with autism. In 2003, NAAR launched the Autism Genome Project (AGP), which brought together a North American and European consortium of researchers in an attempt to improve autism genetic research. In 2005, NAAR merged with a new organisation, Autism Speaks, and in 2006, CAN merged with them as well.

Although the AGP did not match the initial hopes of finding a single-gene mutation, it served as a model to encourage the scientific exchange of data and also encouraged the growth of new theories on the role genes may play in the development of all psychological disorders of childhood. It has also encouraged a greater understanding of the complexity of interactions between genes and the environment in the formation of all developmental disorders.

More recently, genetic studies have considered a broader and more comprehensive range of developmental conditions. This only confirms that autism is a central concept in child development and is associated with many different conditions and dispels the idea that it is a single disorder.

In early 2015 (Yuen et al.), a team of researchers from Canada, the United States, China and Sweden concluded that "there is substantial genetic and clinical heterogeneity in autism and that much more extensive investigations will need to be completed to fully understand the genetic architecture of the autism spectrum".

This is, in fact, the scientific definition of the concept of neurodiversity itself.

Already in 2007, Daniel Geschwind and Pat Levitt had argued that autism was "*analogous to broader syndromes, such as the 'mental retardation syndrome', had multiple causes, and was best thought of as a 'heterogeneous retardation' neurodevelopmental syndrome*". It was, therefore, better to think in terms of 'autism' rather than 'autism'.

According to Bonnie Evans, "*the reason the phenotypic model for autism has been so zealously pursued is because it is the only model that measures social interaction and its atypicalities through genetic phenotypes. It is also the substitute for the psychology of 'human relationships' and so researchers desperately seek the complex architecture of the human organism's genome to support it*".

The approach to children as individual and individual cases, and the approach to the 'data' produced by these children, has changed so radically that the sciences and approaches to human development and human relations have almost completely forgotten it. New epidemiological definitions of autism have redefined the way we think about child development and human relationships. However, these approaches have historical roots. The fact that this particular model of developmental atypicality was formulated through the British epidemiology of the 1960s in the late 1970s is indeed relevant to genetics researchers who are now engaged in the search for diagnostic tools.

In this sense, something has recently changed in the approach to genetic studies.

On 31 August 2021, the University of Cambridge announced the largest genetic study of people with autism to date conducted in the

United Kingdom, Spectrum10K, which was to collect data from 10,000 autistic people and their families over a 10-year period. Ten days later, recruitment for the new study and the study itself were suspended (Baron-Cohen, 2021).

Such a thing had never happened before in the world of research: it was such a surprising event that it garnered major press coverage (i.e. Sanderson, 2021; Chapman, 2021) and produced a storm on social media.

The response of autistic adults to Spectrum10K and its impact represented a turning point in the way autism genetic/genomic research had been conducted up to that point.

As Mitzi Waltz (2023) pointed out, "*this moment also reflects broader societal developments related to genetic research, the use (misuse) of genetic data, privacy, and support for the concept of neurodiversity. To what extent have these changes gone hand in hand with general changes in the ethics of medical research (i.e. institutional or legal requirements for informed consent, patient benefit or privacy), and to what extent have they been responses to the specific demands of autistic research subjects (i.e. autism-specific; support for informed consent, concerns about the use of genetic data in prenatal testing or community research priorities?). The answers to these questions lie not in a sudden reaction to a specific study, but in the opposing changes in practice and views that have developed over decades*".

Autism research has often violated ethical standards. Although genetics and genomics have gradually become the hottest topics in autism research over the past 30 years (Pellicano & Stears, 2011; Sweileh et al., 2016; Quick et al., 2021), this development has not been supported by most autistic people themselves. Their knowledge of past bad ethical practices, concerns about the misuse of genetic data and a desire for research funds to be put to different uses have given rise to discontent (Milton, 2014).

The resulting impasse is a real-life example of the conflict described by medical historian Ian Miller (2018) between the view that conditions such as autism are socially constructed and the predominant medical model of disability. Eugenics has proven to be a persistent and seductive concept, so whenever developments in biomedical research emerge, Miller argues that it is important to pay attention to new forms of eugenics that might be brought in. Biomedical 'progress' is often presented in terms of cost savings instead of quality-of-life improvements, and this attention can cancel out the impact on actual people. All the 'autism awareness' campaigns are not enough to prevent many biomedical researchers from measuring human differences against the norms presented during their medical training, and from making decisions based on assigning lower values to lives lived with differences or impairments. This is a problem that affects all people with disabilities (Wolbring, 2001; Kavanaugh, 2020), but as Waltz argues "*it is particularly problematic when dealing with a socially constructed and subjectively diagnosed category such as autism*".

In Search of the Empathy Gene

In his research on empathy, and the related theory of autism as 'level zero-positive', Baron-Cohen selected a set of candidate genes involved in what he generically termed 'neural growth' because studies on the brains of zero-positive individuals (with autism or Asperger's syndrome) had shown atypical patterns in the connections between neurons and the speed of growth of the autistic brain in the early stages of postnatal development. Of the 68 candidate genes that had been tested, four of them showed a strong and significant association with EQ.

One of these genes was part of the group responsible for steroid sex hormones (cYp11B1). A second gene was linked to the social-emotional behaviour group (wes1). The third and fourth genes associated with EQ were in the neural growth group: NTRK1 and GABRB3.

Thus, summing up, Baron-Cohen claimed to have found four genes for empathy (Chakrabarti, Dudbridge, Kent et al., 2009).

As they admitted in the same study, this was a first step, as there is a big difference between finding genes and understanding how their functioning may impact empathy.

Chakrabarti and Baron-Cohen had asked volunteers chosen from the overall population to fill out not only the questionnaire for EQ but also that for Autism Spectrum Quotient (AQ).

There were four other genes that had been associated with QA alone: one of the genes for neuroligin (NLGN4x), one of the homeobox genes (HOXA1) that control brain formation and ARNT2 (a gene involved in neurogenesis). The fourth was a gene for monoamine oxidase (MAOB).

They then looked to see if any of the candidate genes were strongly associated with zero-positivity (i.e. with a diagnosis of Asperger's syndrome) and found that as many as six genes were. Among these six were three genes for steroid sex hormones: 15F2 (one of the oestrogen receptors), CyP1741 (whose abnormalities cause menstrual cycle irregularities in women) and cYP11B1 (one of the genes that catalyse the production of testosterone from cholesterol and has also been associated with empathy). Then there was the gene for oxytocin (OXT) and also ARNTI and HOXA1 (also associated with empathy).

Their interpretation was that while some genes affect only EQ, others affect either only AQ or only zero-positive states, or more than one of these conditions.

De Novo' Mutations

Despite the heritability, we discussed in the previous chapter, most cases of autism occur sporadically, with no recent evidence of family history. It has been hypothesised that spontaneous de novo mutations in the sperm or ovum contribute to the probability of developing autism (Sebat et al. 2007).[4]

Two pieces of evidence appear to support this hypothesis. Firstly, people diagnosed with autism have a significantly reduced fecundity and are 20 times less likely to have children than average, thus limiting the persistence of mutations in genes over multiple generations in a family (Hodge et al. 2020). Secondly, the likelihood of having a child who develops autism increases as the parents get older, and mutations in the sperm gradually accumulate over a man's lifetime (Kong et al. 2012).

The first genes conclusively shown to contribute to the risk of autism were discovered in the early 1990s by researchers examining human-specific forms of autism caused by mutations on the X chromosome. An expansion of the CGG trinucleotide repeat in the FMR1 gene promoter in boys causes fragile X syndrome, and at least 20 per cent of boys with this mutation exhibit behaviour consistent with autism spectrum disorder (Man et al. 2017; Hatton et al. 2006).

Mutations that inactivate the MECP2 gene cause Rett syndrome, which is associated with autistic behaviour in girls, and in boys, the mutation is lethal at the embryonic level (Amir et al. 1999).

The role of de novo mutations in autism became apparent when DNA microarray technologies achieved a sufficient resolution to allow the detection of copy number variations (CNVs) in the human genome (Sebat et al. 2004; Iafrate et al. 2004).

CNVs are the most common type of structural variation in the genome, consisting of DNA deletions and duplications that vary in size from a kilobase to a few megabases. Microarray analysis has shown that de novo CNVs occur at a significantly higher rate in sporadic cases of autism than in typically developing siblings and unrelated controls. A number of studies have shown that de novo gene-altering CNVs occur approximately four times more frequently in autism than in controls and contribute to approximately 5–10% of cases (Pinto et al. 2014; Levy et al. 2011; Sanders et al. 2011).

It is predicted that there are 350–450 genes that significantly increase susceptibility to autism when affected by de novo inactivating mutations.

Dozens of genes and CNVs have been conclusively identified based on the observation of recurrent mutations in several individuals, and suggestive evidence has been found for more than 100 (Betancur, 2011).

Each mutation is in turn associated with a variety of clinical diagnoses and can also be found in a small percentage of individuals without a clinical diagnosis (Stefansson et al. 2014; Shinawi et al. 2010).

Therefore, the genetic disorders that makeup autism are not specific to autism.

The mutations themselves are characterised by considerable variability in clinical outcome and generally only a subset of mutation carriers meet the criteria for autism. This variable expressivity means that different individuals with the same mutation vary considerably in the severity of the particular trait observed (Brandler, Sebat 2015).

The conclusion of these recent studies on de novo mutations is that the autism spectrum is splitting into as many individual disorders as defined by genetics.

One gene that has been linked to autism is SHANK2. Mutations in this gene act in a dominant manner. Mutations in this gene appear to cause hyperconnectivity between neurons.

A study (Columbia, 2022) of 42,607 autism cases identified 60 new genes, five of which had a more moderate impact on autistic symptoms. The related gene variants were often inherited from the parents of the participants.

The Overall State of Autism Genetic Research

What can we say about the state of current genetic research? First of all, we can answer some seemingly simple questions, always remembering that we are talking about a very recent field of study, which explores territories that we do not know in depth either in terms of functioning or dynamics. We have no definitive theory on the functioning and relationship, correlation and interrelation of genes, the timing of expression and the relationship between gene exploration.

According to Natan Schaefer (2021), only 7 per cent of our DNA is shared exclusively with other modern humans and not with several ancestral cousins. But if we look for the portion of the genome that is unique to our species and also shared by all humans worldwide, regardless of place of origin, then we arrive at an even smaller fraction, 1.5% of the total. This tiny portion of DNA might contain what distinguishes us as a species: it is no coincidence that, according to the authors of the study, part of it consists of mutations involved in neurodevelopment and brain function.

Still a problem of statistics, and how we read the data and the combination of variants.

According to other studies, chimpanzees are genetically the closest to humans with whom we share about 98.6% of our DNA, a percentage that 'stops' at 95% in sharing with pigs.

In these small quantitative differences, there would be 'the whole' that makes us human.

But we must bear in mind that when we talk about DNA, we are talking about the largest known 'data bank' in the universe, and the key lies not only in reading it (which we have, in fact, done), but in understanding the grammar, syntax and dynamics of the relationships and interconnections between sequences (and we are a long way from doing that).

Human imagination got there first—as it often does: it is like being in the library of Babel, a fantastic tale by Jorge Luis Borges from 1941.

Let us see what answers this library has so far given on autism.

Does Autism Have a Hereditary Basis?

The hereditary basis of autism is mainly supported by two observations: identical twins (monozygotic, sharing completely the same genetic inheritance) are more likely to be both autistic than fraternal twins (dizygotic, born from the same birth, but resulting from the fertilisation of two eggs by two different sperm). In the broadest definition of the autism spectrum (i.e. when Asperger's syndrome and developmental disorders not otherwise specified are also taken into account), there is a 92% concordance between monozygotic twins and a 10% concordance between dizygotic twins. In addition, a brother or sister of a person with autism has a higher risk of developing the disorder (approximately 7% for males and 1–2% for females) or of presenting mild forms of developmental problems involving language, sociability or other behavioural symptoms (approximately 4–6%). Finally, certain genetically inherited pathological conditions, such as Fragile X syndrome and Tuberous Sclerosis, often occur in comorbidity with autism.

Are There Autism Genes?

Recognising that autism has a biological origin with a significant hereditary component, research has been directed towards identifying the precise genetic causes, i.e. specific genes underlying autism spectrum disorders. Whole genome analysis and cytogenetic studies carried out on families with at least two members in this condition have made it possible to identify a number of genes most frequently associated with autism, located on different chromosomes such as chromosome 15, 16, 22 and the X chromosome.

In general, to date, genetic analysis has failed to prove the involvement of a single gene associated with autism and there is consensus in pointing to a complex genetic origin, characterised by the involvement of many genes. Some of these genes are involved in the production of important molecules implicated in the development and maintenance of nerve networks: these are (i) proteins involved in the functioning of the nerve synapse (i.e. the structure that allows nerve cells to communicate with each other), (ii) factors that regulate gene expression, (iii) neurotransmitters and their receptors (i.e. the molecules that transmit the nerve signal from one nerve cell to another via the synapse) and (iv) genes involved in brain development.

None of the genes identified so far, in particular, groups of patients and/or their relatives is in itself capable of explaining all the symptoms of this complex condition.

If we combine all the genetic correlations, people diagnosed with autism would have to inherit a particular combination of 'defective' genes to be autistic. However, the presence of this particular combination, although it may be the cause of the disorder, may not be the sufficient condition: it is possible that in some cases this combination of 'defective' genes only gives rise to the disorder as a result of triggering environmental factors, such as exposure to infectious agents during prenatal life, maternal–foetal immunological status, exposure to drugs or toxic agents including through diet during pregnancy.

The role of genes is therefore increasingly interpreted as an important comorbidity factor that interacts with other factors, generically called environmental factors, in determining the onset of the syndrome. For example, the possible involvement of particular conditions that may occur during foetal and neonatal development (*pregnancy progression, infections contracted by the mother, factors linked to childbirth conditions*) is being actively studied. Finally, an important area of research concerns the role of exposure to toxic agents during development, such as mercury or other environmental pollutants such as phosphate or organochlorine pesticides as well as the use of drugs during pregnancy.

We will return to some of these topics in the next chapter, in which I will discuss further aetiological and co-aetiological causes of autism, a discussion we left in Chapter 2.

An international study published in the journal Cell (Satterstrom et al. 2020) identified as many as 102 genes implicated in the risk of autism spectrum disorders, 30 of which are still unknown. The research took advantage of recently developed gene sequencing techniques to study the DNA of more than 35,000 people, almost 12,000 of whom were diagnosed with autism: in particular, the exome, the part of the DNA that codes for proteins, was analysed, thus managing to identify both the role of the genes involved and the type of mutations affecting them. The study was conducted by the Autism Sequencing Consortium (ASC), an international consortium founded and coordinated by Joseph Buxbaum.

The overlaps between genetic brain disorders and neurological and psychiatric phenotypes were included in this table.

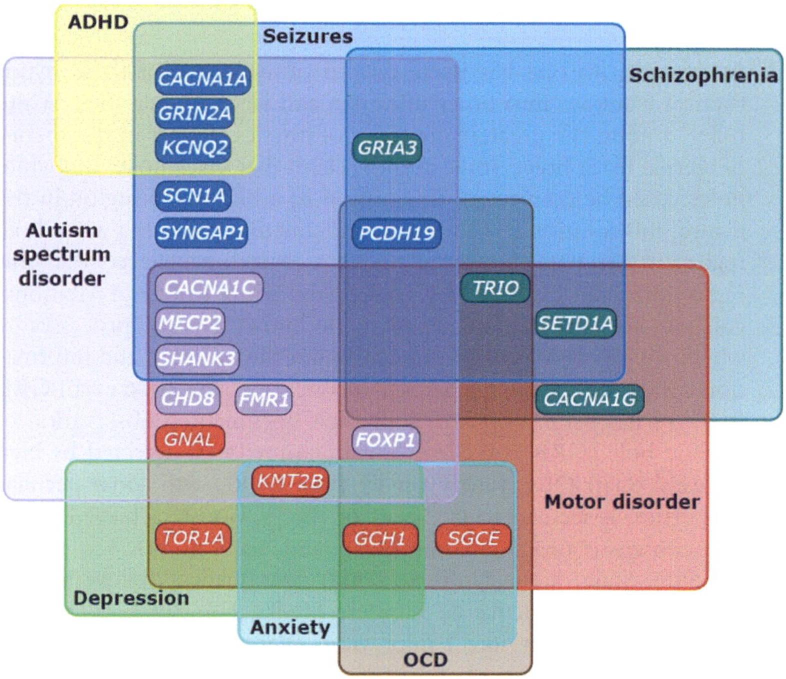

Peall, K.J., Owen, M.J. & Hall, J. Rare genetic brain disorders with overlapping neurological and psychiatric phenotypes. Nat Rev Neurol 20 (2024) [Open Source].

Notes

1. At present, no one has been able to observe one of these hypothetical windows into other universes and large singularities. White holes, among the greatest mysteries of physics, appear in the theory of eternal black holes: some authors claim that black holes and white holes could be connected. In addition to a black hole region in the future, this solution to Einstein's field equation predicts a white hole region in the past. However, Stephen Hawking believed that this region does not exist for black holes originating from a gravitational collapse, and furthermore, there are no known physical processes by which a white hole could form. Although the evidence and information concerning white holes is still inconclusive, the 2006 event GRB 060,614 has been proposed as the first documented observation of a white hole (GRB 060,614 is a gamma-ray burst detected by Neil Gehrels' Swift Observatory on 14 June 2006, with some peculiar properties: it seemed to fall short of the previously achieved scientific consensus on gamma-ray bursts and black holes). See Martin Ratcliffe, State of the Universe 2008: new images, discoveries and events, Springer Science & Business Media, Feb 2008, p. 57. See also what we said in note 1 in previous chapter.
2. The concept is strong, but it is not an exaggeration on my part. Eric R. Kandel in his extraordinary 2016 'In Search of Memory The Emergence of a New Science of Mind' relates this quote from Watson (2003) 'We wanted to do the equivalent of what a word-processing program can do today: cut, paste and copy DNA [...] later we cracked the genetic code. [...] By a heterogenesis of ends, a series of discoveries made in the late sixties and early seventies converged in 1973, giving us the so-called "recombinant DNA" technology, [...] this was no ordinary advancement in laboratory techniques: suddenly scientists were able to package DNA molecules, creating ones never before seen in nature. We could "play god" with the molecular foundations of all life'.
3. In his book "Decisions, Uncertainty, and the Brain: The Science of Neuroeconomics", Paul Glimker comments on training and reflex in a manner not dissimilar to what one might comment—in our case—on the behaviour of an alleged autistic mouse "*what happens when Newsome begins to train a monkey., From the experimenter's point of view, the monkey is learning how to perform a new task, left versus*

right motion discrimination. Only after the monkey has fully learned this new behavior can the experiment begin. From the monkey's point of view, however, the process he participates in looks quite different. The monkey comes to the laboratory already knowing that he wants to maximize the amount of juice he obtains. Slowly he learns to do that by performing the motion discrimination task more and more efficiently. For the monkey there is no new experimental task per se, *just a change in the environment within which he can obtain juice. This may be an important distinction, because we have already seen that the Sherringtonian framework often provides explanatory power only within the limited context of the experimenter's view of the task.*

The risk posed by reflexology is essentially this: Are we developing a theory of individual trees that is testable and verifiable but that does a very poor job of describing an entire forest? If we forced ourselves to employ the much more complicated extrareflexological tools that were available to us, might we be able to extend our descriptive theories of the brain to a broader range of behaviors? Might the identification of simple behaviors as reflexes come to seem an artificial distinction? As Von Holtz said: The sooner we recognize the fact that the [higher functions] which leave the reflex physiologist dumbfounded in fact send roots down to the simplest basal functions of the CNS, the sooner we shall see that the previously terminologically insurmountable barrier between the lower levels of neurophysiology and higher behavioral theory simply dissolves away". (Von Holtz and Mittelstaedt, 1950).

4. Scientists have recently discovered that non-coding DNA in our genome, previously considered 'junk DNA', may play an important role in complex diseases by influencing the activation and silencing of genes. The reason why de novo mutations increase with age is that sperm precursor cells divide every fortnight. This continuous division and copying of DNA leads to errors, and the error rate increases significantly with age. Mothers do not seem to contribute de novo mutations to autism because oocytes, unlike spermatozoa, do not divide and multiply throughout life; they are all generated before the woman is born. The de novo mutations are particularly interesting because the incidence of autism has increased dramatically in recent years. Much of this increase is probably attributable to the fact that we are now much more aware of it and are able to identify it more easily than fifty years ago. But another part of the explanation

is that people are having children at an older age. The genetic mutations that contribute to autism spectrum disorders can be distributed anywhere on our 23 pairs of chromosomes. Regardless of where they are found, these mutations disrupt neuronal circuits in the social brain, and those disorders end up impairing the theory of mind. Some mutations play a key role in the functioning of synapses. In fact, de novo mutations occur most frequently in genes encoding synaptic proteins.

References

Amir RE, Van den Veyver IB, Wan M, Tran CQ, Francke U, Zoghbi HY (1999). "Rett syndrome is caused by mutations in X-linked MECP2, encoding methyl-CpG-binding protein 2". Nature Genetics. 23

Baron-Cohen, S. (2021). *Update on Spectrum10K*. Autism Research Centre. https://docs.autismresearchcentre.com/Spectrum10K/Statement_Sept_10th_2021.pdf

Betancur C (2011) "Etiological heterogeneity in autism spectrum disorders: more than 100 genetic and genomic disorders and still counting" Brain Research. 1380

Brandler WM, Sebat J (2015). "From de novo mutations to personalized therapeutic interventions in autism". Annual Review of Medicine. 66

Chakrabarti B, Dudbridge F, Kent L, Wheelwright S, Hill-Cawthorne G, Allison C, Banerjee-Basu S, Baron-Cohen S. (2009) Genes related to sex steroids, neural growth, and social-emotional behavior are associated with autistic traits, empathy, and Asperger syndrome. Autism Res.

Chapman, A. (2021) Why autistic people are worried by Spectrum10K. *Psychology Today*

Columbia University Irving Medical Center (2022). "60 New Genes Linked to Autism Uncovered". Nature Genetics. SciTech Daily. 54

Cook, E. H., R. Courchesne, C. Lord, N. J. Cox, S. Yan, A. Lincoln, R. Haas, E. Courchesne and B. L. Leventhal. (1997) Evidence of linkage between the serotonin transporter and autistic disorder'. Mol Psychiatry

Evans, B. (2017). The transformation of social life and the transformation of autism in the 1960s. In The metamorphosis of autism in Britain (pp. 189–247). University of Manchester Press.

Geschwind, D. H. and P. Levitt. (2007) 'Autism spectrum disorders: Developmental disconnection syndromes'. *Current Opinion in Neurobiology* 17

Glimcher, P. (2004). Decisions, Uncertainty, and the Brain: The Science of Neuroeconomics—Bradford Books—MIT

Guhathakurta, S., S. Ghosh, S. Sinha, A. Chatterjee, S. Ahmed, S. R. Chowdhury, P. K. Gangopadhyay, S. Ghosh, M. Singh and R. Usha. (2006) 'Serotonin transporter promoter variants: Analysis in Indian autistic and control population'. Brain Res

Hatton DD, Sideris J, Skinner M, Mankowski J, Bailey DB, Roberts J, Mirrett P (2006). "Autistic behavior in children with fragile X syndrome: prevalence, stability, and the impact of FMRP". American Journal of Medical Genetics. Part A. 140A

Hodges H, Fealko C, Soares N (2020). "Autism spectrum disorder: definition, epidemiology, causes, and clinical evaluation". Translational Pediatrics. 9 (Suppl 1)

Iafrate AJ, Feuk L, Rivera MN, Listewnik ML, Donahoe PK, Qi Y, et al. (2004). "Detection of large-scale variation in the human genome". Nature Genetics. 36

Kavanaugh, A. (2020) Disability and public health research in Australia. *Australian and New Zealand Journal of Public Health, 44*

Kong A, Frigge ML, Masson G, Besenbacher S, Sulem P, Magnusson G, et al. (2012). "Rate of de novo mutations and the importance of father's age to disease risk". Nature. 488

LeDoux, J. (2002) Synaptic Self: How Our Brains Become Who We Are—Penguin

Lesch, K. P., D. Bengel, A. Heils, S. Z. Sabol, B. D. Greenberg, S. Petri, J. Benjamin, C. R. Muller, D. H. Hamer, D. L. Murphy. (1996) Association of anxietyrelated traits with a polymorphism in the serotonin transporter gene regulatory region. Science 274

Levy D, Ronemus M, Yamrom B, Lee YH, Leotta A, Kendall J, et al. (2011). "Rare de novo and transmitted copy-number variation in autistic spectrum disorders". Neuron. 70

Man L, Lekovich J, Rosenwaks Z, Gerhardt J (2017). "Fragile X-Associated Diminished Ovarian Reserve and Primary Ovarian Insufficiency from Molecular Mechanisms to Clinical Manifestations". Frontiers in Molecular Neuroscience. 10

Miller, I. (2018). *Medical history (Theory and history, 1)*. Red Globe Press/ Springer Nature

Milton, D. (2014) Autistic expertise: A critical reflection on the production of knowledge in autism studies. Autism, 18

Pellicano, E., & Stears, M. (2011) Bridging autism, science and society: Moving toward an ethically informed approach to autism research. *Autism Research, 4*

Pinto D, Delaby E, Merico D, Barbosa M, Merikangas A, Klei L, et al. (2014). "Convergence of genes and cellular pathways dysregulated in autism spectrum disorders". American Journal of Human Genetics. 94

Quick, V. B., Wang, B., & State, M. W. (2021) Leveraging large genomic datasets to illuminate the pathology of autism spectrum disorders. *Neuropsychopharmacology Reviews, 46*
Robinson, P. D., C. K. Schutz, F. Macciardi, B. N. White and J. J. A. Holden. (2001) 'Genetically determined low maternal serum dopamine beta-hydroxylase levels and the etiology of autism spectrum disorders'. *Am J Med Genet* 100
Rutter, M. (1999). Genetic studies of autism: From the 1970s into the Millennium. Journal of Abnormal Child Psychology, 28
Sanders SJ, Ercan-Sencicek AG, Hus V, Luo R, Murtha MT, Moreno-De-Luca D, et al. (2011). "Multiple recurrent de novo CNVs, including duplications of the 7q11.23 Williams syndrome region, are strongly associated with autism". Neuron. 70
Sanderson, K. (2021) High-profile autism genetics project paused amid backlash. *Nature*, 598
Satterstrom et al., 2020, Cell 180, 568–584
Schaefer NK, Shapiro B, Green RE (2021). An ancestral recombination graph of human, Neanderthal, and Denisovan genomes. Sci Adv.
Sebat J, Lakshmi B, Troge J, Alexander J, Young J, Lundin P, et al. (2004). "Large-scale copy number polymorphism in the human genome". Science. 305
Sebat J, Lakshmi B, Malhotra D, Troge J, Lese-Martin C, Walsh T, et al. (2007). "Strong association of de novo copy number mutations with autism". Science. 316
Shinawi M, Liu P, Kang SH, Shen J, Belmont JW, Scott DA, et al. (2010). "Recurrent reciprocal 16p11.2 rearrangements associated with global developmental delay, behavioural problems, dysmorphism, epilepsy, and abnormal head size". Journal of Medical Genetics. 47
Shuang, M., J. Liu, M. X. Jia, J. Z. Yang, S. P. Wu, X. H. Gong, Y. S. Ling, Y. Ruan, X. L. Yang and D. Zhang. (2004) 'Family-based association study between autism and glutamate receptor 6 gene in Chinese Han trios'. Am J Med Genet B Neuropsychiatr Genet 131B
Stefansson H, Meyer-Lindenberg A, Steinberg S, Magnusdottir B, Morgen K, Arnarsdottir S, et al. (2014). "CNVs conferring risk of autism or schizophrenia affect cognition in controls". Nature. 505
Sweileh, W. M., et al. (2016) Bibliometric profile of the global scientific research on autism spectrum disorders. *SpringerPlus, 5*
Von Holtz, E., and Middelstaedt, H. (1950). Das Reafferenzprinzip. Wechselwirkung zwischen Zentralnervensystem und Peripherie. Naturwissenschaften 37: 464–476. Repr. as The reafference principle. Interaction between the central nervous system and the periphery. In The Behavioral Physiology of

Animals and Man: Selected Papers of E. von Holtz, R. D. Martin (trans.) Miami, FL: University of Miami Press.

Waltz, M. (2023). Genetics, Genomics and the Medical Model of Autism. In: Autism. Palgrave Macmillan, Cham

Watson, J.D. Berry, A. (2003) DNA: The secret of life. Alfred A. Knopf, New York

Wohr, M. Scattoni, M.L. (2013) 'Behavioural methods used in rodent models of autism spectrum disorders: Current standards and new developments'. *Behavioural Brain Research* 251

Wolbring, G. (2001) Where do we draw the line? Surviving eugenics in a technological world. In M. Priestley (Ed.), *Disability and the life course.* Cambridge University Press.

Yuen, R. K., B et al (2015) 'Whole-genome sequencing of quartet families with autism spectrum disorder'. *Nat Med* 21

Zhong, N., L. L. Ye, W. N. Ju, J. Tsiouris, I. Cohen and W. T. Brown. (1999) '5-HTTLPR variants not associated with autistic spectrum disorders'. Neurogenetics 2

CHAPTER 11

Etiological and Co-etiological Factors of Autism

As we have seen in the previous chapter, the role of genes is increasingly interpreted as a factor that interacts with others, generically called environmental, in determining the onset of the syndrome. For example, the possible involvement of particular conditions that may occur during foetal and neonatal development (pregnancy progression, infections contracted by the mother, factors linked to childbirth conditions) is being actively studied. An important area of research concerns the role of exposure to toxic agents during development.

In this chapter, I will attempt to summarise 'the state of the art' on these important and apparently increasingly relevant research fronts.

In this context, it must necessarily be noted that, while it is true that the claim that cases of autism have grown out of all proportion (as we shall see in more detail in the next chapter) is highly critical, it is also true that—if the relationship with broadly understood environmental factors were to be confirmed—at least part of this increase would be attributable to ecological changes over the last century, broadly understood.

For example, the correlations between pollution—particularly chemical and heavy metal pollution—and various forms of cancer are obvious and no longer debatable. It cannot therefore be ruled out a priori that, in the presence of genetic predisposing factors, exposure to toxic agents (also widely understood) during pregnancy cannot intervene as a causal co-element.

M. Di Salvo, *Autism Research between Psychology and Neuroscience*, https://doi.org/10.1007/978-3-031-68338-1_11

However, it is worth considering that, particularly with regard to autism, the statistical data is on the one hand too recent to be historically validated, and on the other hand, lacks a shared metric to avoid the errors typical of other analysis methodologies.

But let us go in order.

At present, many causes of autism have been proposed, but the understanding of the theory of autism causality is incomplete.

As we have seen, attempts have been made to incorporate known genetic and environmental causes into a causal framework. In its most recent definition, autism is a neurodevelopmental disorder characterised by impairments in communication skills and social interaction and restricted/repetitive behaviours, interests or activities not appropriate to the individual's developmental stage, and the severity of symptoms and functional impairment varies from individual to individual (APA 2022).

In recent years, the number of people diagnosed with the disorder has increased dramatically, at least taking into account official statistics. There are several potential reasons for this phenomenon, in particular, changes in diagnostic criteria. Indeed, it is evident that the broader the spectrum, the more open-ended the criteria, the greater the number of cases that may fall within a descriptive case (Salari et al. 2022).

Environmental factors that have been found to contribute to autism or exacerbate its symptoms, or that may be important to consider in future research, include certain foods (Quan et al. 2019), infectious diseases, heavy metals, solvents, exhaust gases, PCBs, phthalates and phenols used in plastic products, pesticides, brominated flame retardants, alcohol, smoking and drugs.

Epigenetic Factors

In the previous chapter, we described genetic studies.

We must now look in more detail at what factors emerge from studies of epigenetics, a branch of genetics that deals with heritable phenotypic changes in a cell or organism in which genotype variation is not observed.

It was defined by Arthur Riggs and colleagues (1996) as "*the study of heritable mitotic and meiotic changes that cannot be explained by changes in the DNA sequence*".

What happens is that a kind of molecular 'fingerprint' is inherited on the genotype that determines the degree of activation of genes, the sequence of which, however, remains identical. This molecular imprint consists of covalent chromatin modifications, both at the DNA and protein level, and is therefore long-lasting, but may be reversible.

These modifications, called epimutations, last for the rest of the cell's life and can be passed on to subsequent generations of cells through cell divisions, without, however, the corresponding DNA sequences being mutated (Bird 2007); they are therefore non-genomic factors that cause the organism's genes to be expressed differently (Hunter 2008).

Epigenetic phenomena underlie, for instance, most cellular differentiation processes (and their alteration, thus also in cancer) (Raman et al. 2009), X chromosome inactivation and contribute to a certain heritable phenotypic plasticity in relation to environmental changes that is reminiscent of Lamarckian inheritance of acquired traits. For example, highly stressful events can leave an epigenetic imprint at the level of DNA methylation (Ryan et al. 2016).

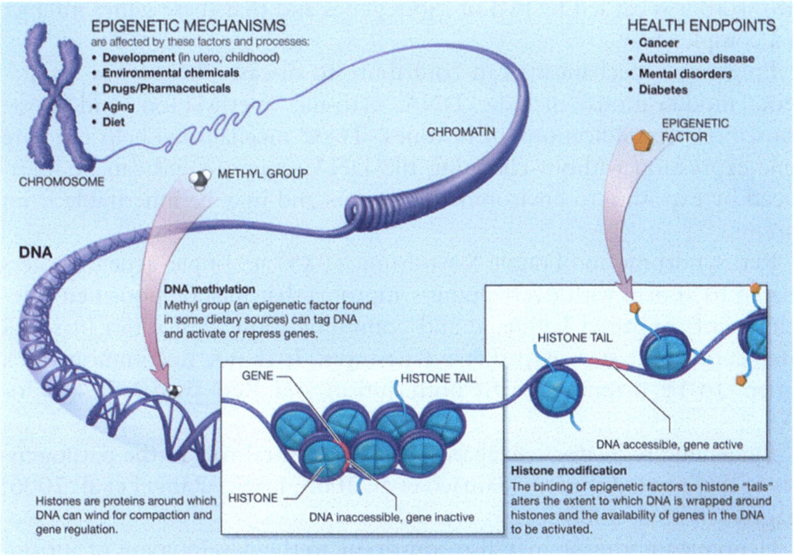

Epigenetic mechanisms are affected by several factors and processes including development in utero *and in childhood, environmental chemicals, drugs and pharmaceuticals, aging and diet. DNA methylation is*

what occurs when methyl groups, an epigenetic factor found in some dietary sources, can tag DNA and activate or repress genes. Histones are proteins around which DNA can wind for compaction and gene regulation. Histone modification occurs when the binding of epigenetic factors to histone "tails" alters the extent to which DNA is wrapped around histones and the availability of genes in the DNA to be activated. All of these factors and processes can have an effect on people's health and influence their health possibly resulting in cancer, autoimmune disease, mental disorders or diabetes among other illnesses. National Institutes of Health.

Epigenetic mechanisms can increase the risk of autism. Epigenetic changes do not result from DNA sequence modifications, but from modifications of chromosomal histones or modifications of DNA bases. Such changes are known to be influenced by environmental factors, including diet, drugs and stress (Miyake et al. 2012).

Most data support a polygenic and epistatic model, which means that the disorder is caused by two or more genes and that these genes interact in a complex way.

Epigenetic mechanisms can contribute to disease phenotypes. Epigenetic modifications include DNA cytosine methylation and post-translational modifications of histones. These mechanisms help regulate gene expression without changing the DNA sequence and can be influenced by exposure to environmental factors and may be inheritable from parents.

Rett syndrome and Fragile X syndrome (FXS) are single-gene disorders related to autism with overlapping symptoms that include poor neurodevelopment, impaired language and communication, difficulties in social interactions and stereotyped hand gestures. It is not uncommon for a patient to be diagnosed with both autism and Rett syndrome and/or FXS.

Epigenetic regulatory mechanisms play a central role in the pathogenesis of these two disorders (Samaco et al. 2005; Lopez-Rangel et al. 2006; Hagerman et al. 2005).

Genomic imprinting may also contribute to the development of autism. Genomic imprinting is another example of epigenetic regulation of gene expression. In this case, epigenetic modifications cause offspring to express either the maternal copy of a gene or the paternal copy of a gene, but not both. The imprinted gene is silenced through epigenetic mechanisms. Candidate genes and susceptibility alleles for autism are identified using a combination of techniques, including genome-wide and targeted analyses of allele sharing in sibling pairs, using association studies and transmission disequilibrium testing (TDT) of functional and/or positional candidate genes and examining novel and recurrent cytogenetic aberrations. The results of numerous studies have identified several genomic regions known to be subject to imprinting, candidate genes and gene-environment interactions. In particular, chromosomes 15q and 7q appear to be epigenetic hotspots in contributing to autism. Furthermore, genes on the X chromosome may play an important role, as in Rett syndrome (Schanen 2006).

An important basis for the causation of autism is also the overproduction or underproduction of permanent brain cells (*neurons, oligodendrocytes and astrocytes*) by neural precursor cells during foetal development (Rutgers University 2022).

The Prenatal Environment

The development of autism is associated with several prenatal risk factors, including the advanced age of either parent, diabetes, haemorrhage and the mother's use of antibiotics and psychotropic drugs during pregnancy (Lee et al. 2019; Morales et al. 2018).

Autism has been linked to birth defects that occur during the first eight weeks after conception, although these cases are rare. If the child's mother has autoimmune conditions or disorders during pregnancy, this can have an effect on the development of autism in the child. All these factors can cause inflammation or impair immune signalling in one way or another (Han et al. 2021).

Obstructive Sleep Apnoea in Pregnancy

Sleep apnoea can cause intermittent hypoxia and its prevalence is also increasing due to the rise in cases of obesity. Known maternal risk factors for autism in offspring are similar to risk factors for sleep apnoea. For example, advanced maternal age, maternal obesity, maternal type 2 diabetes and maternal hypertension are statistically correlated with an increased risk of autism.[1]

One study found that gestational sleep apnoea is associated with low scores on children's reading tests and that this effect may be mediated by an increased risk of the child itself having sleep apnoea (Bin et al. 2017).

Sleep problems in autism have been linked in one study to brain changes, particularly in the hippocampus, although this study does not demonstrate a clear causal link (MacDuffie et al. 2020).

A common presentation of sleep apnoea in children with autism is insomnia (Santapuram et al. 2022).

Infectious Processes

Prenatal viral infection and congenital rubella syndrome are currently considered the main non-genetic infectious causes of autism. Prenatal exposure to rubella or cytomegalovirus activates the mother's immune response and may significantly increase the risk of autism, a study that could obviously only be verified in mice (Libbey 2005).

Immunological events associated with infections in early pregnancy may influence neural development more than infections in late pregnancy, not only for autism but also for psychiatric disorders of presumed neurodevelopmental origin, particularly schizophrenia (Mendelson et al. 2008; Meyer et al. 2007).

Environmental Agents

Teratogens are environmental agents that cause birth defects. Some of them have also been suggested as potential risk factors for autism, although there is little or no scientific evidence to support these claims.

Doubts have also been raised that ethanol increases the risk of autism, as part of foetal alcohol syndrome or alcohol-related birth defects. All known teratogens appear to act in the first eight weeks after conception and, although this does not exclude the possibility that autism may be initiated or influenced later, there is strong evidence that autism arises very early in development (Amdt et al. 2005).

It has been shown that there is a small but significant link between prenatal exposure to air pollutants and the risk of autism. However, this result is not statistically consistent across studies and exposure to pollutants has been measured indirectly (Modabbernia, 2017).

Autoimmune and Inflammatory Diseases

Maternal inflammatory and autoimmune diseases can damage embryonic and foetal tissues, exacerbating a genetic problem or damaging the nervous system (Samsam et al. 2014).

Other Maternity-Related Conditions

It has been hypothesised that thyroid problems, leading to a thyroxine deficiency in the mother in weeks 8–12 of pregnancy, produce changes in the foetal brain that lead to autism. Thyroxine deficiencies can be caused by an inadequate iodine diet and environmental agents that interfere with iodine absorption or act against thyroid hormones. Possible environmental agents include flavonoids in food, tobacco smoke and most herbicides. This hypothesis has not been adequately tested (Román 2007).

The mother's diabetes during pregnancy is a significant risk factor for autism; a 2009 meta-analysis found that gestational diabetes is associated with a twofold increased risk. A 2014 review also found that maternal diabetes was significantly associated with an increased risk of autism (Xu et al. 2014).

Although diabetes causes metabolic and hormonal abnormalities and oxidative stress, no biological mechanism is known for the association between gestational diabetes and autism risk (Gardener et al. 2009).

Maternal diagnosis of polycystic ovary syndrome has been found to be associated with a higher risk of autism (Katsigianni et al. 2019).

Maternal malnutrition during preconception and pregnancy influences foetal neurodevelopment. Intrauterine growth restriction is associated with autism in both term and preterm infants (Vohr et al. 2017).

It has been hypothesised that folic acid taken during pregnancy may play a role in reducing cases of autism by modulating gene expression through an epigenetic mechanism. This hypothesis is supported by several studies (Lyall et al. 2014).

It has been hypothesised that prenatal stress, consisting of exposure to life events or environmental factors that distress the expectant mother, contributes to autism, perhaps as part of a broader, more complex interaction between genes and environment. Autism has been associated with prenatal stress in both retrospective studies that examined stressors and natural experiments that involved prenatal exposure to stress-storms, animal studies have reported that prenatal stress "can disrupt brain development and produce behaviours that resemble the symptoms of autism" (Kinney et al. 2008).

In this context, mention should also be made of the foetal testosterone theory, which, as we have seen, hypothesises that higher levels of testosterone in mothers' amniotic fluid drive brain development towards a better ability to see patterns and analyse complex systems, while decreasing communication and empathy, emphasising 'masculine' over 'feminine' traits or, in the terminology of E-S theory, emphasising 'systemisation' over 'empathy' (Auyeung, Baron-Cohen, 2008; Manson 2008).

In this case, described as 'male extreme' (see Chapter 10) there is a lack of verification to the contrary: would the increase in testosterone be the cause, or is it in turn a response of the (maternal) organism to a factor that is itself the cause? From a strictly epistemological point of view, defining 'male' as correlated with ' foetal testosterone' seems reductive and forced if we consider that testosterone does not directly intervene in sexual determination through gene expression, where the difference between 'male and female' is chromosomal (XX vs. XY).

Some research suggests that maternal exposure to selective serotonin reuptake inhibitors during pregnancy is associated with an increased risk of autism, but it is not yet clear whether there is a causal link between the two.

There is evidence, for example, that this association may be an artefact of confounding with maternal mental illness (Man et al. 2015; Brown et al. 2017).

Perinatal Environment

Autism has also been associated with certain perinatal and obstetrical conditions. Infants born preterm often have various neurodevelopmental disorders related to motor skills, cognition, receptive and expressive language, and social-emotional skills. Preterm infants also have an increased risk of having various neurodevelopmental disorders such as cerebral palsy and autism, as well as psychiatric disorders related to attention, anxiety and impaired social communication (Rogers et al. 2018).

It has also been proposed that hypothalamic–pituitary–adrenal axis function and brain connectivity in preterm infants may be affected by stress related to neonatal intensive care, resulting in deficits in emotional regulation and social-emotional skills. A 2019 analysis of perinatal and neonatal risk factors found that autism was associated with abnormal foetal positioning, umbilical cord complications, low 5-minute Apgar score, low birth weight and duration of gestation and foetal distress, meconium aspiration syndrome, trauma or injury during delivery, maternal haemorrhage, multiple births, feeding disorders, neonatal anaemia, birth defects/malformations, maternal blood group incompatibility and jaundice/hyperbilirubinemia. These associations do not indicate a causal relationship for any individual factor (Gardener et al. 2011).

There is growing evidence that perinatal exposure to air pollution may be a risk factor for autism, although this evidence has methodological limitations, including a small number of studies and failure to control for potential confounding factors (Weisskopf et al. 2015; Flores-Pajot et al. 2016).

Postnatal Environment

A wide range of postnatal factors contributing to autism have been proposed, including gastrointestinal or immune system abnormalities, allergies and children's exposure to drugs, infections, certain foods or heavy metals. Evidence of these risk factors is anecdotal and has not been confirmed by reliable studies (Rutter 2005).

Amygdala Neurons

This theory hypothesises that an early developmental failure involving the amygdala has a cascading effect on the development of cortical areas that mediate social perception in the visual domain. The fusiform area of the ventral stream is involved. The idea is that this area is involved in social knowledge and social cognition and that deficits in this network are decisive in causing autism (Schultz 2005).

Autoimmune Diseases

This theory postulates that autoantibodies targeting the brain or elements of brain metabolism can cause or exacerbate autism. It is related to the maternal infection theory but postulates that the effect is caused by the individual's own antibodies, possibly due to an environmental trigger after birth. It is also related to several other causes, e.g. it has been postulated that viral infections cause autism through an autoimmune mechanism (Ashwood 2004).

Interactions between the immune system and the nervous system begin early during embryogenesis and successful neurodevelopment depends on a balanced immune response. It is possible that aberrant immune activity during critical periods of neurodevelopment is part of the mechanism of some forms of autism (Ashwood et al. 2006).

A small percentage of autism cases are associated with an infection, usually before birth. The results of immune studies have been contradictory. Some abnormalities have been found in specific subgroups and some of these have been replicated. It is not known whether these abnormalities are relevant for autism, e.g. due to infection or autoimmunity, or whether they are secondary to disease processes (Stigler et al. 2009).

As autoantibodies are found in diseases other than autism and are not always present in autism, the relationship between immune disorders and autism remains unclear and controversial (Wills et al. 2007; Schmitz et al. 2008).

A 2015 systematic review and meta-analysis found that children with a family history of autoimmune disease were at higher risk of autism than children without such a history (Wu et al. 2015).

In the presence of an underlying maternal autoimmune disease, circulating antibodies in the foetus could contribute to the development of autism spectrum disorders (Fox et al. 2012).

Intermezzo: The Brain's Immune System and Its Relationship with the Gut

The brain's main immune defence is provided by microglial cells, a type of glia cells that make up the main immune defence of our central nervous system. Microglia cells make up about 20 per cent of the total population of cells within the brain, and they are constantly on the lookout for damaged neurons, plaques and infectious agents (Aloisi 2001).

Henry and colleagues (2020) conducted an experimental study in mice, concluding that acting on the long-lasting inflammation, by cancelling it, leads to good therapeutic results in the treatment of brain trauma (Lull and Block 2010).

A group of researchers from the University of Cambridge (United Kingdom) and the National Institute of Health (US) recently discovered that the meninges host their own immune cells called plasma cells, which secrete antibodies (Fitzpatrick et al. 2020).

These cells are located near the larger blood vessels that run inside the meninges, so they can secrete their antibodies as 'guards' defending the perimeter of the brain. Studying those specific antibodies, the researchers were surprised to discover cells similar to those found in the gut.

Plasmacells are derived from a particular type of immune cells, called B-cells. Each one has on its surface an antibody typical only of that cell. If an antigen (*the part of a bacterium or virus that stimulates an immune response*) binds to the surface of the antibody, the B-cell is activated. And so it divides, multiplying into cells that also recognise the same antigen. Normally, the antibodies found in blood are Immunoglobin G (IgG), produced by the spleen and bone marrow (*the antibodies that protect the body*). In contrast, the antibodies found in the meninges—in animal models and in human samples, in vitro—were Immunoglobulins A (IgA), which are usually produced in the lining of the intestines or the nose and lungs: thus protecting the mucous membranes, the surfaces that interface with the external environment. In other words, the cells that end up in the meninges are those that have selectively multiplied in the gut, where they have learnt to recognise particular pathogens.

In the words of Menna Clatworthy, one of the authors of the study, "*Exactly how the brain defends itself against infection, beyond the physical barrier of the meninges, has been a real mystery until now. But to discover that an important line of defence begins in the gut is a real surprise. Now, however, we can see the point. [...] Even a tiny breach of the intestinal barrier would allow any virus or bacterium to enter the bloodstream with devastating consequences, should it reach the brain. 'Seeding' the meninges with cells that produce antibodies that have already been selected to recognise the microbes in the gut ensures a viable defence against the most likely invaders'. In their experiment with animal models, the researchers showed that when there were no bacteria in their gut, IgA-producing cells were absent from the meninges. This showed that these cells are born in the gut where they are trained to recognise gut microbes, before settling in the meninges. When the researchers removed the plasma cells from the meninges (so no IgA was present to fight the microbes), they entered the brain from the blood. [...] This defence most likely plays an important role in defending us against infections of the central nervous system, such as meningitis and encephalitis*".

According to Stefano Pallanti, associate professor at the University of Florence and professor of psychiatry and humanities at Stanford University, "*in the past, the brain was believed to be outside the world of immunity: its defences were unknown, our professors declared it an impregnable cathedral. The brain was thought to consist basically of neurons and the glia cells, which act as support. The meninges were thought to be just a covering. Now there is this important discovery of an immune system inside the brain, in the meninges. And the immunoglobin is trained in the gut because there the greater the permeability with the outside and, therefore, the more frequent and varied the encounters with foreign agents. The importance of the microbiota is becoming more and more apparent in recent years*".

This relevant discovery also opens up relevant scenarios for psychiatry, which up to now bases its diagnosis and treatment on behavioural symptoms reported by the patient. According to Pallanti, 'We should also look at the results of neuroscience, at the "physical" symptoms they reveal, moving towards neuropsychiatry.

If an autistic child has intestinal problems, treating these with probiotics and anti-inflammatories leads to an improvement in other symptoms.

One has to learn to look outside the head as well. Unfortunately, we don't really know what the healthy normality of the microbiota looks like: we only have indications, not certainties'.

Gastrointestinal Connections

Gastrointestinal problems are one of the medical disorders most commonly associated with autistic people. They are linked to increased social disability, irritability, behavioural and sleep problems, language disorders and mood alterations, so the theory has been put forward that it is an overlap syndrome (Wasilewska 2015).

Studies indicate that gastrointestinal inflammation, food allergies, gluten-related disorders (celiac disease, wheat allergy, non-celiac gluten sensitivity), visceral hypersensitivity, dysautonomia and gastroesophageal reflux are the mechanisms that could link both.

A 2016 review concludes that enteric nervous system abnormalities could play a role in several neurological disorders, including autism. Neural connections and the immune system are a pathway that may allow diseases originating in the gut to spread to the brain (Rao 2016).

A 2018 review suggests that the frequent association of gastrointestinal disorders and autism is due to abnormalities in the gut-brain axis (Israelyan 2018).

The 'leaky gut syndrome' hypothesis developed by Andrew Wakefield is popular among parents of children with autism (Bezawada et al. 2020; Almeida et al. 2020).

It is based on the idea that defects in the intestinal barrier produce an excessive increase in intestinal permeability, allowing substances in the gut (including bacteria, environmental toxins and food antigens) to pass into the bloodstream. Data supporting this theory are limited and contradictory, as both increased intestinal permeability and normal permeability have been documented in people with autism. Studies in mice provide some support for this theory and suggest the importance of gut flora, showing that normalisation of the gut barrier is associated with an improvement in some autism-like behaviours (Rao, 2016).

Studies on subgroups of people diagnosed with autism have shown the presence of elevated plasma levels of zonulin, a protein that regulates permeability by opening the 'pores' of the intestinal wall, as well as intestinal dysbiosis, reduced levels of bifidobacteria and increased abundance of Akkermansia muciniphila, Escherichia coli, Clostridia and Candida fungi that promote the production of proinflammatory cytokines, all producing excessive intestinal permeability (Azhari et al. 2019).

This allows bacterial endotoxins to pass from the gut into the bloodstream, stimulating liver cells to secrete tumour necrosis factor alpha (TNFα), which modulates blood–brain barrier permeability. Studies have shown that TNFα cascades produce proinflammatory cytokines, leading to peripheral inflammation and activation of microglia in the brain, indicating neuroinflammation. Furthermore, neuroactive opioid peptides from digested food have been shown to leak into the bloodstream and permeate the blood–brain barrier, affecting neural cells and causing autistic symptoms (Azhari et al. 2019).

After a 1998 preliminary study of three children with autism treated with secretin infusion reported improved gastrointestinal function and marked improvement in behaviour, many parents requested secretin treatment and a black market for the hormone quickly developed (Johnson 2006).

Subsequent studies have found secretin to be clearly ineffective in the treatment of autism (Krishnaswami 2011).

Endogenous Opiate Precursor Theory

In 1979, a possible association between autism and opiates was proposed; it was observed that injecting small amounts of opiates into young laboratory animals resulted in symptoms similar to those observed in autistic children (Panksepp 1979).

The possibility of a relationship between autism and gluten and casein consumption was first articulated by Kalle Reichelt in 1991 (Millard et al. 2008).

The opioid theory hypothesises that autism is the result of a metabolic disorder in which the opioid peptides gliadorphine (aka gluteomorphine) and casomorphine, produced through the metabolism of gluten (found in wheat and related grains) and casein (found in dairy products), pass through an abnormally permeable gut wall and then proceed to exert an effect on neurotransmission through binding to opioid receptors. The resulting opioid excess has been hypothesised to affect brain maturation and cause autistic symptoms, including: behavioural difficulties, attention problems, impaired communication skills and impaired social and cognitive functioning (Shattock 2002).

Although high levels of these opioids are excreted in the urine, it has been hypothesised that a small portion of them pass through the brain causing interference in signal transmission and disruption of normal activity. Three studies have reported that urine samples from autistic persons show increased excretion of peptides over 24 hours. One study with a control group found no appreciable difference in opioid levels in the urine samples of autistic persons compared to the control group (Azhari et al. 2019).

Theory also states that the removal of opioid precursors from a child's diet may allow for the cessation of these behaviours and normal resumption of neurodevelopment in very young children (Christison and Ivany 2006).

Studies by Aranburu (2021) have shown no benefit of gluten-free diets in the treatment of autism. In the subgroup of people with gluten sensitivity, there is limited evidence to suggest that a gluten-free diet may improve some autistic behaviours.

Nutrition-Related Factors

Multiple attempts have been made to discover a link between various nutritional deficiencies such as vitamin D and folate and the risk of autism. Although many studies have been conducted on the role of vitamin D in the development of autism, most of them are limited by the inability to assess deficiency prior to a diagnosis of autism. One meta-analysis on the association between vitamin D and autism found that autistic people had significantly lower serum levels of 25-hydroxy vitamin D than non-autistic people. Another analysis showed significant differences in zinc levels. Although the studies showed significant differences in protein and calcium intake in autistic persons, the results were limited by their imprecision, inconsistency and indirect nature. Furthermore, low levels of 5-methyltetrahydrofolate (5-MTHF) in the brain can result in brain folate deficiency (CFD), which has been shown to be associated with autism (Modabbernia et al. 2017; Rossignol and Frye 2021).

Toxic Exposure

Numerous studies have attempted to relate toxic exposure and autism.

However, these studies often encountered limitations related to the measurement of toxic exposure, the methods of which were often indirect and cross-sectional. Systematic reviews have been conducted for numerous toxins including air pollution, thimerosal, inorganic mercury and heavy metal levels in hair, nails and body fluids.

Environmental exposure to inorganic mercury may be associated with a higher risk of autism, as high levels of mercury in the body are considered a valid pathogen for autism (Jafari Mohammadabadi et al. 2020).

No significant evidence was found for an association between autism and the concentration of mercury, copper, cadmium, selenium and chromium in hair, nails and body fluids. Lead levels were found to be significantly higher in persons with autism. The accuracy and consistency of the results were not maintained between studies and were affected by an outlier and non-systematic study (Modabbernia et al. 2017).

The atypical eating behaviours of autistic children make it difficult to determine whether increased lead levels are a cause or a consequence of autism (Zafeiriou, 2007), as well as any deficiency or excess.

Locus Coeruleus-Noradrenergic System

This theory hypothesises that autistic behaviour depends at least in part on developmental dysregulation that results in impaired function of the locus coeruleus-noradrenergic (LC-NA) system. The LC-NA system is strongly involved in arousal and attention; for example, it is linked to the brain's acquisition and utilisation of environmental signals (Mehler, 2009).

Oxidative Stress

Oxidative stress, oxidative DNA damage and disruptions in DNA repair have been hypothesised to play a role in the aetiopathology of both autism and schizophrenia (Markkanen et al. 2016).

Physiological factors and mechanisms influenced by oxidative stress are believed to be highly influential on the risk of autism. Interactions between environmental and genetic factors may increase oxidative stress in children with autism. This theory hypothesises that toxicity and oxidative stress may cause autism in some cases. Evidence includes genetic effects on metabolic pathways, reduced antioxidant capacity, enzymatic changes and enhanced biomarkers for oxidative stress (Björklund et al. 2020).

One theory is that stress damages Purkinje cells in the cerebellum after birth and it is possible that glutathione is involved (Kern and Jones 2006).

Polymorphism of genes involved in glutathione metabolism is evidenced by lower levels of total glutathione and higher levels of oxidised glutathione in autistic children. Based on this theory, antioxidants could be a useful treatment for autism (Villagonzalo 2010).

Environmental factors can influence pre-, peri- and postnatal oxidative stress and include heavy metals, infections, certain drugs and toxic exposure from various sources, including cigarette smoke, air pollutants and organophosphate pesticides.

* * *

A very popular story in martial arts schools, with various versions and details, goes something like this: a pupil asks his master what was the most important teaching he had received over many years of practice and the master replies: "*when I began the study of the martial art, to me a punch was just a punch. As I progressed in practice, becoming an expert, mastering the technique, I realised that things were less obvious: a punch was not just a punch. It was something much more complex. As I progressed further, as I continued practising for years I realised a deeper and simpler truth: a punch is just a punch*".

In Chapter 3 and in this chapter, we went through an extensive list of aetiological, causal or co-causal theories of the autism spectrum. In Chapter 4 we described the framework of traits and symptomatology, and in Chapter 6 I tried to describe the cranio-cerebellar features of people diagnosed with autism, while in Chapter 10 I tried to summarise the genetic research on the subject.

I am sure I have not been exhaustive. I hope, however, that I have been respectful of the intense work of researchers at every level and in every part of the world who have worked to research, prove and disprove the results.

But above all, I hope that I have been intellectually honest and that I have provided adequate documentary references for the major claims, where possible in the affirmative and in the negative or critical sense.

Moreover, as Sue Fletcher-Watson and Francesca Happé (2019) suggest, the future of autism research is further complicated, not only by variability in the presentation of autism itself but also by variability in the presence and manifestations of accompanying diagnoses—clinical anxiety, epilepsy, ADHD, etc.—and by the intersectional influence of facets such as race, gender or sexuality.

It is well known that autism often presents together with other medical diagnoses, but new data show that a large percentage of the autistic community also identifies outside the hetero-normative categories in terms of gender and sexual orientation (George and Stokes 2018; Gilmour et al. 2012).

There is no doubt that psychology has an important role to play in probing the experience and impact of co-occurring conditions and intersectional experiences.

However, there is a risk that this complexity is erased by strict inclusion and exclusion criteria for studies that seek a statistically significant result, which is easier when the sample is relatively homogeneous.

One possible solution is to focus less on diagnostic categories and more on trans-diagnostic domains of functioning, as in the Research Domain Criteria proposed as a new research framework by the National Institute of Mental Health in the United States (Insel et al. 2010).

In any case, there is a push for ever larger samples in order to understand the multiple influences on experience and outcomes and to derive recommendations that can be tailored to specific individuals. Of course, the opposite approach is also valid: small groups recruited for qualitative studies performing a deep dive to provide a rich and detailed picture of the multiple influences on life and outcomes.

To return to the metaphor of the fist in martial arts, if Chapter 2 represents the beginning of the story, when a fist is just a fist, we are now in the middle of the middle phase, when progressing in practice, becoming a little more experienced, mastering the technique we realise that things were less obvious: a fist was not just a fist.

In Chapter 8 we saw the evolution of the definition into what Bonnie Evans called the 'second autism' whose diagnostic criteria were referred to in Chapter 7 and in Chapter 9 I tried to highlight the neuroscientific basis of these theories, where they offered confirmations and denials, and to make the reconstructions more understandable I summarised the processes underlying neurogenesis and brain construction in Chapter 5.

Before concluding our journey, however, it is appropriate to anticipate a few considerations. All the aetiological theories and all the possible concauses of the autistic spectrum point to a relevant conceptual synthesis: the damage, if and when it occurred in the very early formation of the child, occurred at the level of the brain system, with integrated repercussions on the peripheral nervous system.

This unity also agrees with all the major psychological and neuroscientific theories, first and foremost of damage to the theory of mind circuit, the empathy circuit and the mirror neuron system.

That this damage may have a genetic basis is tautological: if we are talking about an embryonic phase, it is obvious that something must have happened in modifying the process of embryonic development, most of which is encoded in gene expression.

But genes are a non-blind project, on which the environment, broadly understood, intervenes imponderably. All the more so if this external factor intervenes in a genome that is 'sensitive' to variation and thus 'predisposing'.

In this context, autism can be seen almost more as a symptom of something that has taken a neurodifferent pathway, rather than in itself an autonomous pathology, and in this sense, it makes sense to speak of autism, of the individuality of the autistic person, and in the final analysis to acknowledge definitively that there is no such thing as one autistic person equal to another.

Before proceeding, however, it will be necessary to return to the epidemiology of autism in order to understand how this global spread alarm was generated, we should talk about the role of families, and further consider the categories of the great systemisers.

The story and the research continue.

Note

1. See for example:

Xu G, Jing J, Bowers K, Liu B, Bao W (2014). "Maternal diabetes and the risk of autism spectrum disorders in the offspring: a systematic review and meta-analysis". Journal of Autism and Developmental Disorders. 44

Maher GM, O'Keeffe GW, Kearney PM, Kenny LC, Dinan TG, Mattsson M, Khashan AS (2018). "Association of Hypertensive Disorders of Pregnancy With Risk of Neurodevelopmental Disorders in Offspring: A Systematic Review and Meta-analysis". JAMA Psychiatry. 75

Sandin S, Hultman CM, Kolevzon A, Gross R, MacCabe JH, Reichenberg A (2012). "Advancing maternal age is associated with increasing risk for autism: a review and meta-analysis". Journal of the American Academy of Child and Adolescent Psychiatry. 51

Wang Y, Tang S, Xu S, Weng S, Liu Z (2016). "Maternal Body Mass Index and Risk of Autism Spectrum Disorders in Offspring: A Meta-analysis". Scientific Reports.

Nieto FJ, Young TB, Lind BK, Shahar E, Samet JM, Redline S, et al. (2000). "Association of sleep-disordered breathing, sleep apnea, and hypertension in a large community-based study. Sleep Heart Health Study". JAMA. 283

Muraki I, Wada H, Tanigawa T (2018). "Sleep apnea and type 2 diabetes". Journal of Diabetes Investigation. 9

Punjabi NM (2008). "The epidemiology of adult obstructive sleep apnea". Proceedings of the American Thoracic Society. 5

References

Almeida C, Oliveira R, Soares R, Barata P (2020). "Influence of gut microbiota dysbiosis on brain function: a systematic review". Porto Biomedical Journal. 5

Aloisi, F. (2001). Immune function of microglia. Glia, 36(2), 165-179.

American Psychiatric Association (2022). Diagnostic and statistical manual of mental disorders: DSM-5-TR (5th ed.). American Psychiatric Association Publishing.

Aranburu E, Matias S, Simón E, Larretxi I, Martínez O, Bustamante MÁ, et al. (2021). "Gluten and FODMAPs Relationship with Mental Disorders: Systematic Review". Nutrients. 13

Arndt TL, Stodgell CJ, Rodier PM (2005). "The teratology of autism". International Journal of Developmental Neuroscience (Review). 23

Ashwood P, Van de Water J (2004). "Is autism an autoimmune disease?". Autoimmunity Reviews. 3

Ashwood P, Wills S, Van de Water J (2006). "The immune response in autism: a new frontier for autism research". Journal of Leukocyte Biology (Review). 80

Author National Institutes of Health (2005). http://commonfund.nih.gov/epigenomics/figure.aspx

Auyeung B, Baron-Cohen S (2008). "A Role for Fetal Testosterone in Human Sex Differences". In Zimmerman AW (ed.). Autism. Humana

Azhari A, Azizan F, Esposito G (2019). "A systematic review of gut-immune-brain mechanisms in Autism Spectrum Disorder". Developmental Psychobiology 61

Bezawada N, Phang TH, Hold GL, Hansen R (2020). "Autism Spectrum Disorder and the Gut Microbiota in Children: A Systematic Review". Annals of Nutrition & Metabolism. 76

Bin YS, Cistulli PA, Roberts CL, Ford JB (2017). "Childhood Health and Educational Outcomes Associated With Maternal Sleep Apnea: A Population Record-Linkage Study". Sleep. 40

Bird A. (2007) Perceptions of epigenetics, in Nature, vol. 447

Björklund G, Meguid NA, El-Bana MA, Tinkov AA, Saad K, Dadar M, et al. (2020). "Oxidative Stress in Autism Spectrum Disorder". Molecular Neurobiology. 57

Brown HK, Hussain-Shamsy N, Lunsky Y, Dennis CE, Vigod SN (2017). "The Association Between Antenatal Exposure to Selective Serotonin Reuptake Inhibitors and Autism: A Systematic Review and Meta-Analysis". The Journal of Clinical Psychiatry. 78

Christison GW, Ivany K (2006). "Elimination diets in autism spectrum disorders: any wheat amidst the chaff?". Journal of Developmental and Behavioral Pediatrics. 27 (Suppl 2)

Fitzpatrick, Z., Frazer, G., Ferro, A. et al. (2020) Gut-educated IgA plasma cells defend the meningeal venous sinuses. Nature 587, 472–476

Fletcher-Watson, S., & Happé, F. (2019). Autism: A New Introduction to Psychological Theory and Current Debate (2nd ed.). Routledge.

Flores-Pajot MC, Ofner M, Do MT, Lavigne E, Villeneuve PJ (2016). "Childhood autism spectrum disorders and exposure to nitrogen dioxide, and particulate matter air pollution: A review and meta-analysis". Environmental Research

Fox E, Amaral D, Van de Water J (2012). "Maternal and fetal antibrain antibodies in development and disease". Developmental Neurobiology. 72

Gardener H, Spiegelman D, Buka SL (2009) "Prenatal risk factors for autism: comprehensive meta-analysis". The British Journal of Psychiatry (Review, meta-analysis). 195

Gardener H, Spiegelman D, Buka SL (2011). "Perinatal and neonatal risk factors for autism: a comprehensive meta-analysis". Pediatrics. 128

George, R., & Stokes, M. A. (2018). Sexual orientation in autism spectrum disorder. Autism Research, 11(1), 133–141.

Gilmour, L., Schalomon, P. M., & Smith, V. (2012). Sexuality in a community based sample of adults with autism spectrum disorder. Research in Autism Spectrum Disorders, 6(1), 313–318.

Hagerman RJ, Ono MY, Hagerman PJ (2005). "Recent advances in fragile X: a model for autism and neurodegeneration". Current Opinion in Psychiatry. 18

Han VX, Patel S, Jones HF, Nielsen TC, Mohammad SS, Hofer MJ, et al. (2021). "Maternal acute and chronic inflammation in pregnancy is associated with common neurodevelopmental disorders: a systematic review". Translational Psychiatry. 11

Henry, R. J., Ritzel, R. M., Barrett, J. P., Doran, S. J., Jiao, Y., Leach, J. B., Loane, D. J. et al (2020). Microglial depletion with CSF1R inhibitor during chronic phase of experimental traumatic brain injury reduces neurodegeneration and neurological deficits. Journal of Neuroscience [bioRxiv, 2019, 791871]

Hunter, P. (2008) Special report: 'What genes remember', in Prospect Magazine, n. 146

Insel, T., Cuthbert, B., Garvey, M., Heinssen, R., Pine, D. S., Quinn, K., Sanislow, C., & Wang, P. (2010). Research domain criteria (RdoC): Toward a new classification framework for research on mental disorders. American Journal of Psychiatry, 167(7), 748–751.

Israelyan N, Margolis KG (2018). "Serotonin as a link between the gut-brain-microbiome axis in autism spectrum disorders". Pharmacological Research. 132

Jafari Mohammadabadi H, Rahmatian A, Sayehmiri F, Rafiei M (2020). "The Relationship Between the Level of Copper, Lead, Mercury and Autism Disorders: A Meta-Analysis". Pediatric Health, Medicine and Therapeutics. 11

Johnson TW (2006). "Dietary considerations in autism: identifying a reasonable approach". Top Clin Nutr. 21

Katsigianni M, Karageorgiou V, Lambrinoudaki I, Siristatidis C (2019). "Maternal polycystic ovarian syndrome in autism spectrum disorder: a systematic review and meta-analysis". Molecular Psychiatry. 24

Kern JK, Jones AM (2006). "Evidence of toxicity, oxidative stress, and neuronal insult in autism". Journal of Toxicology and Environmental Health Part B: Critical Reviews. 9

Kinney DK, Munir KM, Crowley DJ, Miller AM (2008). "Prenatal stress and risk for autism". Neuroscience and Biobehavioral Reviews (Review). 32 [in the opposite direction, see: Rai D, Golding J, Magnusson C, Steer C, Lewis G, Dalman C (2012). "Prenatal and early life exposure to stressful life events and risk of autism spectrum disorders: population-based studies in Sweden and England". PLOS ONE]

Krishnaswami S, McPheeters ML, Veenstra-Vanderweele J (2011). "A systematic review of secretin for children with autism spectrum disorders". Pediatrics. 127

Lee E, Cho J, Kim KY (2019). "The Association between Autism Spectrum Disorder and Pre- and Postnatal Antibiotic Exposure in Childhood-A Systematic Review with Meta-Analysis". International Journal of Environmental Research and Public Health. 16

Libbey JE, Sweeten TL, McMahon WM, Fujinami RS (2005). "Autistic disorder and viral infections". Journal of Neurovirology (Review). 11

Lopez-Rangel E, Lewis ME (2006). "Further evidence for pigenetic influence of MECP2 in Rett, autism and Angelman's syndromes". Clinical Genetics. 69

Lull, M. E., & Block, M. L. (2010). Microglial activation and chronic neurodegeneration. Neurotherapeutics, 7(4), 354-365.

Lyall K, Schmidt RJ, Hertz-Picciotto I (2014). "Maternal lifestyle and environmental risk factors for autism spectrum disorders". International Journal of Epidemiology. 43

MacDuffie KE, Shen MD, Dager SR, Styner MA, Kim SH, Paterson S, et al. (2020). "Sleep Onset Problems and Subcortical Development in Infants Later Diagnosed With Autism Spectrum Disorder". The American Journal of Psychiatry. 177

Man KK, Tong HH, Wong LY, Chan EW, Simonoff E, Wong IC (2015). "Exposure to selective serotonin reuptake inhibitors during pregnancy and risk of autism spectrum disorder in children: a systematic review and meta-analysis of observational studies". Neuroscience and Biobehavioral Reviews. 49

Manson, JE (2008). "Prenatal exposure to sex steroid hormones and behavioral/ cognitive outcomes". Metabolism (Review). 57 (Suppl 2)

Markkanen E, Meyer U, Dianov GL (2016). "DNA Damage and Repair in Schizophrenia and Autism: Implications for Cancer Comorbidity and Beyond". International Journal of Molecular Sciences. 17

Mehler MF, Purpura DP (2009). "Autism, fever, epigenetics and the locus coeruleus". Brain Research Reviews. 59

Meyer U, Yee BK, Feldon J (2007). "The neurodevelopmental impact of prenatal infections at different times of pregnancy: the earlier the worse?". The Neuroscientist (Review). 13

Mendelsohn NJ, Schaefer GB (2008). "Genetic evaluation of autism". Seminars in Pediatric Neurology (Review). 15

Miyake K, Hirasawa T, Koide T, Kubota T (2012). "Epigenetics in Autism and Other Neurodevelopmental Diseases". Neurodegenerative Diseases (Review). Advances in Experimental Medicine and Biology. Vol. 724

Millward C, Ferriter M, Calver S, Connell-Jones G (2008). "Gluten- and casein-free diets for autistic spectrum disorder". The Cochrane Database of Systematic Reviews

Modabbernia A, Velthorst E, Reichenberg A (2017). "Environmental risk factors for autism: an evidence-based review of systematic reviews and meta-analyses". Molecular Autism. 8

Morales DR, Slattery J, Evans S, Kurz X (2018). "Antidepressant use during pregnancy and risk of autism spectrum disorder and attention deficit hyperactivity disorder: systematic review of observational studies and methodological considerations". BMC Medicine. 16

Panksepp J (1979). "A neurochemical theory of autism". Trends in Neurosciences. 2

Quan J, Panaccione N, King JA, Underwood F, Windsor JW, Coward S, et al. (2019) A257 Association Between Celiac Disease and Autism Spectrum Disorder: A Systematic Review". Journal of the Canadian Association of Gastroenterology. 2 (Supplement 2)

Raman P. Nagarajan e Joseph F. Costello, (2009) Epigenetic mechanisms in glioblastoma multiforme, in Seminars in Cancer Biology, vol. 19

Rao M, Gershon MD (2016). "The bowel and beyond: the enteric nervous system in neurological disorders". Nature Reviews. Gastroenterology & Hepatology. 13

Riggs AD, Russo VEA, Martienssen RA, (1996) Epigenetic mechanisms of gene regulation, Plainview, N.Y, Cold Spring Harbor Laboratory Press

Rogers CE, Lean RE, Wheelock MD, Smyser CD (2018). "Aberrant structural and functional connectivity and neurodevelopmental impairment in preterm children". Journal of Neurodevelopmental Disorders. 10

Román, GC (2007) Autism: transient in utero hypothyroxinemia related to maternal flavonoid ingestion during pregnancy and to other environmental antithyroid agents. Journal of the Neurological Sciences (Review). 262

Rossignol DA, Frye RE (2021). "Cerebral Folate Deficiency, Folate Receptor Alpha Autoantibodies and Leucovorin (Folinic Acid) Treatment in Autism Spectrum Disorders: A Systematic Review and Meta-Analysis". Journal of Personalized Medicine. 11

Rutgers University (2022). "Scientists Discover That Irregular Production of Brain Cells Could Cause Autism". SciTech Daily.

Rutter M (2005). "Incidence of autism spectrum disorders: changes over time and their meaning". Acta Paediatrica. 94

Ryan, J., Chaudieu, I., Ancelin, M.L. (2016) Biological underpinnings of trauma and post-traumatic stress disorder: focusing on genetics and epigenetics, in Epigenomics, vol. 8

Salari N, Rasoulpoor S, Rasoulpoor S, Shohaimi S, Jafarpour S, Abdoli N, et al. (2022). "The global prevalence of autism spectrum disorder: a comprehensive systematic review and meta-analysis". Italian Journal of Pediatrics. 48

Samaco RC, Hogart A, LaSalle JM (2005). "Epigenetic overlap in autism-spectrum neurodevelopmental disorders: MECP2 deficiency causes reduced expression of UBE3A and GABRB3". Human Molecular Genetics. 14

Samsam M, Ahangari R, Naser SA (2014). "Pathophysiology of autism spectrum disorders: revisiting gastrointestinal involvement and immune imbalance". World Journal of Gastroenterology (Review). 20 (29): 9942–9951.

Santapuram P, Chen H, Weitlauf AS, Ghani MO, Whigham AS (2022) "Investigating differences in symptomatology and age at diagnosis of obstructive sleep apnea in children with and without autism". International Journal of Pediatric Otorhinolaryngology. 158

Schanen NC (2006) "Epigenetics of autism spectrum disorders". Human Molecular Genetics (Review). 15 Spec No 2

Schmitz C, Rezaie P (2008). "The neuropathology of autism: where do we stand?". Neuropathology and Applied Neurobiology (Review). 34

Schultz RT (2005). "Developmental deficits in social perception in autism: the role of the amygdala and fusiform face area". International Journal of Developmental Neuroscience. 23

Shattock P, Whiteley P (2002). "Biochemical aspects in autism spectrum disorders: updating the opioid-excess theory and presenting new opportunities for biomedical intervention". Expert Opinion on Therapeutic Targets (Review). 6

Stigler KA, Sweeten TL, Posey DJ, McDougle CJ (2009). "Autism and immune factors: a comprehensive review". Res Autism Spectr Disord (Review). 3

Villagonzalo KA, Dodd S, Dean O, Gray K, Tonge B, Berk M (2010). "Oxidative pathways as a drug target for the treatment of autism". Expert Opinion on Therapeutic Targets. 14

Vohr BR, Poggi Davis E, Wanke CA, Krebs NF (2017). "Neurodevelopment: The Impact of Nutrition and Inflammation During Preconception and Pregnancy in Low-Resource Settings". Pediatrics (Review). 139 (Suppl 1)

Wasilewska J, Klukowski M (2015). "Gastrointestinal symptoms and autism spectrum disorder: links and risks—a possible new overlap syndrome". Pediatric Health, Medicine and Therapeutics (Review). 6

Weisskopf MG, Kioumourtzoglou MA, Roberts AL (2015). "Air Pollution and Autism Spectrum Disorders: Causal or Confounded?". Current Environmental Health Reports

Wills S, Cabanlit M, Bennett J, Ashwood P, Amaral D, Van de Water J (2007). "Autoantibodies in autism spectrum disorders (ASD)". Annals of the New York Academy of Sciences. 1107

Wu S, Ding Y, Wu F, Li R, Xie G, Hou J, Mao P (2015). "Family history of autoimmune diseases is associated with an increased risk of autism in children: A systematic review and meta-analysis". Neuroscience and Biobehavioral Reviews. 55

Xu G, Jing J, Bowers K, Liu B, Bao W (2014). Maternal diabetes and the risk of autism spectrum disorders in the offspring: a systematic review and meta-analysis. Journal of Autism and Developmental Disorders. 44

Zafeiriou DI, Ververi A, Vargiami E (2007). "Childhood autism and associated comorbidities". Brain & Development. 29

CHAPTER 12

Autism Global Emergency: The New Epidemiology, Parenting, Self-Advocacy and Neurodiversity Movement

The work of Ian Hacking (2009) has shown the relevance of the autobiographical literature of autism, and how it has contributed to the push, together with statistical methods in the social and biological sciences, for the creation of a new 'State Science'.

In scientific thinking, one of the favourite styles of reasoning is the statistical analysis of data and the regularities of populations.

In Chapter 4, when reporting data from scientific reports, we found the following statement, for example: "*it is estimated that from about 0.5% to 10% manifest unusual abilities, ranging from great capacity in specific tasks, such as an extraordinary ability to memorise irrelevant details, to the development of conditions known as 'savant syndrome'*".

Now, if we refer to overall populations, depending on the context, we find that this statement is absolutely neutral: in any population set, it can be estimated that there is a subset "from about 0.5% to 10% that manifests unusual abilities" especially when the set of these "unusual abilities" is not only unspecified but these abilities are made up of an entirely heterogeneous set; above all, this is a tautological definition of average.

The original version of the chapter has been revised: The author names in references have been updated. A correction to this chapter can be found at https://doi.org/10.1007/978-3-031-68338-1_17

M. Di Salvo, *Autism Research between Psychology and Neuroscience*, https://doi.org/10.1007/978-3-031-68338-1_12

If we consider that a statistical set is represented with a bell-shaped graph, with respect to the average peak representing the mean, we will always have a descending curve towards the -10 and -0.5 of any value and on the other side a curve, also descending, with values at + 10 and + 0.5 of the two extremes (*specific inability and hyper ability*) of practically any performance.

And this graph is representative of any distribution and standard deviation of any general ability in a large and adequate set of non-discriminated population considered[1].

Gaussian curve—author's elaboration

Knowledge about child development has certainly been shaped by scientific advances, by new discoveries made possible by technological diagnostic tools that were unthinkable just 60 years ago.

But this knowledge was determined—and in turn contributed to—by factors such as social, political and economic changes, which in turn determined the variation of legal contexts.

Basaglia's international battle (1975) against the institutionalisation of the psychiatric system is part of a movement that, in the wake of May '68, rethought the relationship between state and citizen, redefining the perimeter and the ownership of rights, and this overall and general rethinking could only find its apex expression in the major forms of institutionalisation in which state power is represented: in schools, prisons and psychiatric institutions. Outside Europe, it could only find even more immediate ground in the battles for civil rights (the United States) and the struggles for decolonisation (Africa).

Basaglia (1972) writes "*We, the champions of the great western civilisation that claims the values of the individual, the spirit and reason, find ourselves weakened and destroyed by a system whose logic survives on our weakness, our acquiescence and the manipulation of this weakness and acquiescence. The absolute values that have always been proposed to us (the pride of our civilisation populated by saints and heroes) have acted—in their unattainable and inhuman perfection—as an instrument of domination through the blame game in those who fail to realise them, and as a training in compromise and the acceptance of one's own impotence in the stubborn ones who try to do so. The distance between the absolute and the relative, when the value proposed as the only one is absolute, serves as an instrument of subjection, dependence, manipulation; it serves to make every action absolutely relative (hence empty, useless, meaningless) in the eyes of those who act; it serves to make them supinely and uncritically accept the inhuman condition in which they live*".

It is in this broader international context that new concepts in child development theory and broader changes in approaches to disability in particular are to be found.

Proceeding in order in this chapter, we will try to understand how the diagnosis of autism has expanded and extended globally, how variations in theory have affected the diagnostic criteria, and how and to what extent the movement of families and autistic persons has affected this process.

The English Paradigm

The birth of a branch of science is a child of the (socio-cultural) place where it is born and structured. This was the case for Viennese psychology, and the same for quantum physics, which is Austro-German.

It is with growth and exportation that a true expansion also takes place conceptually: quantum physics in the United States, which meets that social, cultural, economic and political context, will be quite different from great Italian physics in the European context.

And the same goes for psychology, which inevitably could not be the same psychology as Freud's Vienna in the New York of the 1950s or the Los Angeles of the 1980s, or even the London of the 1970s.

As Bonnie Evans (2017) and Miltzi Waltz (2023a) have pointed out, much research on autism during a critical period of its transformation in meaning—from the 1950s to the 1970s—took place at the London Institute of Psychiatry, an institution founded in 1948 as the largest teaching centre for psychiatry in Britain.

Bonnie Evans argues that the intellectual climate created at this institution helped develop a unique concept of autism that has since become the international standard for autism research through international classification systems.

Walz's thesis is that concepts of child development are intimately connected to the legal and social context in which they are formulated.

Contemporary concepts of autism are as strongly linked to our current techniques of modelling, mapping and understanding society as the earlier concepts of '*mental deficiency*' and '*childhood schizophrenia*' were to mapping, modelling and conceptualising British society in the 1930s.

Ian Hacking introduced the idea of '*looping effects*' to demonstrate the interrelation between the two systems: between ideas, on the one hand, and human development, on the other, as both biological and psychological facts.

Although it is usually believed that laws to guarantee the rights of all children were enacted in the late nineteenth century, it was actually only in 1970 that all children in the UK were granted the right to education and the autism category was instrumental in driving this transition.

The 1970 law on the education of disabled children paved the way for autism to be considered a unique impairment requiring specialised training, which led to a transformation in the way schools managed and organised their classrooms through the expansion of staffing to include educational psychologists, speech therapists and teaching assistants.

The 1989 Convention on the Rights of the Child was unanimously adopted by the UN General Council. Many theoretical debates had stalled in the 1960s and 1970s in Europe and the United States over the question of whether children's rights included the 'power of the pupil' to oppose the authority of teachers or complete autonomy from parents, and whether this meant a contrast with 'adult rights'.

It is in the context of these radical changes in the construction of children's rights that the British autism model has been adopted worldwide as part of a legal and psychological framework model for the recognition of disability rights and children's rights in general.

Victor Lotter in Africa and the Globalisation of the Second Autism

Since Lotter's first epidemiological study of autism in 1966, there has been increasing interest in a definition of autism on a global scale.

Moreover, cultural diversity was fast becoming a feature of post-colonial London in the 1960s and 1970s, when immigration from former colonies increased rapidly. As Bonnie Evans reports, in 1981 London's population was 6.6 million, including more than 1 in 6 people born outside the UK, of whom 634,000 were from Asia, Africa and the Caribbean.

The growth of the international community has also attracted the interest of many child neuropsychiatrists due to the influx of doctors from all over the Commonwealth.

In the 1970s, Victor Lotter, the pioneer in the epidemiology of autism, decided it was time to test the durability and stability of the new concept of autism by embarking on an extensive tour of several large cities in Africa.

It was the first attempt to make autism a global category and to use epidemiology to raise questions about the cultural specificity of the condition.

His aim was to test the thesis that all autistic symptoms are universal and to then test to what extent the environment has an impact on the symptomatology.

As he put it, since '*the psychological environment for children is vastly different (in unspecified ways) than in technologically developed countries', a cross-cultural study could reveal the importance of environmental factors. It could also put to rest speculation that autism did not actually 'exist' in Africa, pointing out that his extensive studies have shown that autistic children undoubtedly exist there*'.

Lotter's findings did not explore cultural differences nor did they include post-colonial reflections on identity, raised by scholars such as Frantz Fanon (1961, 1964).

One thing Lotter's study revealed was that the level of services for the treatment of psychiatric and psychological problems in childhood was very limited.

The scarcity of administrative systems also meant that very little was known about the occurrence of all forms of '*mental handicap or behavioural disorders in general*'. Moreover, limited resources meant that priority was often given to what was 'quantitatively more urgent', such as problems like malnutrition and infectious diseases, blindness and deafness.

The English paradigm is ultimately the child of the Anglo-Saxon historical and cultural tradition, within the framework of a hierarchical relationship with traditional authority, being a monarchy, and in the context of the colonial tradition first, and the Commonwealth of Nations later.

The paradigm shift on autism, and in general on the context of institutionalisation and social care, takes place in the context of the pre-Thatcher-post Thatcher cycle, and thus in the general framework of the transformation of the state, and the relationship between state and citizen.

In this process 'numbers' count, and it was relevant that one instance (*e.g. of families in which there were persons diagnosed with autism*) was 'not statistically negligible'.

To some extent, Lorna Wing's 'second autism' and the broadening of diagnostic criteria, as well as the new theses of social and educational inclusion, were moving in this direction. And autism needed to be global, not just a British diagnosis, and to be epidemic.

The American Paradigm

The American context is very different.

Here, the historical and social relationship between state, federal government and citizen, have always been 'typical' in the all-made-in-U.S. debate between more or less federal government, more or less taxes and Washington's laws, in the context of a nation at once economically potentially self-sufficient, but also in full contention with the Soviet bloc for world hegemony: political, cultural, economic and military hegemony.

In the 1960s and 1970s, the American social model was put into crisis.

The generation that won World War II governs an increasingly young, dynamic, debating and confrontational country that unwillingly accepts the police control of opinions dictated by the fight against communism.

The America that had always won began to lose in Vietnam and Cambodia, conflicts of which more and more young Americans are struggling to make sense.

On a continent where slavery was officially abolished, racial segregation persists, African-Americans are practically barred access to university, the right to vote, and even to sit on a bus and in a restaurant except in reserved seats.

The rights of disabled, differently abled, neurodiverse people are part of the broader context of civil rights struggles, but with an American style of associationism, activism and campaigning that is still typical today.

And so in the accounts that we can somewhat define as reference, these typicalities emerge well and differ considerably from the Anglo-Saxon context I mentioned, and from the European context in general.

We can read Gil Eyal (2010), *The autism matrix*, but above all Chloe Silverman (2013), *Understanding Autism*, and Eric Garcia (1990), *We're not broken*, to realise the considerable differences in the path.

Without going into further historical details—for which we refer to these texts and the second edition of Miltzi Waltz's book (2023b)—it should be emphasised here that the history of the autism paradigm in the United States is above all the history of the events surrounding the various associations of autistic parents and persons, as in the tradition of American political participation and activism.

Suffice it to say that the topic of 'autism' has also entered the presidential elections in a big way.

In Garcia's words "*in 2016 the political culture was changing dramatically. The year after the publication of NeuroTribes, former Secretary of State Hillary Clinton became the first presidential candidate to publish an actual policy plan on autism. Autism was nothing new for Clinton, who spent a year working at the Yale Child Study Center when she was in law school. This was when so-called refrigerator parents were accused of autism, but after she married her husband and moved to Arkansas, Clinton said she started spending time with the mother of an autistic child and questioned the veracity of the refrigerator parent theory. In 2007, when she ran against Barack Obama for the Democratic nomination, Clinton adopted much of the common rhetoric about 'fighting' autism and investing 'money in researching causes and cures for diseases from cancer to autism', just as she was talking about her various proposals to assist autistic people. This probably reflected the fact that the most vocal supporters of autism were not autistic. But his political approach changed dramatically in 2016. At that point, his election plan looked like a wish list of issues autists were fighting for, such as a nationwide survey of autistic adults, banning chemical and physical restrictions on autistic children in schools, and promoting an Autism Works initiative*".

And again '*Vaccines have come to dominate much of the political discourse on autism even during the presidential campaign'. In 2008, the*

Republican presidential candidate, Senator John McCain, and the Democratic candidate, Senator Barack Obama, talked about the surge in autism diagnoses and a potential link to vaccines. 'We have seen a skyrocketing rate of autism. Some people suspect it's connected to vaccines, including this person', Obama said, referring to a person in the crowd at a rally in Pennsylvania. 'The science right now is not conclusive, but we need to do research'. At a rally in Texas, Obama's Republican opponent McCain said it was 'indisputable' that autism was 'on the rise' among children. 'And we go back and forth and there's clear evidence that indicates it has to do with a preservative in vaccines'.

Society and Development Models

We have already seen how from Lotter onwards autism—in the definition of the 'British school' transposed in the DSM-5—has become a global diagnosis of an unambiguous spectrum.

The epidemiological approach to autism has allowed researchers and parents to think about the interactions between genes and environment in development in more detail.

The globalisation of Lorna Wing's 'triad of impairments' has allowed the autism spectrum psychology model to be applied worldwide, collecting large amounts of genetic data, and models of 'social disability', which have encouraged researchers to consider how multiple genetic abnormalities can manifest as autism. These studies now seek to analyse the entire autism spectrum, the possible genes involved and the potential environmental factors that could trigger the development of this spectrum condition.

Descriptions of childhood schizophrenia before the 1960s were framed within a legal model that advocated medical interventions, internment and exclusion[2].

Legislative changes from the late 1950s to the early 1990s transformed the diagnosis of autism from a ticket to life in institutionalised settings into a positive description of children's individuality.

After these legislative interventions ensured that the diagnosis of autism was accompanied by rights to education and social acceptance, biological psychiatry changed: it was no longer the repressive diagnosis of a psychiatrist, because the autonomy of children was now protected by law. Autism therefore became an affirmation of the authenticity of children's individuality.

The category 'autism' has been used to label increasing numbers of people and these increases have been largely driven by changes in the administration of institutional care and the expansion of the right to education. Associated with this increase in autism diagnoses has been the adherence to a new model of children's rights.

Through the application of an increasing number of tests for impairments in social engagement, communication and imagination, psychologists have sought to discover the meaning of early child thinking.

The theory of multiple intelligences derived from Gardner's (1987) American cognitive psychology (Di Salvo, 2024) has been replaced by a single definition of intelligence, as a measurable quantity, in the wake, paradoxically, of Skinnerian behaviourism.

The psychology of intelligence testing has become a mainstay of many governments, particularly the US and UK, and has presented a scientific model that could also be used as a powerful administrative tool.

Proponents of the definition of 'second autism' present a crucial challenge to the idea of an 'autism spectrum' as a useful scientific way of thinking about genetic abnormality, emphasising that it embraces individuals who have been excluded due to the limitations of society itself, and who should now be welcomed and appreciated for the skills and abilities they can offer.

This led to support the idea of autism as a biochemical 'difference' rather than a pathology.

Looking at the history of autism as a central concept in child development, it is evident that the assertion of the rights of autistic persons to be autistic and their demands to de-medicalise the entire concept of autism are only a natural progression from the recognition that autism needed to be freed from state control in deficient institutions, into the hands of schools and parents, and finally into the hands of children and adults diagnosed with autism.

The growth of the study of autism as a 'social handicap' has gone hand in hand with the implementation of legislative changes that have defended the rights of children, regardless of their mental impairment.

But the entire foundation of autism rested on the possibility that statistical, epidemiological and genetic studies would be able to define the concept in the future.

The dominant descriptions of autism, such as Baron-Cohen's, state that the potential of 'autistic' thinking lies in its ability to think mathematically, scientifically and logically. This is not a coincidence, but a reflection of the way epidemiological models have been developed to measure the concept and make claims about its scientific accuracy.

The new definition of autism was nothing more than a product of the tools that have been used to measure, contain, control and identify aspects of the child's mind that are simply unknowable.

The way this impenetrable part of the child psyche has been defined has more to say about the society that constructed it than about the child mind itself.

Encountering Neurodiversity

In 2007, the Child Study Center at New York University published advertisements in the form of ransom demands: "*We have your child. We will make sure that he will not be able to take care of himself or interact socially as long as he lives. This is just the beginning*". Signed 'Autism'. Harold Koplewicz, director of the centre, hoped the advertisements would push undiagnosed children to professional diagnosis.

Autism activists led a huge protest. In a note, the Autistic Self-Advocacy Network denounced the fact that the campaign relied on 'the oldest and most offensive stereotypes of disability to scare parents'. People diagnosed with autism and Asperger's have difficulties with social interaction but "*we are not incapable of it and can succeed and thrive on our own terms if supported, accepted and included for who we are*". Autism stereotypes are self-fulfilling, he argued.

Autism advocates are a completely new category: the whole idea did not really exist until the early 2000s. Instead of advocating a cure or trying to research the cause of the much-publicised 'autism epidemic', these activists argued that it was society that needed to change, not autistic people. Koplewicz was stunned. "People kept misinterpreting the advertisements and this caused a contagion. And there was no way to explain", he said. Less than 3 weeks after their publication, the ads were withdrawn. It was a great triumph for the neurodiversity movement.

The first person to articulate the position on autism rights, Jim Sinclair (1993) stated: "*It is not possible to separate autism from the person. Therefore, when parents say: 'I wish my child did not have autism', what they are really saying is: 'I wish my autistic child did not exist and I had a different (non-autistic) child instead*'.

The term neurodiversity was proposed by Judy Singer (1999, 2016), an Australian with a mother and daughter diagnosed with Asperger's and herself on the spectrum. These activists argue that autism is not a disease but an alternative way of being.

Kathleen Seidel runs the website *neurodiversity.com*. According to her, there are two basic principles that need to be understood. The first is the variety of experiences that are grouped under the words autistic spectrum, which is not a simple range from mild to severe symptoms, but rather a three-dimensional universe of behaviour that is as difficult to define as the very notion of human personality. The other is that some people who lack what the majority consider essential for happiness are nevertheless happy.

As Andrew Salomon (2008) reported, "*neurodiversity activists include both people diagnosed with autism and their parents; their opponents also include both groups, with a heavy concentration of parents. There are actually three sides to this debate: those who believe autism is caused by environmental toxins (especially vaccines) and should be cured by addressing these pollutants; those who believe it is genetic and should be addressed through the genome; and the neurodiverse, who believe it is genetic and should be left alone. These camps are blatantly hostile to each other*".

Gerald Fischbach (2010), scientific director of the Simons Foundation, one of the largest private funders of autism research, says: "*I have never seen such an intense and demanding advocacy community. The mercury turns livid when it comes to genetics. Geneticists get angry when it comes to environmental toxins. And these activists get angry at both*".

According to Steven Hyman, former head of the National Institute of Mental Health (NIMH) and former dean of Harvard, it is likely that the disease "*turns out to be a family of related brain problems that do not respond to the same treatments. In the 1970s, the autism rate was said to be about one in 10,000; now it is one in 150. But what do these numbers mean? If an autism epidemic is hitting the country, then enormous resources are appropriate to deal with it: treating it, finding a cause, understanding how to recognise this new population. If we now describe as autism both what was once misdiagnosed as mental retardation and what was once accepted as eccentricity, there is no public health crisis*".

There are also contradictions within the different camps. Vaccine activists tend to dismiss people with articulated neurodiversity as "not really autistic", simply "oddball" individuals who hijack the fate of those most severely affected, they fight for research pointing to the increasing rates of autism that include these very people.

According to Lenny Schafer, editor of the Schafer Autism Report, adoptive father of a severely autistic child, and anti-vaccine activist, "these are *a handful of noisy people who get a lot of media attention but do not represent a large segment of the autism community. They are best ignored. They want to redefine autism as something nice that Einstein and Bill Gates had. They are trivialising what autism really is. It's like stealing money from a blind man's tin cup when you say it's not a disease*".

According to Salomon, "*while those who seek a cure for autism do not want (as their opponents in the neurodiversity movement claim) to erase a vibrant subculture of atypical thinkers, neurodiversity activists are not opposed to helping children either, even if they take exception to some treatments*".

Roy Richard Grinker (2007) author of a book dealing with his daughter's autism, pointed out that vaccine activists and their neurodiversity opponents share one thing, a deep suspicion of scientists' motives: while neurodiversity advocates are furious because scientists want to cure autism, anti-vaccine parents are angry because they are studying the wrong things.

In many ways, the question is whether it is autistic people or their parents who are unhappy and need care. Quite often, neurodiversity activists argue, parents of autistic children confuse their own unhappiness with that of their children. Parents find themselves in a difficult situation to understand how to respond. What should be covered and what should not? Much of the debate hinges on differing ideas about love. Those who believe that vaccines are the cause believe that families who do not accept their hypothesis are neglecting their children; neurodiverse people believe that parents who describe their autistic children as sick are insulting them.

Can neurodiversity bring about change efficiently enough to replace the tyranny of typical neurology? According to Kathleen Seidel "*The word incurable has a rather devastating sound, but it can also be read to mean that autism is an enduring condition. By looking at this 'jewel' through the different facets, I am not trivialising the challenges of people who face tremendous obstacles. I am trying to look at the whole picture, including the beautiful part. Autism is as much a part of our humanity as the ability to dream. This is one of the possibilities in our world. It is part of the human condition, or conditions, as the case may be*".

These considerations lead us to one of the main nodes concerning autism and the other various neurological expressions, which to date have been described in diagnostic manuals and in the public debate in pathological terms: the need to overcome the medical model in favour of the social model.

According to the medical model, disability consists of the physical or mental limitations of a person, who lacks something to be 'normal'. Normality is the benchmark according to which people are diagnosed with an illness or disability, and consequently granted the right to treatment.

The social model, on the other hand, offers a different narrative: disability does not lie in the individual who would be the bearer of it, but is constructed by society. It is the result of attitudes, prejudices and barriers erected by society and not by the individual's deficits.

According to Roberto Mastropasqua, autistic and member of the board of the Neuropeculiar Association—Movement for Neurological Biodiversity—this does not mean ignoring medical knowledge: "*The clinical point of view can be used to put on paper that a person is autistic rather than non-autistic. This makes a difference, of course, because there are implications that have to be taken into account: if I am autistic, I have a different way of relating to others than someone who is not autistic. Because of the way my brain is made and the way my relationship model is constructed, I relate to others in a different way, this must be known, otherwise what has happened so far happens: we do not understand each other*". What is wrong is to conceive of these modes not as differences but as deviations from the norm: "*Normality is a statistical average, and corresponds to the way people relate on average. According to the medical model if you deviate from that norm you have deficits, i.e. you are at fault. But in the specific case of autism, I do not have a deficit: it is not that I behave differently because I lack something, I behave differently because I am different. It's like saying that a pine tree is a pine tree and a fir tree is the wrong pine tree. No! They are two different trees*".

From Parents' Associations to Self-Advocacy

According to historian James Trent (1996), parental organisations dealing with these conditions first came together in the US and UK in the 1950s, mainly in response to a series of articles published after the war on conditions in residential facilities.

When middle- and upper-class parents were told to 'put away' their disabled children, the available institutions proved to be dangerous, even deadly. Poor and working-class parents rarely had a choice.

In 1968, the NAS was not alone. Corresponding societies had sprung up in the United States, Ireland, Australia, France, Norway and Denmark;

Dutch and German parents also came together (National Autistic Society, 1968). These groups undertook to distribute information and provide services that educational and medical institutions did not provide. In many countries, these services included schools, day care centres and residential communities: faced with insurmountable barriers to accessing state services, parents were forced to create, finance and manage their own. The stigma, however, had a huge effect on fundraising.

As Allison (1988) wrote about the NAS, "*the nation did little in response to its appeals: although we had increased our membership from 63 in 1962 to several hundred in 1964, virtually no donations came in*".

Autism has been defined by the medical model as pathological, as have the families of autistic persons, and attempts to question these definitions have met with strong resistance from the outset. For parents, responding to blame has generally meant accepting the most basic narrative of the medical model: the premise that autistic people are abnormal and need to be 'fixed'. This dominant discourse, created by combining the autism-specific narrative with the more general disability narrative, creates and perpetuates a false dichotomy. The choice it presents is between cure or institutionalisation, with the latter equated with a kind of death. As Clara Claiborne Park wrote in a second book describing her daughter's journey to adulthood, "*We could not leave Jessy to her empty serenity. We would not, as was often recommended in those days, have institutionalised her 'for the sake of other children', so that she would spend her days somewhere in a back ward, swinging. We would have kept her with us, we would have enticed, sneaked in where we were not wanted or needed*" (Park, 2001).

In a study of parents' narratives about autism published on the web, Amos Fleischmann (2004) found a common pattern in the way some parents deal with their child's autism diagnosis. For these parents, autism was seen as a challenge to overcome and, as the author writes, these "*parents ... perceived themselves not as victims but rather, as one parent put it, as 'daring mountain climbers'*".

By posing as heroes rather than villains or victims in their personal narratives, these parents strive to 'unveil' their social identity. Firestone's richly illustrated book (2007) is one of many that makes explicit the adoption of this heroic role.

What the role of self-appointed hero might mean for those who seek to save is an open question.

Wolf Wolfensberger (1983) proposed a strategy for dealing with devalued social identity called 'social role enhancement'. His aim is to improve the devalued person's image and individual skills, based on the idea that these steps can increase the chances of obtaining a positive response from society at large and thus help him or her find a valued position in the community.

Although Wolfensberger's work was initially aimed at helping people with intellectual disabilities, it can be used to describe or guide the strategies used by any socially devalued group.

For parents who have experienced devaluation due to their stigmatised association with autism, experiencing social isolation and often even loss of income, one of the few paths to social role enhancement has been to become a parent advocate, someone who fights the system. Parent advocates gain status within the community of parents of children with autism, but being seen as a '*mother from hell*' elicits negative reactions from authority figures, so this can be a problematic choice.

Becoming an effective parent advocate requires a lot of time and commitment. It is necessary to learn the language and processes involved in obtaining a diagnosis and the tactics that unlock access to education and services (Smith-Young et al., 2022).

Once these basic objectives have been achieved, constant vigilance and resilience are required to ensure that services are provided and to communicate effectively with each new professional and system encountered.

Parental advocacy outcomes also depend on socio-economic circumstances, ethnic origin and language competence. Families from ethnic minorities are still much less likely to successfully cope with systems that lead to diagnosis or high-quality educational and therapeutic services (Papoudi et al., 2021).

In the early days these families were almost invisible in the ranks of parental advocacy.

A third possible route towards the enhancement of the social role was that of the parent-researcher. This role was actually crucial for autism research from the late 1950s onwards. Many parents of autistic children became famous researchers, including Lorna Wing and Paul Shattock in the UK and Bernard Rimland in the US. Because major funders were reluctant to do so, much autism research was also funded by parent organisations.

For those who choose to pursue or continue a career in medicine, psychology or education, publicly adopting the role of parent-researcher could lead to being devalued as professionals by their peers.

Expressing one's voice and doing research were acts that could potentially replace negative stigma, at work and in the community, with a sign of courage.

Richard Pollak (1997) described childhood autism as a 'manual resistance' and Rimland as a 'guerrilla leader'.

In the 1970s, both in the United States (through the Education for All Handicapped Children Act of 1975) and in the United Kingdom (through the Education [Handicapped Children] Act of 1970), all children with disabilities gained the right to state education. This did not only affect school attendance. Parents began to use the Special Education Act to gain state-funded access to services such as physical therapy, occupational therapy, sensory integration, speech therapy, augmentative communication devices and ABA for their children. It was, and continued to be, a tough battle. This was particularly important for American parents, who were the only ones then as now in the Western world who did not have a publicly funded national medical or welfare system that could provide at least some limited services for disabled children.

Gil Eyal points out that one social change that has had a key impact on autism has been its gradual (*and still incomplete*) removal from the domain of psychiatry, along with many other developmental conditions, including intellectual disabilities.

In fact, the conjecture that therapies developed for other neurological conditions could also be useful for autistic children came at exactly the time when occupational therapy, speech therapy and physical therapy (*among many other treatment modalities*) became more professionalised and increasingly available in schools. It was in the 1970s and 1980s that these professions began to develop practices specifically geared towards autism.

A new focus on communication as something that happens between people rather than as a function of individual language production has been one of these useful changes in speech therapy (Duchan, 2008).

Using tools such as the Picture Exchange Communication System (PECS) derived from ABA (Bondy, Frost, 1994) and programmes that emphasise pragmatics, speech therapists have been able to offer practical and effective tools to help autistic children communicate. Today, these services are also available to many autistic adults in the North and some in the South.

Geeks and Great Systematisers

The thesis that autistic people are great systematisers and that this characteristic underlies the great abilities they have made available to science and—in recent decades—to STEM and IT has been studied and deepened, especially thanks to the emergence of large specialisation pools (Eindhoven, Boston, Silicon Valley, for example).

The hypothesis, after all, that autistic intelligences could have a usefulness and a use of excellence has always been valued by savants, and found in this context of activism new life and support, particularly from the world of families and the neurodiversity movement.

The literature on the subject is remarkable, and reference texts for an adequate overview can be the following, strictly in chronological order of publication:

Steve Silverman—Neurotribes: The Legacy of Autism and the Future of Neurodiversity—Avery—(2016a);

Simon Baron-Cohen—The Pattern Seekers: How Autism Drives Human Invention Basic Books (2020);

Simon Baron-Cohen—The Pattern Seekers: A New Theory of Human Invention—Penguin Books Ltd, (2022);

Devon Price—Unmasking Autism: Discovering the New Faces of Neurodiversity—Harmony—(2022).

But the first article that opened up this topic to the general public (the first research on the subject was by Baron-Cohen in the mid-1990s) was undoubtedly Steven Silverman's December 2001 article entitled '*The Geek Syndrome—Autism-and its milder cousin Asperger's syndrome-is surging among the children of Silicon Valley. Are math-and-tech genes to blame?*' published in WIRED.

Especially in American areas, studies have led to a veritable epidemiological frenzy.

Silverman wrote "*Over the past decade, there has been a significant increase in the number of children diagnosed with autism throughout California. In August 1993, 4,911 cases of so-called level one autism were recorded in the client management system of the state Department of Developmental Services. This figure does not include children with Asperger's syndrome but only those diagnosed with classic autism. [...] In 1999 the number of clients was more than double the number six years earlier. Then the curve started to steepen. In July 2001, there were 15,441 people in the*

DDS database. Now there are more than seven new level one autism cases—85% of them children—entering the system every day. In the 1990s, cases tripled in California. Anyone who says this is due to better diagnostics has their head in the sand".

He adds "*California is not alone. Rates of both classic autism and Asperger's syndrome are increasing worldwide, which is certainly cause for alarm and urgent mobilisation of research. Autism was once considered a very rare disorder, occurring in one in 10,000 births. Now it is believed to be much more common, perhaps 20 times more. But according to local authorities, the picture in California is particularly bleak in Santa Clara County. Here in Silicon Valley, the family support services provided by the DDS are mediated by the San Andreas Regional Center, one of 21 such centres in the state. SARC distributes needed resources (such as in-home behavioural training, educational aids and respite care) to families in four counties*".

According to SARC Director Santi Rogers, "*While the number of autism cases is increasing in all four cities, the percentage of classic autism cases among the total patient population in Santa Clara County is high enough to be concerning. There is a significant difference and there are no signs that it is decreasing*".

While the diagnostic confusion may contribute to the pervasive feeling that autism is on the rise, Ron Huff, the DDS consultant psychologist who discovered the statistical trend, doesn't believe that all we're seeing now is an increase in children who would have previously been labelled with some other disability, such as mental retardation, or overlooked as perfectly healthy, albeit eccentric, children. "*While we definitely need to do more research I don't think the change in diagnostic criteria will in any way explain all this increase*".

For Rick Rollens, former Secretary of the California Senate and co-founder of the MIND Institute, the idea that there is a frightening increase in autism worldwide is no longer in question. "Anyone who claims that this epidemic is due to better diagnostics", he says, "has their head in the sand".

The one thing almost all researchers in the field agree on is that genetic predisposition plays a crucial role in laying the neurological foundations of autism in most cases.

The Valley is a self-selecting community where passionately brilliant people migrate from all over the world to make smart machines work smarter. The practice of hard work among the components appeals to the predilections of the high-functioning autistic mind. The hidden cost of building enclaves like this, however, may be hidden in the results of almost every major genetic study on autism over the past 10 years. Over and over again, researchers have concluded that DNA copies for autism are likely to be passed down not only from relatives who are classically autistic, but from those who exhibit only a few typically autistic behaviours (*geneticists call those who do not fit into the diagnostic schema 'broad autistic phenotypes'*).

Silverman writes "*The old phrase 'madness is hereditary, you get it from your children' has a twist in the autistic world. It has become commonplace for parents to diagnose themselves as having Asperger's syndrome, or to identify other relatives living on the spectrum, only after their children have been diagnosed with Asperger's syndrome. High-tech hotspots like the Valley and Route 128 outside Boston are a curious oxymoron: they are fraternal associations of loners. In these places, if you are a geek living in the high-functioning regions of the spectrum, your chances of meeting someone who shares your persistent obsession (Linux or Star Trek) are greatly increased. As more and more women enter the world of IT work, guys who could never have hoped to find a kindred spirit suddenly discover that 'she's hacking Perl scripts in the next cubicle'*".

Genetic hypotheses such as these do not exclude that environmental factors play a role in the increase in numbers. Autism is almost certainly not caused by the action of a single gene, but by an orchestration of several genes that may make the developing child more susceptible to a trigger in the environment. One consequence of increased reproduction among people carrying some of these genes could be an increased 'genetic load' in subsequent generations, leaving them more vulnerable to threats posed by the environment.

In Valley clinics and schools, the observation that most parents of autistic children are engineers and programmers who themselves exhibit autistic behaviour is nothing new. And it may not be news to other communities either. Microsoft has become the first major American company to offer its employees insurance benefits to cover the costs of behavioural training for their autistic children.

At least a third of CAN's funding comes from donors in the Valley, who now want to get the maximum return on investment: better treatments, smarter assistive technologies and, eventually, a cure.

Shestack states "*Now we need brute force processing power. We need high-density SNP mapping and microarray analysis so we can design pharmaceutical interventions. We need Big Pharma to realise that although*

450,000 people in America may not be that many. For cholesterol drugs, we are talking about a demand for new products that will take 2–70 years. We need new technologies that measure perception patterns and tools for neural retraining. And we need a website where families with a newly diagnosed child can connect to a network of therapists in their city who have been rated by users, just like eBay".

One of the things that Dan Geschwind, director of UCLA's neurogenetics lab, finds fascinating about dyslexia and autism is what they suggest about human intelligence: that certain types of excellence might require not just different ways of thinking, but different types of brains: "*Autism addresses fundamental questions about the way we view talents and disabilities. The flip side of the coin of dyslexia is increased abilities in mathematics and architecture. There may be an aspect of this phenomenon with autism and assortative mating in places like Silicon Valley. In parents, who carry a few genes, they are a good thing. In children, who carry too many, it's very bad*".

The UK Brain Types Study was an ambitious attempt to measure levels of empathy and systemisation in 600,000 people. In this vast research, the first of its kind, Baron-Cohen and colleagues (Greenberg et al., 2018) asked people to complete short versions of two questionnaires: the Systematisation Quotient (sQ) and the Empathy Quotient (eQ). We discussed these in Chapter 7.

The sQ investigates the level of interest in systems, systems that follow 'if–then' rules. If you score high on the sQ test, you are the type of person who pays a lot of attention to detail.

In contrast, the eQ test measures the ease with which one can imagine what another person might think or feel. Cognitive empathy is the ability to imagine what another person might believe, know, desire, perceive or feel. This is the capacity that allows human beings to engage in flexible deception, flexible communication, teaching, cooperation around shared values and even in the spiritual realm.

The first relevant finding from the UK Brain Types Study was that both empathy and systemisation trace the bell curves of the population.

Baron-Cohen and colleagues examined, out of the five brain types determined as divisional categories in the experiment (Wakabayashi, A. et al., 2007; Groen, Y. et al., 2015) two extreme brain types: the extreme Type E, those with extremely high levels of empathy but are below average in systematisation, and their polar opposite, the extreme Type S, those with extremely high levels of systematisation but are below average in empathy. The type of brain one has depends on the tuning of the Empathy Circuit and the Systematisation Mechanism, which in turn determines where one is on the bell curves of empathy and systematisation.

The two 'extreme' types can be described as follows:

Extreme E-type—People with this brain type have highly developed empathy, but are below average in systemisation. Extreme E-type people are rare in the population, and this brain type is more common in women than in men (3% of women have it but only 1% of men). These individuals are super empathic: they empathise effortlessly and instinctively because their empathy circuit is set very high.

Extreme S-type—In the opposite case, there are individuals who are very good at systematising, but are below average in empathy. Like the extreme E-type, this brain type is also quite rare in the population. The extreme S-type is the mirror image of the extreme E-type in that it is twice as common in men (4%) as in women (2%). They are super-systematisers as their minds are perpetually looking for patterns. Autistic people also belong to this group.

The fact that the majority of the population (two-thirds) is specialised to be either empathic (type E) or systematiser (type S) may also suggest that these brain types evolved under the pressure of natural selection, with some survival advantages in having a specialised brain. This may make sense if empathisers do well in one environment (the world of people, intuitively knowing what others may think or feel) and systemisers do well in a different environment (the world of objects, understanding how things work).

Psychologist Laurent Mottron et al., (2009) found that autistic people were 40% faster than typical individuals in detecting patterns in a non-verbal visual intelligence test, exhibiting greater activity in the visual areas of the brain while achieving this result. A study (Wei et al., 2013) conducted in Silicon Valley found that, compared to any other type of disability, more autistic students were studying a STEM subject at university, again suggesting that autistic people have super-systematic minds.

Baron-Cohen's group (Baron-Cohen S. et al., 2001, 2007; Wheelwright S. et al., 2006) found that students in STEM disciplines have more autistic traits than students in the humanities, and that mathematics students are more likely to be diagnosed with autism than humanities students.

To extend this concept, he conducted the Big AQ Study, at the time the largest study on autistic traits, which involved half a million people worldwide and found—as predicted—that those working in STEM disciplines had higher AQ scores than those working in non-STEM professions. This finding was confirmed in the UK Brain Types Study, this time with 600,000 people.

They found three times as many females as males among those who were extreme E-type, and twice as many males as females among those who were extreme S-type. One of the candidate biological factors determining this difference was the testosterone level to which a foetus' brain is exposed in the womb.

During prenatal life, when the brain is developing, male foetuses produce at least twice as much of this hormone as female foetuses, and animal research shows how this hormone changes (or 'masculinises') the brain (Baron-Cohen S. et al., 2004, 2011; McCarthy M. M. et al., 2015).

We now know that there are, on average, some key differences between the sexes in the structure of the human brain, differences that emerge when large-scale brain scans are performed.

On average, there are also differences between the sexes in the number of neurons, as found in post-mortem studies of male and female brains. Volume differences are also expressed in different volumes of white mass and grey matter, but also structural differences: females have many more connections between the two hemispheres than males. Many of these differences between the sexes have been shown to be present at birth, leading to the hypothesis that, whatever the subsequent role of culture, a prenatal biological factor is also involved (Gilmore. et al., 2018; Knickmeyer et al., 2000).

Various studies (Chapman et al., 2006; Auyeung et al.,2006, 2012; Lombardo et al., 2012a, 2012b) have linked this difference between the sexes to prenatal testosterone exposure and found that the higher the prenatal testosterone to which a child is exposed, the greater its autistic traits.

According to Baron-Cohen (2005), "*autistic children on average had undoubtedly higher prenatal testosterone levels than typically developing children and this also had knock-on effects on their prenatal oestrogen levels, which on average were also elevated in children later diagnosed as autistic, which is not surprising, since testosterone is then converted to oestrogen*".

To find out whether super-systemic couples are more likely to have an autistic child, Baron-Cohen and colleagues (Baron-Cohen et Hammer,

1997; Windham et al., 2015) conducted the Parents' Occupations Study in 1997.

The study on parents' occupations was retrospective: starting with a child whose autism diagnosis was known, and then going backwards and checking whether the parents were super-systematisers.

From anecdotal accounts, autism seemed much more common in Silicon Valley and that it was more common among children of Massachusetts Institute of Technology alumni.

'Like marries like' is what biologists call 'assortative mating', whereby, for example, tall people are more likely to marry tall people.

There are a few ways in which assortative mating between super-systematisers can occur. It is possible that people who share similar interests end up living in the same place, as they have followed similar paths in their educational and career choices. Or perhaps the super-systematisers are more likely to become a couple, as the more socially competent individuals form couples first, while the less socially competent ones, years later, are still available and looking for a mate, and find each other because they are both still single. A third possibility is that the assortative mating between the super-systematisers is the result of two people being attracted to each other because their minds are similar.

Eindhoven was a 'natural' experiment to observe what happens when evolution is accelerated to see how the offspring of super-systematisers develop. The Eindhoven Study again provides evidence consistent with the idea that genes for systemisation overlap with genes for autism (Roelfsema et al., 2012).

The Risk of Self-Self-Enforcing-Diagnosis and White Males in Alabama

Research has made it clear that in certain professions and areas of specialisation, more people are diagnosed with autism "than the average" in other professions.

However, we need to clarify the reverse relationship, i.e. whether it is people with certain intelligences, interests, or professional applications who display characteristics that may be common with autism diagnoses.

According to Rosenhan (1973) "*From Bleuler, down to Kretschmer, and the authors of the American Psychiatric Association's Diagnostic and Statistical Manual there has always been an ingrained belief that patients present symptoms, that these symptoms can be classified, and, implicitly, that the healthy are distinguishable from the insane. In recent times, however, this belief has been challenged. Based in part on theoretical and anthropological considerations, but also on philosophical, legal and therapeutic ones, the view has developed that the classification of mental illness into psychological categories is at best useless, while at worst it is downright harmful, misleading and pejorative. According to this view, psychiatric diagnoses are in the minds of observers and are not a valid condensation of the characteristics displayed by the observed*".

I think it is worth re-reading that experiment 50 years later, which poses a further inverse problem, namely, as Watzlawick (1981) puts it: "*the traditional psychiatric criterion of a person's health or illness is the degree of his or her 'adaptation to reality'. This criterion tacitly assumes that there is a real, objective reality that is open to our examination and understanding*". In other words, it posits the need to understand 'the spectrum of the norm' before defining what lies outside it.

In this process, we must also clearly distinguish what is 'a type of behaviour', even if it is atypical, from 'pathological behaviour', with criteria that do not self-advertise the a priori definitions of textbooks.

Take for example the symptomatology of 'oppositional defiant disorder': adolescents who defy adults, refuse to obey rules, 'deliberately' annoy people, blame others for their own mistakes or bad behaviour. Moreover, if they admit the diagnosis, the diagnosis is confirmed; if they deny the diagnosis, this denial is confirmation of the correctness of the diagnosis.

What is the line between a rebellious teenager (who among us has not been rebellious at least at one stage of our adolescence?) and one who suffers from 'oppositional defiant disorder'? The risk is an entirely individual, subjective metric of who makes the diagnosis and how they interpret and classify data and behaviour.

People will tell me that autism is quite different, and that there are clearer protocols, but I—provocatively—ask myself, for example in this context of studies on STEM professions, what is the 'borderline with pretensions to scientificity' between the behaviour of an engineer, who is particularly meticulous and fixated on precision, and a diagnosis of tendential Asperger's? Because the risk is to classify a behaviour, however atypical, as a symptom (which affects the number of diagnoses).

That the super-systematisers are autistic is not an objective fact, but the diagnosis of autism provides that the description may include this characteristic.

However, it does not seem strange that NASA engineers or code programmers are meticulous, fussy, almost maniacal about precision, tend to be super-systematisers. It is a behavioural characteristic.

The risk is that the two will end up overlapping and self-fulfilling.

That a 'high-functioning' autistic person then finds a professional field that enhances his or her characteristics 'congenial to him or her' is also a statistically normal phenomenon, without any necessary correlation.

For example, it is entirely plausible—without this constituting a pathological symptom—that 'more women with a strong maternal sense' teach kindergarten than 'women who cannot stand children'.

Let us go deeper. One looks for the autism gene, searching for it in the context of, for example, super-systematisers. I will now give an even more provocative example.

No one (I would hope) would look for the racism gene in 'white Alabama males of the 1950s'. Yet it is really hard to find 'white Alabama males of the 1950s' who were not members of the KKK. All racists? Is racism genetic? First of all, racism is not genetic, indeed in nature it is counterintuitive that racism is an evolutionary disadvantage. The selfish gene (Dawkins 1976) could not survive long in a racist environment.

From a scientific point of view, then, races do not exist: Richard Lewontin (1972) was the first geneticist to unequivocally disprove the myth of the existence of different human races. It is biologically impossible to identify an individual on the basis of any genetic marker and thus categorise humanity into 'races'. It is therefore simply incorrect for science today to speak of 'human races'.

It is clear that racism is an environmental and cultural phenomenon, as are consequential behaviours. These are the things we should consciously ask of science.

So while it is true that the vast majority of 'white Alabama males of the 1950s' were white primes, this does not mean that there is a racism gene.

Furthermore, the fact that a behaviour falls within the symptomatology or characteristics of a definition does not mean that it constitutes a diagnosis.

If this were the case, we would all be liver and carotid artery patients, because typical diseases of these organs are essentially asymptomatic, and the absence of symptoms could be a symptom.

The fact that a behaviour is more frequent in a professional category does not necessarily mean an inverse relationship. A behaviour can be typical of a profession and also of a diagnosis without necessarily overlapping.

White males in Alabama may or may not be racist (we hope less and less) but this cannot make them fit into a pathological diagnosis (however convenient that would be).

Is There an Autism Emergency?

Higher diagnostic rates have often been confused with higher prevalence rates, but this is the result of improved diagnostic services and measures and diagnostic substitution. In the past, most people who now meet the diagnostic criteria for autism were not diagnosed. The current criteria are broader. Moreover, where a person once could only carry an intellectual disability label, he or she can now receive an autism label in place of or in addition to this (Eyal, 2010).

Diagnostic substitution is influenced not only by increased medical knowledge about autism but also by mediated cultural beliefs. Parents today are encouraged to see autism as something potentially treatable, while most accept that intellectual disability is not and is instead something to be embraced. In a socio-political environment where families are often expected to bear the costs and practicalities of ongoing care, this difference is crucial.

An autism label may attract additional funding for special education, for example, giving families and educators the hope of providing additional help at school or specific services that promise to 'normalise' autistic children.

Of course, the possibility that some recent environmental factor may be contributing to the current increased prevalence cannot be ruled out at all, and new candidates emerge frequently. Potential causes proposed range from selective mating via the Silicon Valley effect (Silverman, 2016) to endocrine disrupting chemicals in common plastics (Miodovnik et al., 2011).

However, the language of 'epidemic' and 'global health crisis' disseminated by parents' groups, charities and some doctors has proved powerful in obtaining significant research funding.

As Waltz reports in the United States for example—also driven by the strong activism we have seen in the political campaigns up to the presidential elections—the Combating Autism Act of 2006 authorised an expenditure of USD 950 million over 5 years for epidemiological surveillance and basic research, as well as the funding of state screening programmes for developmental disorders: it represented the largest investment in autism research in history. The Combating Autism Act was re-funded in 2011, and in 2019, it was renamed the Autism Collaboration, Accountability, Research, Education and Support (CARES) Act, which authorised $369.4 million in federal spending each year until 2024 (Interagency Autism Coordinating Committee, 2019). The Autism CARES Act expanded its focus to include the needs and well-being of autistic adults (Diament, 2019).

This funding has had an impact on education, care and support, but the lion's share of research has gone to genetic research, and genomics in particular. Genetics is the study of genes and how they are passed on from one generation to the next, while genomics refers to the study of all the genes in a type of person and how they relate to each other and the environment. On the autobiographical narrative of autism, this author fully subscribes to the work of Hacking (2009).

Daniel Kahneman (2011) was the first psychologist to win the Nobel Prize in economics. His studies with Amos Tversky were very important in highlighting the processes of choice and decision-making.

But among many studies, it has also highlighted many fallacies in science, particularly in the ability to read statistics and trends.

Following his reasoning, "*the algebra of a regression equation is less important than the grand idea exhibited by its form: events have multiple causes, all statistical. This sounds like an elementary principle, but it is regularly ignored in public discourse. Too often it is written as if every outcome has only one, undoubted cause: if it is proven that A affects B, this proves that C cannot affect it. […] An expert spends ten thousand hours practising his craft; which, it is said, proves that it takes practice to become an expert and talent has nothing to do with it. Men today cry twice as often as their fathers; this shows that the difference in tears between men and women is social, not biological. The possibility of multiple causes—nature and culture, talent and practice—is inconceivable. […] Even more elusive is the idea of interacting causes: the possibility that the effect of one cause may also depend on another cause. Perhaps everyone benefits from practice, but talented people benefit more. What we need is a vocabulary to talk and think in terms of multiple causes: yet another area where a few simple concepts from statistics can make everyone smarter. The revealing concepts are main effect and interaction. […] It is not only inexperienced individuals who misinterpret the concept of chance. A study on statistical intuition conducted on a sample of experienced research psychologists revealed how ingrained even among them was the belief in what we might call the 'law of small numbers', according to which even small samples are highly representative of the population from which they are drawn. From their answers, it was clear that these psychologists thought that a valid hypothesis about a population was the statistically significant result of a sample almost regardless of the sample size. Consequently, they relied too much on the results of small samples and severely overestimated their replicability. In the actual conduct of re-search, this bias leads to choosing inadequate sample sizes and placing undue weight on findings*".

There is not much to add to these considerations.

To say that a genetic malformation is present in '10% more' cases in people diagnosed with autism may seem an enormity, but if that gene already appears in, say, 10% of the general population, it means that it is expressed that way in 10 out of 100 people considered normal and 11 times out of 100 in people diagnosed with autism. That sounds quite different. Especially if the sample of people diagnosed with autism is much smaller than those considered 'normal'.

Furthermore, if autism is present in, say, 1% of the total population, then in order to have a true comparison we would have to analyse the data in the weighted form: for example, 100 people with a definite diagnosis of autism and, proportionally then, 10,000 people considered normal. But this never happens.

This is not a new mistake, nor can we seriously claim that no one has noticed it before. It is just that certain lessons need to be remembered often because they are easily forgotten.

Richard Feynman (1963) stated:

"Statistics and probability are mathematical subjects. The general idea is quite obvious: to know how many people are taller than six feet, we can take one hundred people at random, see that perhaps forty are taller, and conclude that the same holds true over the whole population. It sounds like a no-brainer. Well, it is and it isn't. If we choose hundred people from among those who pass through a slightly low doorway, the conclusion will surely be wrong. If we choose them in a group of friends we will again be wrong, because they may all be located in one area. But if we choose them absolutely independent of their stature, then, if forty of them are over six feet, we can expect that, out of a hundred million people, more or less forty million will have this characteristic. How much more and how much less can be calculated with great accuracy: for an error of no more than 1% we would have to take a sample of ten thousand people.

We often do not realise how difficult it is to achieve high precision; ten thousand attempts to get one or 2%.

Anyone who does statistics knows how crucial the choice of a reliable sample is. Statistical investigation is a scientifically sound thing, provided you do it seriously.

This business of statistical sampling and its use to determine the characteristics of a population is very serious, because this method is used all the time, and one should be very careful.

The most serious thing, apart from the sampling problem, is that in practice only measurable characteristics are considered for selection purposes. A person's spirit, his or her attitude towards life, are hardly so; and this is generally made up for by the personal interview, which is a good thing. But it is much easier to take more tests than to waste time with interviews, and the result is that the only qualities that count are those that can be measured, or that one believes can be measured. And so many interesting characteristics are never taken into account and many good students are discarded. So the system is dangerous, and one should seriously think about it.

To be honest, it should be done like this: now that a test has been designed, and there is a method to assign points to each question ... you have to redo the test [on more and many samples] and only then will we know if the test works. The reason why nobody does this 'test of the test' is, first: because it takes too long, and second: because the few times they tried it, it was seen that the test did not work.

If we look closely at all the countless unscientific attitudes that cause us so much trouble, we will see that very often they have nothing to do with inability to reason, but are due to simple ignorance, to lack of information".

Chapman and Chapman (1967, 1969) described an interesting bias in the judgement of the frequency with which two events occur concurrently. They provided inexperienced subjects with the information on several hypothetical mentally ill persons.

The data on each patient consisted of the clinical diagnosis and a human figure drawn by the patient. Subsequently, the subjects estimated the frequency with which each diagnosis (such as paranoia or suspiciousness) was accompanied by various features of the drawing (such as strange eyes). The subjects greatly overestimated the frequency of spontaneous correlations, such as that between suspiciousness and strange eyes.

This effect was called ‘illusory correlation’. With their misjudgments about the data to which they had been exposed, the volunteers ‘rediscovered’ much of the widespread clinical doctrine that draws unfounded interpretations from the human figure test. The illusory correlation effect proved extremely resistant to data that contradicted it. This effect persisted even when the correlation between symptoms and diagnosis was in fact negative, and prevented the subjects from identifying relationships that were in fact present.

Availability easily explains the illusory correlation effect.

When judging the frequency with which two events occur concurrently, one relies on the strength of the associative link between them. If the association is strong, one is likely to conclude that the events were concurrent and thus judge that the strong correlations often occurred together. According to this view, the illusory correlation between suspiciousness and the drawing of two strange eyes, for example, is due to the fact that suspiciousness is more easily associated with eyes than with any other part of the body.

Long life experience has taught us that, in general, examples of large classes are more easily recalled to memory than those of less frequent classes, that probable events are easier to imagine than improbable ones, and that associative connections between events are strengthened when the events themselves often occur in conjunction.

Consequently, man has a procedure (the availability heuristic) that allows him to estimate the magnitude of a class, the probability of an event or the frequency of a concurrence on the basis of the ease with which the relevant mental operations of retrieval, construction and association can be performed. However, this very useful estimation procedure produces systematic errors.

If we put all this information together, we have a broader—not necessarily clearer—picture to consider.

The diagnosis of autism has changed, as have the general diagnostic framework, assessment criteria and perspectives.

By opening up to social forms of care and acceptance, instead of closure in institutionalised structures, and by offering services, it became easier to accept the diagnosis, and the change in the perception of the role of parents in society—from neglected, ineffective, guilty to protagonists—helped facilitate their approach to the diagnosis.

Participatory mobilisation—of both parent associations and self-advocacy—has helped to build a perception of the autism diagnosis as something common, addressable, manageable, and not antisocial.

In the latter sense, publicity—from films to books, from interviews to public speaking—has made the world of neurodiversity visible and acceptable to society.

The vision of autistic people as inventors, geniuses, progress-drivers engaged in STEM specialisations has come full circle.

The more visible autism became, the more it became a widespread social emergency—and not a niche problem—the more research funds were channelled, and the more researchers had a 'vested interest' in ensuring that attention did not wane (science needs funds, but also time!), and in addition to promising cures, they intensified the effort to expand diagnostic criteria.

Finally, there is medical liability recognised in many western countries, whereby in case of doubt a doctor has the responsibility to make a diagnosis—especially if not accompanied by an invasive or strong drug therapy—rather than not making it, and thus answer for the possible damage of the omitted diagnosis.

At the same time, there is also a 'lack' of diagnosis.

For example, in some countries, families do not always have sufficient insurance or financial resources for an autism diagnosis, or there are not enough social services available in that country to handle an autism diagnosis.

In some countries, it is easier to obtain a diagnosis of autism if the person lives in a large capital city than if he or she lives in a rural town, which of course also concerns access to specialised school services and public assistance.

Finally, there are cultural and social factors. In some countries and cultures, autism may be unknown, or little or poorly known, and therefore one may not consider seeking a diagnosis.

The concept of neurodiversity (Singer, 1999) is fundamental in bridging the social divide and in constructing new inclusive words: it seeks to place the autism spectrum within the human spectrum, alongside other areas of human diversity such as race and gender and related discourses on rights, freedoms and cultural choices.

It is the writer's opinion that recognising and working from the strengths and valuing the rich and varied lived experiences of persons currently designated as being 'on the autism spectrum' are the first steps into the current autism territory.

Notes

1. The specific error of statistical computation in the social sciences is often underestimated, with the belief instead that what is significant is a certain degree of approximation that essentially allows a trend to be drawn.

   Behind this apparent justification for the lack of mathematical rigour is often concealed either the lack of conviction of mathematical statisticians in research, or a simple method of justifying bias and remorse discussed in Chapter 2.

   Such errors are unfortunately so frequent that even 'exact sciences' are affected.

   As Dyson (2004) recounts, Enrico Fermi once clarified "that [his] friend Johnny von Neumann used to say: with four parameters I can make the data describe an elephant, and with five I can make it shake its trunk".

   Fermi himself can give us—as he used to do—a simple and suitably precise method of solving the problem. This time we draw it from an anecdote told by Stanislaw M. Ulam.

   Ulam once won a tennis match with Fermi 6–4, but the latter refused to admit defeat, noting that the difference between the two scores was less than the square root of the sum of the scores themselves, 3.16.

   This is a practical method used by statisticians to determine whether a result is significant or within the limits of measurement error.

   Dyson, F. (2004) A Meeting with Fermi, in "Nature", 427, no. 6972 p. 297.

   Ulam, S. M. (1995) Avventure di un matematico, trans. it. by Tecla R. Sportelli, Sellerio, Palermo 1995, p. 215.
2. The internment solution for the 'mentally handicapped' will be challenged by Franco Basaglia's movement against forced institutionalisation in the early 1970s. However, this in turn can also be seen as an improvement.

References

Allison, H. G. (1988). Prospettive su un pezzo di puzzle. *Communication, 22*

Auyeung B. et al. (2006) "Fetal testosterone and the Child Systemizing Quotient (sQ-C)" European Journal of Endocrinology, 155

Auyeung B. et al. (2012) "Effects of fetal testosterone on visuospatial ability" Archives of Sexual Bebavior, 41

Baron-Cohen, S., Hammer, J., (1997) "Parents of children with Asperger syndrome: What is the cognitive phenotype?" Journal of Cognitive Neuroscience, 9

Baron-Cohen S. et al. (2001) "The Autism Spectrum Quotient (AQ): Evidence from Asperger syndrome/high-functioning autism, males and females, scientists, and mathematicians" Journal of Autism and Developmental Disorders, 31

Baron-Cohen S. et al. (2004), Prenatal Testosterone in Mind: Amniotic Fluid Studies, MIT Press/Bradford Books, Cambridge

Baron-Cohen S. et al (2005), "Elevated fetal steroidogenic activity in mo, vedi Molecular Psychiatry, 20

Baron-Cohen S. et al. (2007) "Mathematical talent is linked to autism" Human Nature, 18

Baron-Cohen S. et al.(2011) "Why are autism spectrum conditions more prevalent in males?" PLOS Biology, 9

Baron-Cohen, S. (2020) The Pattern Seekers: How Autism Drives Human Invention Basic Books

Baron-Cohen, S. (2022) The Pattern Seekers: A New Theory of Human Invention—Penguin Books

Basaglia, Franca Ongaro (a cura di), Crimini di pace, Michel Foucault, Erving Goffman, Ronald David Laing e Noam Chomsky, Torino, Einaudi, 1975

Basaglia, F., (1972) Introduzione a *La salute mentale in Cina*, di Gregorio Bermann, Einaudi

Bondy, A. S., & Frost, L. A. (1994). The Picture Exchange Communication System. *Focus on Autism and Other Developmental Disabilities*

Chapman E. et al. (2006) "Fetal testosterone and empathy: Evi-dence from the Empathy Quotient (EQ) and the 'Reading the Mind in the Eyes' test" Social Neuroscience, 1

Chapman, L.J., Chapman, J.P., (1967), Genesis of Popular but Erroneous Psychodiagnostic Observations, Journal of Abnormal Psychology, 73

Chapman, L.J., Chapman, J.P., (1969) Illusory Correlation as an Obstacle to the Use of Valid Psychodiagnostic Signs, Journal of Abnormal Psychology, 74

Dawkins, R. (1976) Selfish Gene

Diament, M. (2019). Autism Act clears Senate. Disability Scoop.

Di Salvo, M. (2024) Multiple intelligences and specialization. A return to Gardner and the wisdom of rethinking psychology in Hillmanian transparency.

- Revista De Estudios Socioeducativos. ReSed, 1(12). doi.org/10.25267/Rev_estud_socioeducativos.2024.i12.18
Duchan, J. (2008). *A history of speech-language pathology.*
Eyal, G., Hart, B., Oncular, E., Oren, N., & Rossi, N. (2010). *The autism matrix: The social origins of the autism epidemic.* Polity Press.
Evans, B. (2017). The transformation of social life and the transformation of autism in the 1960s. In The metamorphosis of autism in Britain (pp. 189–247). University of Manchester Press.
Fanon, F., (1961)Les Damnés de la terre, tr. it. I dannati della terra, Einaudi, 2000
Fanon, F., (1964) Pour la révolution africaine, tr. it. Scritti politici—volume I. Per la rivoluzione africana (Prefazione di Miguel Mellino; Postfazione di Paul Gilroy), Roma, Derive Approdi, 2006
Feynman, R. (1963, 1998) The Meaning of it All—a collection of three lectures in April 1963 at the University of Washington as part of the John Danz Lectures Series.
Firestone, B. (2007). *Autism Heroes: Portraits of Families Meeting the Challenge* Jessica Kingsley Publishers.
Fischbach, G. D., & Lord, C. (2010). The Simons Simplex Collection: A resource for identification of autism genetic risk factors. *Neuron, 68*
Fleischmann, A. (2004). Narrazioni pubblicate su Internet da genitori di bambini con autismo: Cosa rivelano e perché sono importanti? *Focus on Autism and Other Developmental Disabilities*
Garcia, E., 1990—We're not broken : changing the autism conversation. Boston : Houghton Mifflin Harcourt,
Gardner, H. *The Mind's New Science.* New York: Basic, 1987
Gilmore J. et al. (2018) "Imaging structural and functional brain development in early childhood" Nature Reviews Neuroscience, 19
Greenberg et al. (2018) "Testing the Empathizing-Systemizing (E-S) theory of sex differences and the Extreme Male Brain (EMB) theory of autism in more than half a million people", in Proceedings of the Natio-nal Academy of Sciences
Grinker, R.R., (2007). *Unstrange minds: Remapping the world of autism.* Basic Books.
Groen, Y. et al. (2015) "The Empathy and systemizing Quotient: The psychometric properties of the Dutch version and a review of tre crs, 45,9, 2001, abil 2, Journal of Autismand Developmental Disorders, 45
Hacking, I. (2009). Autistic autobiography. *Philosophical Transactions of the Royal Society of Biological Sciences, 364*
Kahneman, D., (2011). *Thinking, Fast and Slow.* Farrar, Straus and Giroux.[*also in Steven Pinker Rationality, Penguin Books Ltd, 2022]*

Knickmeyer R. et al. (2000) "Impact of sex and gonadal steroids on neonatal brain structure" Cerebral Cortex, 24

Interagency Autism Coordinating Committee. (2019). Autism CARES Act of 2019. IACC. https://iacc.hhs.gov/about-iacc/legislation/autism/cares-act-2019/

Lewontin, R (1972) "The Apportionment of Human Diversity," Evolutionary Biology, vol. 6

Lombardo M. et al. (2012a) "Fetal testosterone influences sexually dimorphic gray matter in the human brain" Journal of Neuroscience, 32

Lombardo M. et al. (2012b) "Fetal programming effects of testosterone on the reward system and behavioral approach tendencies in humans" Biological Psychiatry, 72

Lotter, V.(1966) 'Epidemiology of autistic conditions in young children'. *Social Psychiatry* 1

McCarthy M.M. et al (2015) "Surprising origins of sex differences in the brain", Hormones and Behavior, 76

Miodovnik, A., et al. (2011) Endocrine disruptors and childhood social impairment. *Neurotoxicology, 32*

Mottron L. et al. (2009) "Enhanced perception in savant syndrome: Patterns, structure, and creati-vity", Philosophical Transactions of the Royal Society of London

Papoudi, D., Jørgensen, C. R., Guldberg, K., & Meadan, H. (2021). Perceptions, experiences, and needs of parents of culturally and linguistically diverse children with autism: A scoping review. *Review Journal of Autism and Developmental Disorders*

Park, C. C. (2001). *Uscire dal nirvana:* La *vita di una figlia con l'autismo*. Little, Brown and Company.

Pollak, R. (1997). La creazione del Dr. B: una biografia di Bruno Bettelheim. Simon e Schuster.

Price, D. (2022) Unmasking Autism: Discovering the New Faces of Neurodiversity—Harmony

Roelfsema M. et al. (2012) "Are autism spectrum conditions more prevalent in an information-technology region? A school-based study of three regions in the Netherlands" Journal of Autism and Developmental Disorders, 42

Rosenhan, D. L. (1973). On being sane in insane places. *Science, 179*

Salomon, A., (2008)—The Autism Rights Movement—NYMagazine

Silverman, C. (2013). Understanding autism: Parents, doctors and the history of a disorder. Princeton University Press.

Silverman, S. (2016a) Neurotribes: The Legacy of Autism and the Future of Neurodiversity—Avery

Silverman, S. (2001) "The Geek Syndrome—Autism—and its milder cousin Asperger's syndrome—is surging among the children of Silicon Valley. Are math-and-tech genes to blame?" WIRED

Sinclair, J. (1993). Don't mourn for us. *Our Voice*, http://web.archive.org/web/20090329155049/http://web.syr.edu/~jisincla/dontmourn.htm

Singer, J. (1999). 'Why can't you just be normal for once': From a 'problem with no name' to a new category of difference. In M. Corker & S. French (Eds.), *Disability discourse*. McGraw-Hill Education

Singer, J. (2016) Neurodiversity the birth of an idea

Smith-Young, J., Chafe, R., Audas, R., & Gustafson, D. L. (2022). 'I know how to advocate': Parents' experiences in advocating for children and youth diagnosed with Autistic Spectrum Disorder. *Health Service Insights, 15*

Trent, J. W. (1996). *Inventare la mente debole: Una storia del ritardo mentale negli Stati Uniti*. University of California Press.

Wakabayashi, A. et al. (2007) "Empathizing and systemizing in adults with and without autism spectrum conditions: Crosscultural stability", Journal of Autism and Developmental Disorders, 37

Waltz, M. (2023). The Social Construction of Autism. In: Autism. Palgrave Macmillan, Cham

Waltz, M. (2023). Parent-Blaming, Parent Power and Research. In: Autism. Palgrave Macmillan, Cham

Watzlawick, P. (1981) Die Erfundene Wirklichkeit—R. Piper & Co. Verlag, München

Wei, X. et al. (2013) "Science, Technlogy, Engineering, and Mathematics (STEM) participation among college students with an autism spectrum disorder", Journal of Autism and Developmental Disorders, 43

Wheelwright S. et al. (2006) "Predicting Autism Spectrum Quotient (AQ) from the Systemizing Quotient-Revised (sQ-R) and Empathy Quotient (EQ)" Brain Research

Windham G. et al. (2015), "Autism spectrum disorders in relation to parental occupation in technical fields" Autism Research

Wolfensberger, W. (1983). Social role valorization: A proposed new term for the principle of normalization. *Mental Retardation, 21*

CHAPTER 13

Therapeutic and Supportive Treatments and Interventions for People Diagnosed with Autism

As Uta Frith (1989) wrote "*What is the cause of autism and what can be done about it? This is the most common question asked by anyone confronted with this enigmatic disorder for the first time. In theory, there should be an answer that can explain autism and at the same time prevent it by pointing to its cure. Unfortunately, such a simple answer does not exist. Scientists are sceptical about the profusion of hypotheses about the physical and chemical causes of autism that are constantly being produced. They can live with the abstract idea that the causes of autism are not yet known. For parents it is much more difficult. When they are given no explanation as to why their healthy, beautiful child is autistic, they are driven to find a reason. The human mind automatically detects cause and effect relationships between events that occur together, and this can generate powerful illusions. But correlation is not the same thing as causation. The scientific method provides tools to determine whether events perceived as causal really are. To do this takes time. […] When we have identified the specific causes and mechanisms of action, will our attitude towards autism change? It is possible. In other pathologies, our attitude has often changed when the cause has become known. Parents of children with Down syndrome today know that the cause is a particular chromosomal abnormality, and accept the permanent nature of the condition. They are less likely to fall prey to charlatans offering cures. […] The question of the causes of autism will remain a subject of intense speculation for some time to come. We have to think of a long causal chain*

M. Di Salvo, *Autism Research between Psychology and Neuroscience*,
https://doi.org/10.1007/978-3-031-68338-1_13

with many distinct links. As a reminder, we can summarise that there is a risk, followed by an attack, followed by harm. The risk can be of many types, including genetic defects, chromosomal abnormalities, metabolic disorders, viral agents, immune intolerances and anoxia resulting from perinatal problems. We can assume that each of these risks has the potential to produce very serious damage to neurodevelopment. Due to these severe interferences, permanent damage can be produced in the development of specific brain systems involved in higher mental processes. The damage may be slight or severe, but often involves the shutdown of one or more crucial systems at a critical stage".

Uta Frith's hypothesis is that only in such a case will autism occur, and with the necessary adjustments we shall see this author agrees with this thesis.

In this chapter, I will attempt to describe the therapies currently adopted with some success to improve the quality of life for families and help give greater independence and autonomy to people diagnosed with autism.

I will state at the outset that I will not discuss drug therapies.

These depend, country by country, on the criteria and authorisations of the local authorities (*for example, melatonin, widely used in Europe as a sleep aid, is banned in many Arab countries, and in others it is not recommended for paediatric use*).

Above all, drug therapies are aimed not at autism per se but at alleviating some specific, often related characteristics, such as anxiety and hyperactivity. Discussing this is beyond the scope of this paper, and is often a topic to be left to the general sensitivity and consideration of the physician.

It is the general opinion of this author that "*pharmacological intervention has a relevant value, especially in cases where the symptomatic component represents a source of pain and discomfort for the person. This is in all cases in which soothing the symptom at the same time does not make the pathological source (and thus the source of the symptom) inaccessible, and keeps the patient, his self and his person accessible, and especially when it does not claim, by eliminating the symptoms, to have put an end to the pathological situation, which could find other, far more dramatic ways of expression*". (Di Salvo, 2024).

Nor will I discuss clearly and unanimously discredited therapeutic approaches (*chelation therapies for example*) or those without evidence (*secretin for example*) or vitamin supplements.

That there is damage to the central nervous system in autistic persons is beyond dispute, and therefore a clear negative effect of potentially inflammatory foods in general is logical and to be avoided. As well as vitamin supplementation, mineral salts, and probiotics are part of the normal treatment of any person who is, on clinical analysis, deficient and in need of them.

In this regard, see the very important and comprehensive work by Marta Herbert (2013) of Harvard Medical School.

An investigation and evaluation that cannot be 'by standard reference tables' but must necessarily always be individualised (Moalem, 2014).

Nor will I deal here with therapies that are so new that they do not have sufficient studies to support them, in one sense or another, and that require years, if not decades, of further research in order to be able to indicate in which direction they may be useful at least in diminishing the most dramatic aspects of an overall condition (*for example stem cell research*).

Here I will discuss—and try to document with studies for and against and general verification studies—the most accredited therapies in general use, which as we shall see are essentially psychological and psychotherapeutic in nature.

It is important to state at the outset that there is no 'one therapy', nor is there a therapy that is valid for everyone.

In itself, the psychological and psychotherapeutic approach depends very much on the schools in each individual country and the psychotherapeutic approach.

In addition to this difference, there is an ontological one: therapy is personal and individualised; this is true and ontologically unavoidable for neuronormal people, all the more so for neurodiverse people.

What is true for psychological support is even more true for integrated and multimodal therapies.

In this overall context, it is the writer's opinion that—precisely because of the peculiar characteristics of the autistic spectrum—psychological, psychomotor, mind–body relationship and even bioenergetic support approaches are all avenues to be pursued in the search for the most suitable, functional and useful form of support for the person.

I will not divide this chapter into paragraphs, but I will highlight the individual interventions I will discuss in bold type, precisely to give an idea of a continuum and not an alternative between the various types of intervention.

I hope this approach will also be graphically effective.

Autism therapies include a wide variety of interventions that help people diagnosed with autism and often their families. These therapeutic methods seek to help people cope with difficulties and increase their functional independence.

Many therapies claim results that have not been supported by LOE 1.[1]

Autism is a neurodevelopmental condition characterised by differences in social interaction and mutual communication as well as restricted and repetitive interests, behaviours or activities.

There is no therapy to eliminate autism.

Treatment is generally tailored to the person's needs.

Treatments fall into two main categories: educational interventions and medical management. Training and support are also provided to families of persons diagnosed with autism spectrum disorders (Myers, Johnson, 2007).

Studies on interventions present some methodological problems that prevent definitive conclusions on effectiveness from being drawn (Ospina et al. 2008).

Although many **psychosocial interventions** have some positive evidence suggesting that some form of treatment is preferable to no treatment, systematic reviews have reported that the quality of these studies has generally been poor, their clinical results are mostly provisional and there is little evidence for the relative effectiveness of treatment options (Seida et al. 2009).

Intensive and prolonged **special educational programmes** and behavioural therapy early in life can help autistic children acquire self-care, social and work skills, and can often improve functioning and decrease the severity of observed maladaptive signs and behaviours (Rogers and Vismara 2008).

Available approaches include **applied behavioural analysis** (ABA), developmental models, structured teaching, speech and language therapy, social skills therapy and occupational therapy.

Occupational therapists work with autistic children by creating interventions that promote social interaction such as sharing and cooperation. Educational interventions have some effectiveness in children: intensive ABA treatment has demonstrated efficacy in improving global functioning in preschool children, and is well-established for improving intellectual performance in young children (Eikeseth, 2009).

Historically, 'conventional' **pharmacotherapy** has been used to reduce behaviour and sensitivity associated with autism (Leskovec et al. 2008).

Many of these treatments were prescribed off-label to target specific symptoms. Today, drugs are mainly prescribed to autistic adults.

Therapeutic treatments such as behavioural or immersive therapies are gaining popularity in treatment plans for autistic children. Depending on the symptomatology, one or more psychotropic drugs may be prescribed, i.e. antidepressants, anticonvulsants and antipsychotics (Popow et al. 2021).

Educational interventions attempt to help children not only learn school subjects and acquire traditional readiness skills, but also to improve functional communication and spontaneity, enhance social skills such as joint attention, develop cognitive skills such as symbolic play, reduce disruptive behaviour and generalise what they have learnt by applying the acquired skills to new situations.

Intervention studies have methodological flaws that prevent definitive conclusions on effectiveness from being drawn. Concerns about outcome measures, such as their inconsistent use, most influence how the results of scientific studies are interpreted (Wheeler et al. 2008).

A 2009 study in Minnesota found that parents follow behavioural treatment recommendations far less often than they follow medical recommendations and that they adhere more often to reinforcement recommendations than to punishment recommendations (Moore and Symons 2009).

Currently, claims that **early intervention** by the age of three is crucial are unsubstantiated although it seems plausible, based on neuroscientific evidence, that any early intervention can maximise neuroplastic potential (Howlin et al., 2009).

Mind–body therapies are often used by people with autism spectrum disorders. However, a comprehensive examination of the specific types of mind–body therapies and their expected outcomes is lacking, which means that further research and exploration in this area is needed. However, even in this case, it seems plausible and consistent that an intervention that is not only psychological but also physical may yield greater results (Hourston and Atchley, 2017).

Applied behaviour analysis (ABA) is the applied research field of the science of behaviour analysis and forms the basis of a wide range of techniques used to treat autism and many other behaviours and diagnoses (Dillenburger and Keenan 2009).

ABA-based interventions focus on teaching one-to-one tasks using behaviourist principles of stimulus, response and reward, and reliable measurement and objective evaluation of observed behaviour (Howard et al. 2005).

There is wide variation in the professional practice of behavioural analysis and among the assessments and interventions used in school-based ABA programmes (Steege et al. 2007).

Technology has started to be used to a considerable extent in ABA therapy for the treatment of autism. Robots, gamification, image processing, storyboards, augmented reality and web systems have proven to be useful in the treatment of autism. These technologies are used to teach autistic children skill acquisition. Web programmes are designed to address skills such as attention, social behaviour, communication and/or reading (Alves et al. 2020).

ABA has faced much criticism over the years. Recently, studies have shown that ABA can create abuse and can increase symptoms of post-traumatic stress disorder in patients (Kupferstein, 2018; Sandoval-Norton et al., 2019; Wilkenfeld and McCarthy 2020; Bottema-Beutel et al., 2021; McGill and Robinson, 2021; Shkedy et al., 2021). For example, the Autistic Self Advocacy Network has campaigned against the use of ABA (DeVita-Raeburn 2016).

Many intensive behavioural interventions are based on **discrete trial teaching methods** (DTT), which use stimulus–response-reward techniques to teach fundamental skills such as attention, compliance and imitation (Ryan and Hemmes 2005).

However, children have often been found to have problems using the skills taught by **DTT in natural environments**. In functional assessment, a common technique, the teacher formulates a clear description of a problem behaviour, identifies the antecedents, consequences and other environmental factors that influence and maintain the behaviour, develops hypotheses about what causes and maintains the behaviour, and collects observations to support the hypotheses. Some more comprehensive ABA programmes use multiple methods of assessment and intervention in an individual and dynamic manner (Steege e al., 2007).

ABA-based techniques have demonstrated efficacy in several controlled studies: children have been shown to achieve lasting improvements in school performance, adaptive behaviour and language, with significantly better results than control groups.

A 2009 comprehensive synthesis of **Early Intensive Behavioural Intervention (EIBI)**, a form of ABA treatment, found that EIBI produces strong effects, suggesting that it may be effective for some children with autism; it also noted that the large effects may be an artefact of comparison groups with treatments that have yet to be empirically validated and that no comparison between EIBI and other widely recognised treatment programmes has been published (Reichow and Wolery, 2009).

A 2009 meta-analysis concluded that EIBI has a large effect on intelligence on a large scale and a moderate effect on adaptive behaviour (Eldevik et al. 2009).

However, a 2009 systematic review and meta-analysis found that **Applied Behavioural Intervention (ABI)**, another name for EIBI, did not significantly improve outcomes compared to standard care for preschool children with autism in the areas of cognitive achievement, expressive language, receptive language and adaptive behaviour (Spreckley et al., 2009).

Pivotal response therapy (Koegel et al., 1999; Koegel and Koegel 2006) is a naturalistic intervention derived from ABA principles. Instead of individual behaviour, it targets crucial areas of a child's development, such as motivation, responsiveness to multiple cues, self-management and social initiations; it aims for widespread improvements in areas that are not specifically targeted. The child determines the activities and objects that will be used in a PRT exchange. Planned attempts to achieve the target behaviour are rewarded with a natural reinforcer: for example, if a child attempts to request a stuffed animal, the child receives the animal, not a piece of candy or another unrelated reinforcer.

The inability to communicate, verbally or non-verbally, is a fundamental deficit in autism. Communication interventions fall into two broad categories. Firstly, many autistic children do not speak, or speak little, or have difficulty in the effective use of language. Social skills have been shown to be effective in the treatment of children with autism (Gillis and Butler, 2007).

Interventions that attempt to improve communication are commonly conducted by speech therapists and work on joint attention, communicative intent and **augmentative and alternative communication** (AAC) methods such as visual methods. A 2006 study reported benefits for both the joint attention intervention and the symbolic play intervention (Kasari et al. 2006), and a 2007 study found that the joint attention intervention was more likely than the symbolic play intervention to induce children to subsequently engage in shared interactions (Gulsrud et al. 2007).

Social skills treatment attempts to increase social and communication skills, addressing a fundamental deficit in autism. A wide range of intervention approaches are available, including modelling and reinforcement, adult-peer mediation strategies, peer tutoring, social games and stories, self-management, fundamental response therapy, video modelling, direct instruction, visual cues, circle of friends and social skills groups (Matson et al. 2007).

The **SCERTS model** (Prizant et al. 2005) is an educational model for working with autistic children. It is designed to help families, educators and therapists work together cooperatively to maximise progress in supporting the child.

The acronym refers to the focus on:

SC—social communication—the development of functional communication and emotional expression.

ER—emotional regulation—the development of well-regulated emotions and the ability to cope with stress.

TS—transactional support—the implementation of support to help families, educators and therapists respond to children's needs, adapt the environment and provide tools to enhance learning.

Relationship-based models emphasise relationships that help children reach and master early developmental milestones.

Relationship development intervention (Gutstein et al., 2007) is a family-based treatment programme for children with autism. This programme is based on the belief that the development of dynamic intelligence (the ability to think flexibly, take on different perspectives, deal with change and process information simultaneously) is the key to improving the quality of life of children with autism.

The **Treatment and Education of Autistic and related Communication-handicapped CHildren (TEACCH)**, which has been called 'structured teaching', emphasises structure by using organised physical environments, predictably sequenced activities, visual programmes and visually structured work/activity systems. where each child can practise various tasks. Parents are taught to implement the treatment at home. A 1998 controlled study found that children treated with a TEACCH-based home programme improved significantly more than a control group (Ozonoff and Cathcart, 1998).

A 2013 meta-analysis that reviewed all clinical studies on TEACCH indicated that it has little or no effects on perceptual, motor, verbal, cognitive and motor functioning, communication skills and activities of daily living. Positive effects were found in social and maladaptive behaviour, but these required further replication due to the methodological limitations of the pool of studies analysed (Virues-Ortega et al., 2013).

Unusual responses to sensory stimuli are more common and prominent in children with autism, although there is no concrete evidence that sensory symptoms differentiate autism from other developmental disorders. Several therapies have been developed to treat **sensory processing disorder (SPD)** (Rogers and Ozonoff, 2005).

The term **multisensory integration** in simple words means the ability to use all one's senses to perform a task. Occupational therapists sometimes prescribe sensory treatments for children with autism, however, there is generally little or no scientific evidence of effectiveness (Baranek, 2002).

Animal-assisted therapy, in which an animal such as a dog or horse becomes a key part of a person's treatment, was a controversial treatment. A 2007 meta-analysis found that animal-assisted therapy was associated with "moderate improvement in autism spectrum symptoms" (Nimer and Lundahl, 2007).

Reviews of published studies on dolphin-assisted therapy (DAT) have found major methodological flaws and concluded that there is no convincing scientific evidence that DAT is a legitimate therapy or that it offers anything other than temporary improvements in mood (Marino, 2007).

Modern animal-assisted therapy for autism is not about 'autistic symptom control', but a natural way to foster socialisation (bridging the 'double empathy gap') and also for stress reduction (Ávila-Álvarez, 2020; Wijker et al., 2019, 2021).

The remarkable adherence to the therapeutic programme by participants in the 2020 study and the clinically relevant effects of the programme indicate that TAA with dogs can be used to reduce perceived stress and agoraphobia symptoms and to improve sociality, awareness and communication in adults with autism with normal to high intelligence" (Wijker et al., 2020).

Neurofeedback attempts to train individuals to regulate their own brainwave patterns by allowing them to observe their own brain activity more directly. In its most traditional form, EEG electrode output is fed into a computer that controls a game-like audiovisual display. Neurofeedback has been evaluated 'with positive results for autism', but the studies lacked random assignment to controls. This research has been ongoing since 2019, although it now focuses on "improving attention" and "reducing anxiety" (Coben et al., 2010; Mercado et al., 2019).

Environmental enrichment is concerned with how the brain is affected by the stimulation of information processing provided by its environment (including the opportunity for social interaction). Brains living in richer, more stimulating environments have more synapses and the dendrite structures on which they reside are more complex. This effect occurs mainly during neurodevelopment, but also to a lesser extent in adulthood.

Research on humans suggests that lack of stimulation (deprivation, as in old-style orphanages) delays and impairs cognitive development. Research also finds that higher levels of education (which is cognitively stimulating in itself and associates people with more cognitively demanding activities) result in greater resilience (cognitive reserve) to the effects of ageing and dementia.

In all interventions for autistic children, the main strategy is to focus on improving sensitivity in all senses. Autistic children may lack the ability to name or even feel their emotions. This can also have an impact on the relationship with other people's emotions and the deduction of other people's moods. Many autistic children also live with a sensory processing disorder. In sensory-based interventions, there were signs of progress in children who exhibited an appropriate response when receiving a stimulus after being in sensory-based therapies for a period of time. Autism spectrum disorder varies from child to child, which can make it difficult for clinicians to assess and know which therapies to apply.

The aim of these differentiated interventions is to intervene at the neurological level of the brain in the hope of developing appropriate responses to different sensations in the body and also to external stimuli in the environment. Scientists have used music therapy, massage therapy, occupational therapy and more. Because the autism spectrum is so diverse and widespread, each case or scenario is different.

Emerging evidence in favour of mindfulness-based interventions to improve mental health in adults with autism is supported by a recent systematic review. This includes evidence for reductions in stress, anxiety, ruminative thoughts, anger and aggression (Benevides et al. 2020).

Drugs, supplements or diets are often used to alter physiology in an attempt to alleviate common autistic symptoms such as seizures, sleep disturbances, irritability and hyperactivity that may interfere with education or social adjustment. There is plenty of anecdotal evidence to support medical treatment; many parents who try one or more therapies report some progress, and there are some well-publicised reports of children being able to return to mainstream education after treatment, with marked improvements in health and well-being. However, this evidence may be confounded by the improvements observed in autistic children growing up without treatment, the difficulty of verifying reports of improvements, and the lack of reporting of negative outcomes of treatments. Only very few medical treatments are well supported by scientific evidence using controlled experiments (Levy and Hyman, 2005; Schreibman, 2005).

Transcranial magnetic stimulation, which is a rather established treatment for depression, has been proposed and used as a treatment for autism (Novella 2013). A review published in 2015 found insufficient evidence to support its widespread use for autism spectrum disorders (Oberman et al., 2015).

A 2015 review found tentative but insufficient evidence to justify its use outside of clinical trials (Oberman et al., 2016).

New findings show that TMS can positively influence gamma brainwave oscillations and help improve performance accuracy (Ospina et al. 2008, Nevada Autism Center, 2020).

It is now known that "children with autism are at risk for ***eating tract disorders*** above all, they are at increased risk for general gastrointestinal (GI) problems, constipation, diarrhoea and abdominal pain" (Wasilewska and Klukowski, 2015; Abdelrahman et al. 2015; Gorrindo et al. 2012).

Currently, there is no 'diet for autism', only dietary suggestions.

As we have seen for each therapy or intervention there are conflicting statistical findings in relation to the degree of effectiveness.

From the point of view of the evaluation of results in terms of statistical effectiveness, much has already been said in the second part of Chapter 2.

From my point of view, starting from the consideration that no 'one autism is the same as another', the value—and thus the effectiveness—of any intervention must be assessed in individual terms.

If one assumes that for a therapy to be considered effective, it must benefit a certain percentage threshold of people, it is the measure adopted for measurement that is in itself wrong, because it is like measuring a distance using the kilogramme.

What should be evaluated is the individual impact, the degree of benefit and improvement of the person's condition, individually understood.

If we expect, for example, that animal-assisted therapy is valid for all or most autistic persons, it is our approach that is wrong, because it leads back to the logic of 'therapy for all'.

What has to be established is whether for that person, individually, animal therapy is effective; if even one person benefits then that intervention is—indivudually—useful, valid and effective and should be pursued. In this context we might also find that for one person interaction with a dolphin is more effective than interaction with a horse. And this does not exclude two different degrees of effectiveness for two different people.

There are other therapies currently being tested related to the latest neuroscientific research and will be the subject of the last chapter.

Before concluding our research and this narrative, we must return to Chapter 9 where we left the topic of music.

Music is a typical human specialisation, which goes beyond the emission of sounds by other animals.

Together with mathematics, music represents perhaps the only universal human language, belonging to all humans, everywhere, and regardless of specific culture, capable of conveying emotions beyond language.

Anyone who has experience relating to an autistic person cannot have failed to grasp the absolutely special relationship of autistic people with music.

The next chapter is devoted to this report.

Note

1. LOE - Levels of Evidence. Winona State University: Darrell W. Krueger Library—the highest level assigned based on methodological quality, validity and applicability to patient care—Level 1 research includes evidence from a systematic review or meta-analysis of all relevant RCTs (randomised controlled trials) or evidence-based clinical practice guidelines and systematic reviews of RCTs with comparable results.

References

Abdelrahman HM, Sherief LM, Alghobashy AA, Abdel Salam SM, Hashim HM, Abdel Fattah NR, Mohamed RH (2015). "Association of 5-HT2A receptor gene polymorphisms with gastrointestinal disorders in Egyptian children with autistic disorder". Research in Developmental Disabilities

Alves FJ, De Carvalho EA, Aguilar J, De Brito LL, Bastos GS (2020). "Applied Behavior Analysis for the Treatment of Autism: A Systematic Review of Assistive Technologies". *IEEE Access*. 8

Ávila-Álvarez A, Alonso-Bidegain M, De-Rosende-Celeiro I, Vizcaíno-Cela M, Larrañeta-Alcalde L, Torres-Tobío G (2020) "Improving social participation of children with autism spectrum disorder: Pilot testing of an early animal-assisted intervention in Spain". *Health & Social Care in the Community*. 28

Baranek GT (2002) "Efficacy of sensory and motor interventions for children with autism". *Journal of Autism and Developmental Disorders*. 32

Benevides TW, Shore SM, Andresen ML, Caplan R, Cook B, Gassner DL, et al. (2020). "Interventions to address health outcomes among autistic adults: A systematic review". Autism. 24

Bottema-Beutel K, Crowley S, Sandbank M, Woynaroski TG (2021). "Research Review: Conflicts of Interest (COIs) in autism early intervention research—a meta-analysis of COI influences on intervention effects". *Journal of Child Psychology and Psychiatry, and Allied Disciplines*. 62

Nevada Autism Center—"How Gamma Oscillations Can Be Used as a Vital Biometric for Autism Spectrum Disorder (ASD) Diagnosis". August 2020

Coben R, Linden M, Myers TE (2010). "Neurofeedback for autistic spectrum disorder: a review of the literature". Applied Psychophysiology and Biofeedback. 35

DeVita-Raeburn E (2016). "Is the Most Common Therapy for Autism Cruel?". *The Atlantic*.

Di Salvo, M. (2024) Psicoterapia e plasticità cognitiva e neuronale. Cervello sociale e disturbo post traumatico da stress.—Ricerca Psicoanalitica Journal of Relations in Psychoanalysis

Dillenburger K, Keenan M (2009) "None of the As in ABA stand for autism: dispelling the myths". *Journal of Intellectual & Developmental Disability.* 34

Eikeseth S (2009). "Outcome of comprehensive psycho-educational interventions for young children with autism". *Research in Developmental Disabilities.* 30

Eldevik S, Hastings RP, Hughes JC, Jahr E, Eikeseth S, Cross S (2009). "Meta-analysis of Early Intensive Behavioral Intervention for children with autism". *Journal of Clinical Child and Adolescent Psychology.* 38

Ospina MB, Krebs Seida J, Clark B, Karkhaneh M, Hartling L, Tjosvold L, et al. (2008). "Behavioural and developmental interventions for autism spectrum disorder: a clinical systematic review". *PLOS ONE.*

Novella S (2013). "Transcranial Magnetic Stimulation for Autism". Neurologica Blog

Frith, U. (1989). Autism and "Theory of Mind". In: Gillberg, C. (eds) Diagnosis and Treatment of Autism. Springer, Boston, MA.

Gillis JM, Butler RC (2007). "Social skills interventions for preschoolers with Autism Spectrum Disorder: A description of single—subject design studies". *Journal of Early and Intensive Behavior Intervention.* 4

Gorrindo P, Williams KC, Lee EB, Walker LS, McGrew SG, Levitt P (2012). "Gastrointestinal dysfunction in autism: parental report, clinical evaluation, and associated factors". Autism Research. 5

Gulsrud AC, Kasari C, Freeman S, Paparella T (2007). "Children with autism's response to novel stimuli while participating in interventions targeting joint attention or symbolic play skills". *Autism.* 11

Gutstein SE, Burgess AF, Montfort K (2007). "Evaluation of the relationship development intervention program". *Autism.* 11

Herbert, M. (2013) The Autism Revolution: Whole-Body Strategies for Making Life All It Can Be—Harvard Health Publications

Hourston S, Atchley R (2017). "Autism and Mind-Body Therapies: A Systematic Review". *Journal of Alternative and Complementary Medicine.* 23

Howard JS, Sparkman CR, Cohen HG, Green G, Stanislaw H (2005) "A comparison of intensive behavior analytic and eclectic treatments for young children with autism". *Research in Developmental Disabilities.* 26

Howlin P, Magiati I, Charman T (2009). "Systematic review of early intensive behavioral interventions for children with autism". *American Journal on Intellectual and Developmental Disabilities.* 114

Kasari C, Freeman S, Paparella T (2006). "Joint attention and symbolic play in young children with autism: a randomized controlled intervention study". *Journal of Child Psychology and Psychiatry, and Allied Disciplines.* 47

Koegel RL, Koegel LK (2006). *Pivotal Response Treatments for Autism: Communication, Social, & Academic Development*. Brookes

Koegel LK, Koegel RL, Harrower JK, Carter CM (1999). "Pivotal response intervention I: overview of approach". *Journal of the Association for Persons with Severe Handicaps*. 24

Kupferstein H (2018). "Evidence of increased PTSD symptoms in autistics exposed to applied behavior analysis". *Advances in Autism*. 4

Leskovec TJ, Rowles BM, Findling RL (2008). "Pharmacological treatment options for autism spectrum disorders in children and adolescents". *Harvard Review of Psychiatry*. 16

Levy SE, Hyman SL (2005). "Novel treatments for autistic spectrum disorders". Mental Retardation and Developmental Disabilities Research Reviews. 11

Marino L, Lilienfeld SO (2007) "Dolphin-Assisted Therapy: more flawed data and more flawed conclusions" *Anthrozoös*. 20

Matson JL, Matson ML, Rivet TT (2007). "Social-skills treatments for children with autism spectrum disorders: an overview". *Behavior Modification*. 31

McGill O, Robinson A (2021). ""Recalling hidden harms": autistic experiences of childhood applied behavioural analysis (ABA)". *Advances in Autism*. 7

Mercado J, Espinosa-Curiel I, Escobedo L, Tentori M (2019). "Developing and evaluating a BCI video game for neurofeedback training: the case of autism". Multimedia Tools and Applications. 78

Moalem, S. (2014) Inheritance: How Our Genes Change Our Lives, and Our Lives Change Our Genes—Sceptre

Moore TR, Symons FJ (2009). "Adherence to behavioral and medical treatment recommendations by parents of children with autism spectrum disorders". *Journal of Autism and Developmental Disorders*. 39

Myers S.M. Johnson C.P. (2007). "Management of children with autism spectrum disorders" *Pediatrics*. 120

Nimer J, Lundahl B (2007). "Animal-assisted therapy: a meta-analysis". *Anthrozoös*. 20

Oberman LM, Rotenberg A, Pascual-Leone A (2015). "Use of transcranial magnetic stimulation in autism spectrum disorders". Journal of Autism and Developmental Disorders. 45

Oberman LM, Enticott PG, Casanova MF, Rotenberg A, Pascual-Leone A, McCracken JT (2016). "Transcranial magnetic stimulation in autism spectrum disorder: Challenges, promise, and roadmap for future research". Autism Research. 9

Ozonoff S, Cathcart K (1998). "Effectiveness of a home program intervention for young children with autism". *Journal of Autism and Developmental Disorders*. 28

Popow C, Ohmann S, Plener P (2021). "Practitioner's review: medication for children and adolescents with autism spectrum disorder and comorbid conditions". *Neuropsychiatrie.* 35

Prizant BM, Wetherby AM, Rubin E, Laurent AC, Rydell PJ (2005). *The SCERTS manual: a comprehensive educational approach for young children with autism spectrum disorders.* Baltimore, Md.: Paul H. Brookes

Reichow B, Wolery M (2009). "Comprehensive synthesis of early intensive behavioral interventions for young children with autism based on the UCLA young autism project model". *Journal of Autism and Developmental Disorders.* 39

Rogers SJ, Ozonoff S (2005) "Annotation: what do we know about sensory dysfunction in autism? A critical review of the empirical evidence". *Journal of Child Psychology and Psychiatry, and Allied Disciplines.* 46

Rogers SJ, Vismara LA (2008). "Evidence-based comprehensive treatments for early autism". *Journal of Clinical Child and Adolescent Psychology.* 37

Ryan CS, Hemmes NS (2005). "Post-training Discrete-Trial Teaching Performance by Instructors of Young Children with Autism in Early Intensive Behavioral Intervention". *The Behavior Analyst Today.* 6

Sandoval-Norton AH, Shkedy G, Shkedy D (2019). Rushby JA (ed.). "How much compliance is too much compliance: Is long-term ABA therapy abuse?". *Cogent Psychology.* 6

Schreibman L (2005). "Critical evaluation of issues in autism". The Science and Fiction of Autism. Harvard University Press

Seida JK, Ospina MB, Karkhaneh M, Hartling L, Smith V, Clark B (2009). "Systematic reviews of psychosocial interventions for autism: an umbrella review". *Developmental Medicine and Child Neurology.* 51

Shkedy G, Shkedy D, Sandoval-Norton AH (2021). "Long-term ABA Therapy Is Abusive: A Response to Gorycki, Ruppel, and Zane". *Advances in Neurodevelopmental Disorders.* 5

Spreckley M, Boyd R (March 2009). "Efficacy of applied behavioral intervention in preschool children with autism for improving cognitive, language, and adaptive behavior: a systematic review and meta-analysis". *The Journal of Pediatrics.* 154

Steege MW, Mace FC, Perry L, Longenecker H (2007). "Applied behavior analysis: beyond discrete trial teaching". *Psychology in the Schools.* 44

Virues-Ortega J, Julio FM, Pastor-Barriuso R (2013) "The TEACCH program for children and adults with autism: a meta-analysis of intervention studies". *Clinical Psychology Review.* 33

Wasilewska J, Klukowski M (2015) "Gastrointestinal symptoms and autism spectrum disorder: links and risks—a possible new overlap syndrome". Pediatric Health, Medicine and Therapeutics. 6

Wheeler D, Williams K, Seida J, Ospina M (2008). "The Cochrane Library and Autism Spectrum Disorder: an overview of reviews". *Evidence-Based Child Health*. 3

Wijker C, Leontjevas R, Spek A, Enders-Slegers MJ (2019). "Process Evaluation of Animal-Assisted Therapy: Feasibility and Relevance of a Dog-Assisted Therapy Program in Adults with Autism Spectrum Disorder". *Animals*. 9

Wijker C, Leontjevas R, Spek A, Enders-Slegers MJ (2020). "Effects of Dog Assisted Therapy for Adults with Autism Spectrum Disorder: An Exploratory Randomized Controlled Trial". *Journal of Autism and Developmental Disorders*. 50

Wijker C, Kupper N, Leontjevas R, Spek A, Enders-Slegers MJ (2021). "The effects of Animal Assisted Therapy on autonomic and endocrine activity in adults with autism spectrum disorder: A randomized controlled trial". *General Hospital Psychiatry*. 72

Wilkenfeld DA, McCarthy AM (2020). "Ethical Concerns with Applied Behavior Analysis for Autism Spectrum Disorder". *Kennedy Institute of Ethics Journal*. 30

CHAPTER 14

Music and Autism: New Ways of Research for Neuroscience

In Chapter 9 regarding the relationship between mind reading and central coherence, I mentioned a reference to Francesca Happé's (2001) research that lesions of the right hemisphere often lead to a decrease in global perception. A fragmented style has also been identified in children with Williams syndrome, and this finding may be a clue to its brain basis.

This mention of Uta Frith immediately reminded me of a chapter in a wonderful book by Oliver Sacks (2007): Musicophilia.

Ursula Bellugi—a cognitive neuroscientist who had done pioneering work on deafness and sign language—was fascinated by Williams syndrome.[1]

Ursula Bellugi is a linguist, although she focuses her attention on the emotional powers of language and all its poetic uses no less than on its formal linguistic character. She was attracted by the wide vocabulary and unusual words used—in spite of their low IQ (see references note 6 in Chapter 8) - by very youngsters with Williams syndrome.

What seemed to be so highly developed in these children was not only the vast and unusual vocabulary, but also all communicative faculties. In particular, children with Williams syndrome showed a special penchant for stories. In Bellugi's eyes, it was increasingly evident how this storytelling ability went hand in hand with these children's marked sociability.

However, they seemed strangely indifferent to the non-human in their environment. In some cases, children with Williams syndrome

M. Di Salvo, *Autism Research between Psychology and Neuroscience*,
https://doi.org/10.1007/978-3-031-68338-1_14

were incapable of tying their shoes, of assessing obstacles and steps, of 'understanding' the arrangement of objects in the house.

This was in stark contrast to what is observed in autistic children, who can fixate on inanimate objects while appearing indifferent to the emotions of others. Somehow, people with Williams syndrome seemed the exact opposite of those with severe autism.

Observant parents, while noticing their children's problems and difficulties, also noticed their unusual social attention, their friendly disposition and their openness to others. Many were struck by the way in which, even at a few months old, their children listened very attentively to music and began to accurately reproduce melodies by singing in an open or closed-mouthed voice, even before they learned to speak. Some parents observed that their children were so completely absorbed by the music that they were unable to pay attention to anything else; other children were extremely sensitive to the emotions expressed in the music and could burst into tears listening to a sad song. Still others played their instrument every day for hours, or—if the melody and rhythm were to their taste—learned songs in three or four languages.

Bellugi and Levitin (1998) wrote: "*All those people with Williams syndrome [...] had a good, if implicit, understanding of rhythm and its role in grammar and musical form*".

Not only rhythm, but all aspects of musical intelligence seemed to be greatly and often precociously developed in them.

According to Oliver Sacks (2007), "*The fact that the entire panoply of musical talents could be so impressively developed in persons so affected in general intelligence demonstrated—just as the isolated faculties of musical savants also demonstrate—that it is indeed possible to speak of a specific 'musical intelligence', as Howard Gardner had at one time postulated in his theory of multiple intelligences*" (Di Salvo, 2024).

The three inclinations so accentuated in people with Williams syndrome—that for music, for stories and for the social dimension—seem to proceed in parallel—they are all distinct, yet intimately associated, elements of that ardent motivation to express oneself and to communicate, absolutely central to Williams syndrome.

Seeing such an extraordinary constellation of talents and cognitive deficits, Ursula Bellugi and others began to research its possible brain basis. Neuroimaging techniques—together with rare autopsy reports—revealed considerable deviations from the norm. The brains of persons with Williams syndrome were on average 20% smaller and had a highly unusual morphology, as the reduction in volume and weight seemed to affect only the posterior part of the organ, more precisely at the level of the occipital lobe and parietal lobes; the temporal lobes, on the other hand, were of normal size and sometimes larger than normal. This coincided with the very clear picture of the unequal cognitive abilities of people with Williams syndrome: the devastating impairments in the visuospatial sense could in fact be attributed to the underdevelopment of the parietal and occipital areas, while the strong abilities in the auditory, verbal and musical fields could, in general terms, be traced back to the large size and rich neural networks of the temporal lobes. In people with Williams syndrome, the primary auditory cortex was larger and significant changes were found in the planum temporale, a structure known to be important in language and music perception, as well as in the absolute ear (Bellugi et al. 1999).

Levitin, Bellugi et al. (1998, 2006) finally studied the functional correlates of musicality found in individuals with Williams syndrome. They wondered whether musicality and the emotional response to music in the latter were based on the same type of neurofunctional architecture as in normal subjects or professional musicians. To find out, they had all three groups listen to different types of music—from Bach cantatas to Strauss waltzes—and neuroimaging techniques clearly showed that people with Williams syndrome processed music very differently from others. In order to perceive music and react to it, they employed a much wider range of neural structures, including regions of the cerebellum, brainstem and amygdala, which are scarcely activated in normal subjects (musicians and non-musicians). This extensive brain activation, particularly at the level of the amygdala, seemed to be linked, in people with Williams syndrome, to the irresistible attraction to music and the sometimes overwhelming emotional reactions it aroused in them (Levitin and Cook, 1996; Levitin and Bellugi, 2006).

According to Ursula Bellugi, all these studies indicate that 'the brains of individuals with Williams syndrome are organised differently from those of normal individuals, both at the macroscopic and at the microscopic level. The mental and emotional characteristics, so typical, of people with Williams syndrome are very accurately reflected in the peculiarities of their brains.

As Sacks stated "*we do not know how the deletion of those genes could give rise to the particular talents that these people have. It is not even certain that those talents have a direct genetic basis: it is possible, for example, that some of those abilities are merely spared by the particular vicissitudes of brain development in Williams syndrome; or, that they arise as a kind of compensation for the relative lack of other functions*".

And this reasoning applies to all savants.

In the words of Oliver Sacks "*Williams syndrome certainly offers an extraordinarily rich and precise insight into how a particular genetic endowment shapes the anatomy of a brain and how this, in turn, can define particular cognitive strengths and weaknesses, as well as personality traits and perhaps even creativity. Nevertheless, behind the superficial similarities that exist between people with Williams syndrome, there is an individuality that, as with all of us, is to a large extent determined by experience*".

And once again, this is a statement that this author feels able to endorse with regard to autism.

Matthew State (Sanders et al. 2011) at the University of California, San Francisco (UCSF), found that having an extra copy of a segment of chromosome 7 significantly increases the risk of developing an autism spectrum disorder. In contrast, when that same region is lost, the result is Williams syndrome: is in a way a mirror image of autism. According to Thomas Insel (2010), former director of the National Institute of Mental Health, the contrast between autism and Williams syndrome suggests that our brains use specific networks for specific types of functions, such as social interaction. Deficits in social network functioning may lead the brain to compensate by developing skills in a non-social network, the result of which are the types of unusual abilities we see in autistic savants.

Alfred Tomatis and Audiopsychophonology

For Alfred Tomatis (1986),[2] autism is a sort of "multiform generic" for which there is certainly a common trunk but also a thousand and one details that give rise to as many clinical pictures as there are autistic children and that highlight as many relational disorganisations in the lives of those involved. Tomatis avoids entering into the discussion as to whether or not this syndrome is organic, just as he does not talk about possible genetic factors. He immediately goes to the essentials by highlighting the major element of autism that is closely related to his work. This element, this essential sign is the following: the autistic child does not listen.

In his 1986 paper on autism, Tomatis explains his view of the syndrome:

> "He has good hearing certainly, so at least he shows, but somehow, somewhere there lurks a refusal to listen. Unintentional refusal, mind you. The consequences of this, however, are not long in coming. By refusing to listen, i.e. to integrate language, the child will live in a universe in which speech will be excluded. [...] In this way, the autistic person lives intensely and paradoxically in an acoustic-sound world in which all semantic meaning is eliminated. Although he can hear a fly fly, he is unable to perceive its name when it is spoken even when standing close to him. Things happen as if there were a disconnect between hearing and perceiving. Passive action, that which leaves him immersed in sound, subsists at times even excessively, while the transition to vigilance, from which consciousness will emerge, does not take place. [...] Autistics are assailed by so many stimulations without any of them entering into categorical structuring. They hear but do not listen. They look but do not see. They make sounds but do not speak. Their world around them is paradoxical, without threads, without a synthesis in the transitions from one event to another. These children reassure themselves in their own way through stereotypes that fix them on themselves while simultaneously giving them a semblance of behavioural reality. The world is presented to them without unity, without development. Their vision is made up of a succession of separate clichés. The sounds manifest themselves as repetitive sequences without any reference to the word. If even one word were perceived semantically, the syndrome would disappear. It is not a question, as we unfortunately often see, of training a child to pronounce words".

Clinical experience with a large number of autistic children has led Tomatis to think that what needs to be stimulated is the desire to communicate. And this desire, so deeply rooted in humans, latches onto one of the ear, one of the ear's earliest organs in their development.

The work with the Tomatis method consists of trying to arouse the desire to communicate through the awakening of listening. By sending specially designed sounds through the Electronic Ear, the child is acoustically invited to retrace the stages of his or her psycho-affective development in order to encourage the re-weaving of the communicative texture.

Among the reactions observed, those of the neurovegetative life are generally the first to manifest themselves and in a sensitive manner:

- sleep becomes calmer, with nightmares lessening and often disappearing;

- appetite normalises; the child who used to eat little begins to eat more while the one who used to devour demands less food. He also tends to be less selective about food.

Other things change in terms of behaviour: the child in general becomes more affectionate, primarily with his mother. He seeks contact, loves to sit on her lap, caress her. He accepts being touched more easily.

Autistic people often suffer pain because of their multiple hypersensitivities. Many of them are hypersensitive to sounds. The intensity of pain can sometimes be unbearable. Some indicators of this hypersensitivity are:

- covering one's ears with one's hands;
- outbursts of anger, due to the frustration of constantly having to fight with sounds;
- repetition of the same words, phrases or statements. Probably a way of calming or stabilising oneself in the face of a mass of sounds perceived as intense or confusing.

Tomatis explains this hypersensitivity by the fact that these children are more sensitive to sound through bone and skin conduction than through the tympanic (airway). The sounds picked up by the body, however, reach the brain without being filtered out, thus creating a 'noise', which, without the tympanic counterpart, becomes irrelevant and adds to the background noise that attacks the child. Thus, when working with people with autistic syndrome who are hypersensitive to sounds, the first thing we try to do is to balance bone conduction against tympanic perception. In this way sounds can be processed more effectively.

Since all our senses are interrelated, it often happens that by reducing hypersensitivity to sounds, other hypersensitivities such as that exhibited by the tactile defence, ocular avoidance and aversion to foods with different nuances of taste are alleviated.

It thus becomes clear that the reduction of hypersensitivity and the regulation of sensory integration are key steps in helping the autistic child reconnect with his family and his environment, inducing him to gradually come out of his protective shell.

This improvement in sensory integration will also enable the child to respond better to other therapies that were previously ineffective precisely because of the lack of contact through the senses.

Among the scientific research on the effectiveness of the Tomatis method in autism problems is that carried out by Joan Neysmith-Roy (2001) of the Department of Psychology at the University of Regina (Canada).

This was a double-blind study with six (male) children diagnosed with profound autism. The study confirmed the clinical evidence that autistic children benefit from the Tomatis method. At the end of the treatment, three (50%) of these children showed significant changes in behaviour. Of these, two went from a profound form of autism to a milder form of autism. The other three children showed no significant changes. Of particular interest was the fact that five of the six treated children also showed significant improvement in behaviour in the prelinguistic area, such as adaptability to change, listening skills, non-verbal communication and emotional responsiveness.

The Neuroscientific Basis of the Music-Brain Relationship

For over a century, most neuroscientists have thought of the brain as an organ consisting of an 'upper' part and a 'lower' part. Although there was no unanimous agreement on exactly where to draw the line between the 'top' and 'bottom', almost everyone agreed that the front part of the thin outer layer of the brain, the so-called frontal cortex, was 'the most superior of all'. It was thought that it was the frontal cortex that processed the superior human qualities, such as the ability to reason, plan, control impulses, maintain concentration for long periods, resort to abstract thinking, make decisions, and imagine what other people think and feel.

This idea was based on the observation that a lesion in those areas of the brain impaired all the functions listed above.

Since many childhood psychiatric disorders affect these 'higher' faculties, therapies for these types of problems are designed to stimulate the frontal cortical structures. They are usually aimed at controlling or alleviating symptoms, but none of them heal the brain or make disorders disappear once and for all.

One of the reasons why sound therapy has not attracted more attention is that little has been known so far about the structure on which it acts most strongly—the subcortical brain. We speak of subcortical structures because they are located below the outermost thin layer of the brain, the cortex. Thus, anatomically speaking, they are located deeper, i.e. towards the lower part of the brain.

Unfortunately, the subcortical structures of the brain have also sometimes been regarded as much less evolved than they actually are. There are several reasons for this.

Firstly, because they are housed deep in the brain, it has often been difficult to access them with the tools and technologies that have been available for much of the last century. Therefore, it was difficult both to observe them and to fully appreciate their functions.

Secondly, the subcortical brain is the only brain structure present in many less evolved animals, and since these animals do not possess the same 'sophisticated' intellectual capacities as humans, by deduction it was assumed to be a more rudimentary brain.

During evolution, a thin outer layer of cortex would have developed that enveloped the subcortex, and the outer cortex was regarded as something that was 'added' to the subcortical brain. Since animals with a more recent evolutionary history, endowed with a cortex, appeared more intelligent, it was hypothesised that their intellectual superiority derived precisely from the cortex, the supreme achievement of evolution. The human being is among animals the one in which the cerebral cortex is most developed.

Because of the rigid locationalist model in vogue at the time, it was believed that all higher intellectual faculties took place exclusively in the cortex. If a person had difficulty performing a complex cognitive task, the cause of their problem had to lie in cortical damage.

The error in this reasoning lay in the assumption that each new structure that developed with evolution simply added to the older structure, functioning independently of it.

In fact, as a new structure developed, the more ancestral ones adapted and integrated. The presence of a new structure modified the old, and old and new now work together holistically.

Magnificent proof of this phenomenon can be found in recent human and animal studies: these show that while the cortex evolved and increased in size, the subcortical structures underwent significant growth and modification. Once again, the moral is that localisation theory, however instructive, lends itself to exaggeration, and in fact the correct measure is often overstepped. Because of our cortico-centric viewpoint, the contribution of the subcortical brain has not been sufficiently taken into account. Just how important the subcortical cortex is shown to us by the fact that, if it is stimulated with sounds, the 'higher' mental abilities in children suffering from the most common psychiatric disorders of childhood undergo astounding improvements (Herculano-Houzel, 2010).

Paul Madaule (1994) found that listening therapy can help the two-thirds of autistic children who come to the Listening Centre in Toronto. By 'help', he means improvements that enable the child to make better use of available therapies and to be more involved in school, social and family life, so that he has more controlled behaviour and is more present to himself and more autonomous. Generally, the earlier one starts with therapy, the better.

Marta Herbert's Thesis

Although autism has so far been considered incurable, Dr Martha Herbert (2012, 2014), a paediatric neurologist and researcher at Harvard Medical School, also documents cases of children with autism who have experienced extraordinary improvements. "*For decades, most doctors told parents that autism depended on a genetic problem in their children's brains", writes Herbert",and that [...] they had to prepare themselves for the fact that those disorders would accompany their children throughout their lives?" However, Herbert shows, autism is a dynamic process. It is not just genetic in origin, it is not just a brain problem, it does not have a single trigger and it is not always 'completely intractable' especially if therapy starts when the child is very young. [...] In some cases autism is already present at birth or manifests itself immediately afterwards; in 'regressive autism', on the other hand, the child initially appears to have absolutely normal mental development, after which—usually between the second and third year of life—the first symptoms appear*".

According to Herbert: "*Today there are hundreds of genes that are associated with autism. The effects of these genes, however, are for the most part negligible. In most cases they probably only result in a modest vulnerability (...). But even genes that have a more direct correlation with autism (...) only affect a fraction of the total number of autistic people (...), and some carriers of that gene do not become autistic*".

It is genetics that determine a child's autistic risk, but environmental factors are usually essential for the risk to develop into a full-blown syndrome. Many of these factors excite the child's immune system, causing it to release antibodies that produce chronic inflammation that affects the brain. Many autistic children have immune abnormalities and overactive immune systems (Goines et al. 2010).

In these cases, there is a high incidence of gastrointestinal infections and inflammations (Parracho et al. 2005), food intolerances (often to cereals, gluten, dairy products and sugar), asthma (also an inflammatory disease) and skin inflammations. Herbert gives many examples of children who experienced dramatic improvements after undergoing anti-inflammatory therapies.

Another stressor is toxins, which in turn can irritate the brain and cause inflammation (Valicenti-Mcdermott et al. 2006).

These toxins, being foreign to the body, trigger immune reactions.

Herbert shows that autism is an expression of an organic pathology, involving the entire body, which also affects the health of the brain. Chronic inflammation in the body can have negative consequences on all organs, including the brain. In 2005, a study conducted at the Johns Hopkins University School of Medicine showed that the brains of autistic people are often inflamed. Autopsies revealed the presence of inflammatory states in the cerebral cortex (the outermost layer of the brain) and at the level of the axons; the inflammation was 'particularly evident in the cerebellum', the subcortical area that has close links to the vestibular apparatus (on which sound therapy is used). The cerebellum is involved in the refinement of certain cognitive and motor functions; moreover, it is one of the brain structures stimulated by modern versions of sound therapy.

Since 2008, five studies have shown that a significant number of autistic children have antibodies from their mothers and that their nerve cells were attacked when they were still in the womb (Vargas et al. 2005; Enstrom et al. 2009; Braunschweig et al. 2013; Bauman et al. 2013).

It is likely that the mother was exposed to an infection or toxin that caused an alteration in her immune system. By injecting this type of antibody into pregnant monkeys, it was found that their pups exhibited behaviour similar to that of autistic children. Autistic children also have a high level of antibodies in their blood (Lee et al. 2010).

Herbert's hypothesis is that the sum of all these stresses and inflammation affects the nervous system by damaging neurons.

Chronic inflammation interferes with the development of neuronal circuits.

Brain scans show that in autistic children, many neuronal circuits are 'hypoconnected' and that neurons in the frontal areas of the brain (involved in planning and intentionality) are poorly connected with those in the posterior areas (responsible for sensations). In other areas of the brain, however, there is a state of 'hyperconnectivity', often the cause of epileptic seizures, which are also common in autistic children. It is likely that this combination of hypo- and hyperconnectivity hinders the synchronisation of brain activity between the different areas. In short, autism is said to be the product of genetic risk factors and numerous environmental triggers that sometimes affect the child in the foetal state, other times after birth, and in this immune reactions and inflammatory states play a major role. The combination of all these factors has a disruptive effect on the developing brain and results in neurons not connecting properly and not being able to communicate well with one another (Just et al. 2004; Coben et al. 2008; Schipul et al. 2011).

Recently, neuroscientists have made important discoveries about 'wiring problems' related to autism (Porges et al. 2014).

These findings offer an explanation for the fact that autism negatively affects listening ability. A team of Stanford University scientists led by Daniel A. Abrams in 2013 and Vinod Menon demonstrated that in autistic children, the area of the auditory cortex responsible for processing the human voice is hypoconnected to the subcortical structure of the brain known as the reward centre.

When a person completes a task, the reward centre is activated and produces dopamine, inducing positive emotions and thus reinforcing the motivation to repeat the same actions. The research, which was conducted using a special type of magnetic resonance imaging that makes it possible to see the connections between different areas of the brain, showed that the language areas in the left hemisphere (*responsible for the symbolic and abstract components of language*) and the language areas in the right hemisphere (*involved in the musical and emotional components of speech, i.e. so-called prosody*) were in a state of hypocontact with the brain's reward centre.

What is the consequence of this hypoconnexion? That a child unable to connect the areas of the brain that control speech functions with the reward centre experiences no pleasure in language exchange.

Marta Herbert's idea is that when 'the entire organism is challenged by a combination of bad food, toxins, microbes and stress, and some genetic vulnerability to complicate it all', the cellular system that supports the brain becomes overwhelmed.

Inflammation produces waste. The brain, like all other organs in the body, must constantly dispose of waste products and dead cells, and then rebuild itself and supply the neurons with new nutrients. This task falls to the glial cells.

When they succumb, the glial cells swell and can no longer adequately support the neurons; the neurons get less blood and their mitochondria (*the cells' energy generators*) go under stress.

Eventually, some neurons, no longer receiving support from the glial cells, become 'numb' to the point of stopping normal signalling activity.

Dysfunctional or injured neurons continue to activate, producing 'noise', or they become overexcitable and dysregulated.

When the glio-neuronal system goes haywire, Herbert explains, an enzyme, glutamate, which normally has an excitatory action on neurons, is released in quantity. This contributes to the excessive excitability of neurons and possibly to the onset of hypersensitivity and what he calls a noisy brain.

Coming back to listening therapy applied to cases of autism, it is possible that the loss of the pleasure of linguistic communication has devastating effects on the child's ability to bond with parents or any other individual.

Leo Kanner (1943) had noted that autistic children seemed indifferent to the human voice and made no attempt to speak; in one of his patients "*there was no change in expression when spoken to*". Today, the fundamental role of the voice in the attachment process (*bonding*) between child and parent is considered an unequivocal fact, and indifference to the voice has consequences for this bond.

A 2010 study (Salizer et al. 2010) shows that if a stressed non-autistic child is made to hear his mother's or father's voice, his brain begins to produce oxytocin (Modahl et al. 1998).

Oxytocin is a neurotransmitter that induces calmness and emotional closeness, which strengthens feelings of tenderness and trust, enabling parents and child to bond with each other. The parents' voice has a calming effect on the child and encourages the development of communication. In autistic children, however, oxytocin levels are significantly lower.[3]

Whatever the cause, what one might call 'vocal bonding' does not take place.

While a large number of autistic children are indifferent to the pleasure of the human voice, they are not indifferent to noise.

In many autistics, there is a hypersensitivity to sounds and noises—which explains their very frequent gesture of plugging their ears in the sign of great discomfort—and their nervous system is prone to go into 'fight or flight' mode.

Neuroscientist Stephen Porges (2011, 2014)[4] found that certain ranges of sound frequencies have a relationship with our sense of safety or danger. Every species has its predators, and the sounds produced by those predators trigger the fight or flight mechanism in the prey. There is a direct link between the auditory cortex and the brain's threat-protection system, which explains, among other things, why a sudden loud noise jolts us and triggers an immediate reaction of intense anxiety. Evolution has also meant that individuals of the same species communicate with each other using sound frequencies that predators are unable to hear (*reptiles, which for millions of years have preyed on medium-sized mammals such as humans, are unable to perceive human voice frequencies*).

When we feel calm and safe, the myelinated parasympathetic nervous system deactivates the 'fight or flight' reaction.

As Porges has shown in his research, the parasympathetic system activates a 'social engagement system', as well as the muscles of the middle ear, and this enables the individual to listen to others, to communicate and to connect with others. The parasympathetic system helps us to bond with other people precisely because it regulates the brain areas that control the middle ear muscles, which are used to tune into the high frequencies of human speech and to activate the muscles that serve vocal and facial expressions.

Being in 'myelinated parasympathetic mode' means being calm, controlled and connected.

Tomatis shown that many children with autism, learning disabilities and delayed language development—as well as those who have suffered repeated ear infections—are unable to tune in to human speech frequencies because they are unable to use their middle ear muscles to attenuate low-frequency sounds.

If the volume of the lower frequencies is very high, they overpower the higher frequencies of spoken language, causing the autistic child to be hypersensitive to sounds and noises, especially continuous ones such as those produced by a hoover or an alarm. In us humans, moreover, low-frequency sounds generate anxiety because they remind us of those of ancient predators.

Dazed by noises, these children remain in a 'fight or flight' state and fail to activate the social engagement system. Working on the circuit that controls the middle ear muscles can reduce hypersensitivity and encourage social involvement (as Tomatis put it) so that attachment to others becomes a source of pleasure.

The findings of Porges, Tomatis and other scholars tell us that it is time to rethink the theory that the core of autism is the inability to feel empathy and recognise the existence of other minds.

Perhaps this is not always the case. Children who find themselves constantly dazed by sensations and in a constant state of 'fight or flight' are unable to activate their social involvement system or realise the existence of other thinking beings.

In many cases, their inability to recognise the existence of other people may be a consequence of sensory processing problems in the brain. As Paul Madaule (1978) argues, the purpose of our sensory systems is to "*project us out into the world and, at the same time, protect us from a universe of sensations. When we are too sensitive, we create mechanisms that isolate us from the world*".

To understand whether a child suffers from hypersensitivity, Porges points out, one only has to look at him or her.

The facial nerve, which regulates the middle ear muscle, the stapedium, also innervates the muscles of the eyelids and those that control facial expressions.

When we are interested in what a person is saying, the muscles of our middle ear contract, allowing us to tune in to the speech frequencies of our interlocutor and keep our eyes wide open: that is, we assume an interested expression.

From a pupil's facial expressions a teacher can tell if the child is listening to the lesson. In many autistic children this circuit does not work, which is why they have an absent expression; their facial muscles, and therefore their facial expressions, are flat, expressionless.

Music, Neuroscience and Mirror Neurons

Remember Ramachandran's theory of mirror neuron system dysfunction?

According to him, the discovery of mirror neuron dysfunction opens the way to new therapeutic methods.

"The failure to suppress mu waves could, for example, become a valuable tool for screening and diagnosing the disorder in early childhood and adopting the behavioural therapies currently available long before other, more 'overt' symptoms occur. [...] A second and more fascinating possibility would be the use of biofeedback as a therapy. In biofeedback, a physiological signal from the subject's body or brain is detected by a machine and displayed on a monitor. The objective is to decrease or increase, through concentration, the signal and thus exert some conscious control over the phenomenon one wishes to modify. A biofeedback system, for example, can show someone his or her heart rate, represented as a dot bouncing on the monitor emitting a characteristic beep: with practice, most people are able to use such feedback to learn how to slow down their heart rate at will. Brain waves can also be modified with biofeedback".

A third possibility, discussed in an article written with Lindsay Oberman and published in 'Scientific American' (2006), would be "*to experiment with certain drugs that enhance empathy, probably by increasing the amount of neurotransmitters called 'empathogens', which are naturally present in the brains of highly social individuals such as primates. If administered early enough, cocktails of these drugs could perhaps help overcome some initial symptoms sufficiently to minimise the subsequent cascade of events leading to the full spectrum of symptoms*".

One last method to revive the lethargic mirror neurons of autism would be to exploit the great fun that it is for all human beings, autistic included, to dance to the rhythm of music.

"It has already been tried to make autistic children dance to the sound of rhythmic music, but no attempt has ever been made to exploit the known properties of the mirror neuron system. One idea would be to have several dancers dance in unison as models and tell the child to imitate their dance in synchrony. Perhaps it would also help the child to place the dancers in a room lined with mirrors that would reflect the various images and multiply the effect".

Remember the martial arts metaphor and the words of the master?

"*When I began studying martial arts, to me a punch was just a punch. As I progressed in practice, became an expert, mastered the technique, I realised that things were less obvious: a punch was not just a punch. It was something much more complex. As I progressed further, as I continued practising for years, I realised a deeper and simpler truth: a punch is just a punch*".

We are approaching the conclusion of our story, and with the next chapter we will try to find a unity in this journey that has had many stops and offered many different views.

A famous line by T.S. Eliot (1943) is "We shall not cease to explore. And at the end of all our going we shall return to the starting point to know it for the first time".

My wish is certainly that we will never stop exploring.

As I pointed out at the beginning, homo sapiens has always done two things—in addition to all the other things in common with all other living beings—research and telling.

And if you have come this far, it is because your desire to research is similar to mine, and I hope to have been a good companion on the journey so far.

If I have aroused any doubts, any questions, any desire to 'go and search and verify', half of what I have set out to do has been done.

I wish that at the end of all our going, returning to the starting point, we would really recognise it as new, as known for the first time.

It means that I will have been able to offer you new, hopefully broader, even different views.

A great psychotherapist wrote:

> "Every observation is partial, every opinion is a counterfeit. When you examine the totality you break it down into individual fields of observation, and to form an opinion you have to break it down into parts. The flaws in observation can be eliminated to a certain extent by looking at the world from several sides; so can the flaws in opinion be eliminated by choosing appropriate subdivisions. What I have to say is everyone's experience and each of you sees it, probably better than I do, though certainly in a very different light. But it is worth considering, for once, things universally known through the eyes of an outsider, and even if the result were only to provoke a more lively clash of opinions, I would already be satisfied". (Groddeck, 1920).

With this quote, which I fully endorse, I offer you a final stage of this journey, hoping that it is just, for everyone, a new starting point.

Notes

1. Williams syndrome is very rare, perhaps affecting one child in ten thousand, and to find it described in the medical literature we had to wait until 1961, when J.C.P. Williams, a New Zealand cardiologist, published an article on it; the following year, it was also described independently in Europe by Alois J. Beuren and his colleagues. (In Europe, therefore, we tend to speak of Williams-Beuren syndrome, while in the United States it is generally known as Williams syndrome.) Both Williams and Beuren described a syndrome characterized by defects of the heart and great vessels, unusual facial conformation, and mental retardation.

   The term "mental retardation" suggests a global or general intellectual defect, such as to compromise the ability to speak together with all other cognitive faculties. In 1964, however, G. von Arnim and P. Engel observed the elevated calcium levels that seemed associated with Williams syndrome and—along with those—also a curiously unequal profile of ability and disability. They spoke of the children's "sociable and talkative personalities" and "their unusual command of language"—the very last thing one would expect from a "retarded" child. (In that publication, the authors also noted, albeit in passing, that these children appeared to have a strong attachment to music.).

   Their parents were often struck by their children's unusual combination of intellectual strengths and disabilities and had great difficulty finding appropriate environments or schools, since they were not "retarded" in the usual sense of the term. At the beginning of the 1980s, in California, some of these parents met and joined together to form the initial nucleus of what would become the Williams Syndrome Association.

   A path very similar to what happened for autism.
2. In the 1950s, starting from studies on the professional deafness of aviators and on vocal emission disorders in singers, he hypothesized that there was a link between the development of personality and the listening experience, giving rise to "audiopsychophonology", discipline of Human Sciences, having as its object of study and application the re-education of the ear, which would have had the aim of increasing an individual's ability to listen. According to Tomatis, listening would be a process that would invest the organism in its psychocorporeal globality, to be understood as a voluntary act aimed

at finding what we want to hear. The ear, therefore, the object of study and clinical application of audiopsychophonology, represents a "psychocorporeal functional unit responsible for listening".

3. An explanation for this phenomenon has not yet been found, but it is possible that in many cases it is a secondary effect: in many children it could be the consequence of sensitivity at the auditory level which makes listening painful and which produces a hypoconnectivity between the auditory areas and the brain reward center.
4. Porges S.W. et al. (2014) «Reducing Auditory Hypersensitivities in Autistic Spectrum Disorders: Preliminary Findings Evaluating the Listening Project Protocol» Frontiers in Pediatrics.

   His intervention was aimed at training the cerebral regulation of the middle ear muscles. By listening to filtered music, autistic children began to show more expressive facial expressions and no longer looked away in the presence of other people; in half of the subjects a decrease in sensitivity to sounds was found; emotional regulation improved in 22% of them, compared to 1% in the control group. Evidently this depended on the activation of the "social involvement system". Porges' program, however, is not yet available to the public. Music works because, as Porges points out, it "duplicates the spectrum of the human voice".

   Porges S.W. (2011) The Polyvagal Theory: Neurophysiological Foundations of Emotions, Attachment, Communication, and Self-Regulation—W W Norton & Co Inc.

References

Abrams D.A.et al. (2013) Underconnectivity Between Voice-Selective Cortex and Reward Circuitry in Children with Autism, Proceedings of the National Academy of Sciences

Bauman M.D., et al. (2013) Maternal Antibodies from Mothers of Children with Autism Alter Brain Growth and Social Behavior Development in the Rhesus Monkey, Translational Psychiatry

Bellugi, U., Levitin, D. J. (1998) Musical abilities in individuals with Williams syndrome, Music Perception: and Interdisciplinary Journal, 15

Bellugi, U., Lichtenberger, L., Mills, D., Galaburda A., Korenberg, J. R. (1999) Bridging cognition, the brain and molecular genetics: evidence from Williams syndrome, Trends in Neuroscience, 22

Braunschweig, D., et al., (2013) Autism-specific Maternal Autoantibodies Recognize Critical Proteins in Developing Brain, Translational Psychiatry

Coben, R., Myers, T.E., (2008) Connectivity Theory of Autism: Use of Connectivity Measures in Assessing and Treating Autistic Disorders, Journal of Neurotherapy

Di Salvo, M. (2024) Multiple intelligences and specialization. A return to Gardner and the wisdom of rethinking psychology in Hillmanian transparency. - Revista De Estudios Socioeducativos. ReSed, 1(12). doi.org/10.25267/Rev_estud_socioeducativos.2024.i12.18

Eliot, T. S. (1943) *Four Quartets*

Enstrom A., et al., (2009) Increased IgG4 Levels in Children with Autism Disorder, Brain, Behavior, and Immunity

Goines, P., Van De Water, J., (2010) The Immune System's Role in the Biology of Autism, Current Opinion in Neurology 23

Groddeck, G. (1920) Psychoanalytische Schriften zur Literatur und Kunst

Happé, F. (2001) Social and non-social development in autism: Where are the links?, in J.A. Burack, T. Charman, N. Yirmiya, P.R. Zelazo (a cura di), The development of Autism: Perspectives from Theory and Research, Erlbaum, Mahwah (NJ)

Herbert, M., Weintraub, K. (2012) The Autism Revolution, Ballantine Books, New York

Herbert, M., (2014) Translational Implications of a Whole-Body Approach to Brain Health in Autism: How Transduc-tion Between Metabolism and Electrophysiology Points to Mechanisms for Neuroplasticity, in V.W. Hu, Frontiers in Autism Research: New Horizons for Diagnosis and Treatment, World Scientific, Hackensack.

Herculano-Houzel, S. (2010) Coordinated Scaling of Cortical and Cerebellar Numbers of Neurons, Frontiers in Neuroanatomy

Insel, Th.R. (2010) "The challenge of translation in social neuroscience: A review of oxytocin, vasopressin, and affiliative behavior". In Neuron, 65, 6, pp. 768–779.

Just et al., (2004) Cortical Activation and Synchronization During Sentence Comprehension in High-Functioning Aútism: Evidence of Underconnectivity, Brain: A Journal of Neurology

Kanner, L. (1943) Autistic Disturbances of Affective Contact, Nervous Child

Lee R.H. et al. (2010) Neurodevelopmental Effects of Chronic Exposure to Elevated Levels of Pro-Inflammatory Cytokines in a Developing Visual System, Neural Development

Levitin, Daniel J. (2006) This Is Your Brain on Music, Dutton, New York

Levitin, Daniel J., Perry R. Cook, (1996) Memory for musical tempo: Additional evidence that auditory memory is absolute, Perception and Psychophysics,

Levitin, Daniel J., Ursula Bellugi (2006) Rhythm, timbre and hyperacusis in Williams-Beuren syn-drome, in Williams-Beuren Syndrome: Research and Clinical Perspectives, a cura di C. Morris, H. Lenhoff e P. Wang, Johns Hopkins University Press, Baltimore,

Madaule, P., (1978), The Dyslexified World, testo originale dell'intervento presso il convegno Listening and Learning, Toronto – published inT.M. Gimour, P. Madaule, B. Thompson (a cura di), About the Tomatis Method, Listening Centre Press, Toronto, 1989

Madaule, P. (1994) When Listening Comes Alive: A Guide to Effective Learning and Communication

Modahl, C. et al. (1998) Plasma Oxytocin Levels in Autistic Children, Biolo-gical Psychiatry

Parracho H., et al. (2005) Differences Between the Gut Microflora of Children with Autistic Spectrum Disorders and That of Healthy Children, Journal of Medical Microbiology

Porges S.W. (2011) The Polyvagal Theory: Neurophysiological Foundations of Emotions, Attachment, Communication, and Self-Regulation - W W Norton & Co Inc

Porges, S.W. et al. (2014) Reducing Auditory Hypersensitivities in Autistic Spec-trum Disorders: Preliminary Findings Evaluating the Listening Project Protocol, Frontiers in Pediatrics

Ramachandran VS, Oberman LM. (2006) Broken mirrors: a theory of autism. Sci Am.

Neysmith Roy, Joan M (2001) The Tomatis Method with severely autistic boys. Individual case studies of behavioral changes, South African Journal of Psychology, 31

Sacks, O., (2007) - Musicophilia - Tales of Music and the Brain.

Salizer, L.J. et al. (2010) Social Vocalizations Can Release Oxytocin in Humans, Proceedings of the Royal Society: Biology

Sanders, S.J., Ercan-Sencicek, A.G., Hus, V., Luo, R., Murtha, M.T., et al. (2011) "Multiple recurrent de novo CNVs. Including duplications of the 7q11.23 Williams syndrome region, are strongly associated with autism". In Neuron, 70, 5, pp. 863–885.

Schipul et al. (2011) Inter-regional Brain Communication and Its Disturbance in Autism, Frontiers in Systems Neuroscience

Tomatis, A., (1986) Considérations sur l'autisme, Paris, Centre Tomatis

Valicenti-Mcdermott M. et al., (2006) Frequency of Gastrointestinal Symptoms in Children with Autistic Spectrum Disorders and Association with Family History of Autoimmune Disease, Developmental and Behavioral Pediatrics

Vargas D.L. et al. (2005), Neurological Activation and Neuroinflammation in the Brain of Patients with Autism, Annals of Neurology

CHAPTER 15

The Energy That Makes the Brain Grow from NGF to BDNF the New Frontiers on the Origins of Autism Towards a General Theory and a New Definition

The human brain is the most complex structure we know.

Actually, we don't know it at all.

We have finally realised that body-soul-brain are one, although this discovery is not new. While the West specialised by dividing and belittling in the search for the 'infinitely small place', traditional Chinese medicine proceeded holistically, and the Australian aborigines, bereft of all Western medical-scientific knowledge, lived in 'one' immersed in the whole.

The greatest mystery of the universe, the greatest immensity, is within us.

Weighing just under one and a half kilogrammes, the human brain is made up of more than 150 billion cells, of 3000 different types (Sanders, 2023) including neurons, glial cells, interneurons, and each of the approximately 85 billion neurons has 100 to 20,000 axonal and dendritic connections with the others. An integrated, integral, discrete and redundant system in which each neuron does not exchange simple 1/0 messages, but complex messages with a language made up of different discharge intensities, different frequencies and repetitions, different chemical messages. A language of which we barely understand the basic alphabet and very few elementary rules, which do not yet allow us to know the symphony of its music and the works of art of its literature.

M. Di Salvo, *Autism Research between Psychology and Neuroscience*,
https://doi.org/10.1007/978-3-031-68338-1_15

We perceive something in dreams, in intuitions, in desires, in inventions, in memories.

Immensity appears violent and grandiose to us, presenting us with extraordinary phenomena.

Williams syndrome, which we learned about in the previous chapter, Capgras syndrome (in which a person is firmly convinced that a friend, spouse and other members of his or her family have been replaced by impostors or impersonators) or proposagnosia, the inability to recognise faces, the phantom limb syndrome (in which one does not recognise an arm or leg as one's own even though it is attached to the body), have enabled us to understand the human brain more than anything else, just as on another front, that of injury, we have finally really understood the complex systems of language thanks to the study of strokes in Broca's and Wernicke's areas.

Penfield's very useful and revolutionary maps allowed us to discover certain specialised areas for certain functions, and Sperry's revolutionary split-brain interventions and Gazzaniga's studies taught us about specialisation between the two hemispheres, but we had to make a titanic effort against ourselves and our omnipotence to get out of the locationalist misunderstanding and admit that things were more complex, and that there is no 'one place for one function', but that the brain operates by circuits and systems integrated in an infinitely complex structure of relations.

And then there is autism: '*here, where nature is greater than man, no one can escape it*' (Jung, 1929).

The journey into the brain, to the engine that builds this extraordinarily complex structure, which I propose in this chapter, proceeds in stages and begins in Italy, with Rita, a young woman with an extraordinary passion for research.

First Stage: Turin-Brussels-Turin-(mountains)-St.Louis

Rita Levi Montalcini was born in 1909 in Turin, where she studied medicine.

In 1938, when with the racial laws the fascist regime forbade Jews access to many professions, including medical practice and scientific research, Rita fled to Brussels to continue her studies.

When the Nazis invaded Belgium, she returned to Turin and built a laboratory in her bedroom, where she studied the formation of the nervous system, making her own microsurgical instruments from sewing needles.

When the Allies bombed the city in 1940, he fled to the Piedmont countryside.

Travelling in a cattle car, as she recounted, sitting on the floor, she read a scientific article by Viktor Hamburger, who had conducted pioneering research on neuronal development in chicken embryos.

He decided to repeat and extend these experiments, working on a table in a mountain hut with eggs provided by a local farmer (when he finished an experiment, he ate the eggs, there was always war and with it hunger).

After the war, Hamburger invited her to join him and his researchers in St. Louis to work on their findings, the results of which indicated that nerve fibres in chicks grew faster in the presence of tumours. Rita hypothesised that the tumour probably induced the secretion of a substance that promoted nerve growth.

Together with biochemist Stanley Cohen, she isolated the protein responsible for this phenomenon and called it 'nervous system growth factor' or NGF (nerve growth factor). For this discovery Rita and Stanley Cohen received the Nobel Prize in 1986.

Stage Two: The Growth Factor Within Us

Rita Levi Montalcini's work led to the discovery of several growth factors, one of which, the 'brain-derived neurotropic factor', or BDNF, has recently caught the episodic and spotty attention of some scientists working on the autism and XFragile syndrome front.

BDNF plays a key role in reinforcing the plastic changes produced in the brain during the critical period (Lai, 2013).

When we perform an activity that requires specific neurons to fire together, they release BDNF. This growth factor consolidates the connections between neurons so that they will activate reliably in the future. BDNF also promotes the growth of the myelin sheath that surrounds each neuron, which accelerates the transmission of electrical signals.

During the critical period, BDNF activates the nucleus basalis, the brain region that allows us to focus our attention and maintains it for the duration of the critical period. Once activated, the nucleus basalis also enables us to remember our experiences. It enables the differentiation of maps and makes changes occur effortlessly.

In this sense, we can consider the nucleus basalis as a 'modulatory plasticity control system', i.e. the neurochemical system that, once activated, puts the brain in a flexible condition.

The fourth function performed by BDNF—once the consolidation of the core connections is complete—is to help close the critical period.

When the basic neuronal connections have been established, greater stability and thus less plasticity of the system is required. When BDNF is released in sufficient quantities, this deactivates the nucleus basalis and ends the wonderful epoch of spontaneous learning. From this moment on, the nucleus basalis can only be reactivated when something important, new or surprising occurs, or if we make an effort to focus our attention.

Third Stage: A First Explanation Within Ourselves

A dysfunctional theory of the neurogrowth factor regulatory system now allows us to develop a theory that explains why so many different problems can fall within the characteristic and symptomatic autistic spectrum.

During the critical period, certain environmental circumstances overexcite the neurons of children who have a genetic configuration susceptible to autism, leading them to a premature and massive release of BDNF. Thus, all connections are consolidated, and not only the most important ones.

Such a release of BDNF causes the critical period to end too prematurely, 'sealing' all these connections: the child is left with an excessive amount of undifferentiated brain maps, i.e. with a pervasive developmental disorder.

The brains of these children are hyperexcitable and hypersensitive. When they hear a certain frequency, for example, the entire auditory cortex is activated. Other autistic children have a hypersensitive touch and are tormented by clothing labels touching their skin.

The theory also explains the high frequency of epilepsy in autism: due to the release of BDNF, the brain maps are poorly differentiated, and since so many connections in the brain have been indiscriminately reinforced, when even a few neurons are activated, the brain risks bursting.

This also clarifies why autistic children have larger brains: BDNF increases the sheath surrounding the neurons.

The interesting thing about this theory is that it integrates perfectly with all the major psychological and neuroscientific theoretical evidence.

A quantitative and qualitative dysfunction of BDNF in critical periods explains the malfunctioning of the circuitry involved in the theory of mind, its components being affected, and for the same reason explains the extreme empathy-systematisation.

It also agrees perfectly with the malfunctioning of the mirror neuron circuit highlighted by Ramachandran, where even in the presence of 'singularly intact' neurons, what becomes dysfunctional is the integrated system of relationship and correlation.

It explains physiologically the cranio-cerebellar features we described in Chapter 6 and accords with practically explaining the entire symptomatological picture we described in Chapter 3.

And it indirectly explains the marked difference in the incidence of autism between males and females (certainly better than the extreme male brain theory based on intrauterine testosterone alone).

A recent study just on BDNF and its incidence in autism cases found that "*preferential expression of exon 1 and exon 4-Bdnf transcripts in females may contribute to sex differences in autism cases by protecting females from the adverse effects of genetic variants or environmental factors such as VPA on the developing brain* " (Konopko et al. 2017).

Step Four: An Initial Explanation Outside of Us

It has now been made clear (Chapter 10) that autism has a genetic basis of susceptibility. Initially, theories on the causes of autism could be divided between those that attributed the major responsibility to a genetic predisposition, and those that gave weight mainly to environmental factors. The question is whether prolonged damage, whatever its origin, prevents the normal development of a specific neural system at a critical stage.

After establishing the genetic basis of autism, research to locate potentially predisposing genes entered a phase of rapid development. This research, however, has proven to be very arduous. While genetic studies are being intensively pursued, researchers need to investigate how the consequences of genetic risk should be conceptualised.

There is a big gap between what genes do and how they influence brain development. To explain autism, we need to understand how genetic defects impair critical brain systems.

Basically, any environmental risk factor that can lead to early developmental brain damage must be considered a potential non-genetic cause of autism. Prenatal conditions and birth complications are risk factors that can lead to brain damage.

Both viral infections and autoimmune disorders have been reported in cases of autism. Viral disorders are characterised by sudden attacks and can infect the central nervous system and lead to permanent brain damage. In rare cases, if the central nervous system is attacked by an infection at a critical time, before or after birth, autism can occur.

The immune system itself, which protects us from viral damage, can be subject to dysfunction for genetic reasons. Severe forms of immunological intolerance in the mother lead to foetal death; mild forms can interfere with normal growth processes and thus lead to developmental disorders. Once again, whether a dysfunction of the immune system is the cause or effect of a brain abnormality remains an open question.

As Uta Frith (1989) wrote "*we must think of a long causal chain with many distinct links. As a reminder, we can summarise that there is a risk, followed by an attack, followed by an injury. The risk can be of many types, including genetic defects, chromosomal abnormalities, metabolic disorders, viral agents, immune intolerances and anoxia resulting from perinatal problems. We can assume that each of these risks has the potential to produce very serious damage to neurodevelopment. Due to these severe interferences, permanent damage can be produced in the development of specific brain systems involved in higher mental processes. The damage may be slight or severe, but often involves the shutdown of one or more crucial systems at a critical stage. Our hypothesis is that only in such a case will autism occur*".

If we see it this way, the theory of a BDNF dysfunction fits perfectly with all the evidence related to immune system dysfunction, environmental causes, pollutants, and is perfectly adequate to provide the neurobiological basis for Marta Herbert's research evidence (which we met in Chapter 14).

Fifth Stage: A First Synthesis

In Chapter 5 I attempted to summarise, as far as it is useful in the context of this publication, the process of neurodevelopment.

In the context of research into cell differentiation by chemical substances, molecules have been identified that play a fundamental role in the growth and trophism of the nerve cell. The prototype of such substances is the nerve growth factor (NGF) isolated in the early 1950s by Rita Levi Montalcini and Stanley Cohen.

Disciplines such as developmental biology and genetics analyse the mechanisms that preside over the formation of nerve cells and glial cells from the first sketches of the nervous system.

The central problem of these studies is to assess which genetic components and which epigenetic influences preside over the progressive functional specialisation of neurons, and the formation of the complex network of nerve fibres that constitutes a hitherto unsurpassed example of miniaturisation, forming a set of circuits with approximately 1015 synaptic contacts. What emerges from all these studies is that, broadly speaking, the circuits and their realisation are encoded in the genome, while environmental stimuli play a fundamental role in the ultimate realisation of synaptic connections.

Neurons, organised in networks, ganglia, centres and other highly complex structures, process nerve impulses, store them and emit behavioural responses such as general movement, foraging, mating, fleeing in the face of danger, etc. The basic units of these structures at the level of the cerebral cortex are the columnar structures, which are present in the mammalian cortex in quantities proportional to the number of neurons.

It is estimated that each of these structures consists of a minimum of 500 to a maximum of 10,000 neurons. A column is connected with other adjacent columns and with distant columns located in other cortical areas. Through this organisation, the brain processes information simultaneously in series and in parallel.

Columns are present in all cortical areas dedicated to the perception of stimuli, their processing and the emission of a response. Connected to the process of processing is that of the memory of events that are to be recorded for a longer or shorter period. Presumably all neurons are equipped with molecular elements that allow synaptic events to be fixed for a certain period of time.

What 'goes wrong' in this process—which begins in the intrauterine environment and continues during the first three years of extra-uterine life—directly affects the structural and functional formation of the brain and the nervous system as a whole.

But mind-brain-body are a single, non-separable entity. And what happens in the brain has a relationship and correlation that manifests itself throughout the body, cascading on the functioning and relationship of the latter with all organs, whether through the nervous system, through hormonal communication, or through the bloodstream.

Genes are certainly the basis, and having a genetic susceptibility is certainly an implicit risk factor.

A number of triggers intervene on this basis, and there are many things that can go wrong. Paradoxically, the need for a genetic basis and the low incidence of the objective number of people diagnosed with autism demonstrate the extraordinary strength of our bodies to 'resist' external agents.

At any time a toxic agent, a hormonal overload, a pollutant, a problem in gestation, can cause an atypical functioning of NGF and BDNF mechanisms, cascading into a degeneration of neuronal circuit development.

It is likely that the evidently different severity of the final characteristics in which autism is expressed will depend on the cause itself that has intervened in the neurodevelopment process.

Evidence shows, for example, that a birth accompanied by a phase of anoxia rather than a particularly 'intoxicated' amniotic picture correlates with autism in its most severe and debilitating expressions.

The picture is thus complicated by the fact that symptoms that are already serious in themselves (*e.g. the limb-motor problems linked to anoxia at birth*), even when considered individually, overlap in the same person.

It is also evident from these considerations that the thesis is consistent with the evidence of the autistic spectrum with particularly differentiated characteristics.

Sixth Stage: In the Laboratory

As I said in the preface, by basic training I am culturally bound to the concept of "fact" and evidence.

I will summarise here the results of the major published scientific studies on the relationship between BDNF and autism spectrum disorder. I will attempt to do so without further comment, in roughly chronological order and providing bibliographical references to the articles.

Out of more than 76,000 English-language articles on autism catalogued on PubMed just 200 address the relationship between BDNF and autism (many are notes and reviews and commentaries) and all within the last fifteen years.

Let us now pretend we are in court, and listen to the testimony of the 'expert witnesses'.

One of the earliest studies is that of Nishimura and colleagues from 2007 "*In this study, we examined the SNP and haplotype association of BDNF with autism in a trii (Autism Genetic Resource Exchange) based association study. We also examined BDNF mRNA expression in peripheral blood lymphocytes of drug-naïve autistic patients and control subjects. In the TDT of autism trios, SNP haplotype combinations showed significant associations in the autism group. BDNF expression in the drug-naïve autism group was significantly higher than in the control group. We suggest that BDNF has a possible role in the pathogenesis of autism through its neurotrophic effects on the serotonergic system*".

According to Raznahan and colleagues (2009), "*alterations in the relationship between BDNF val66met genotype and surface area in autism may drive the findings for volume. If correct, this suggests that autism is associated with a distorted relationship between BDNF val66met genotype and regional cortical surface area determinants—gyrification and/or furrow positioning*".

Sheikh and colleagues (2010) state that "*these findings suggest that down-regulation of the BDNF-Akt-Bcl2 anti-apoptotic signalling pathway in the autistic brain could be one of the underlying mechanisms responsible for the pathogenesis of autism*".

According to Correia and colleagues (2010), "*By substantiating the previously observed increase in BDNF levels in autistic children in a larger group of patients and suggesting a genetic association between NTRK2 and autism, this study integrates evidence from multiple levels supporting the hypothesis that alterations in BDNF/tyrosine kinase B (TrkB) signalling contribute to an increased vulnerability to autism*".

Oxidative stress was the focus of Al-Ayadhi's (2012) study according to which "*serum levels of Sonic HedgeHog (SHH) protein and brain-derived neurotrophic factor (BDNF) could be linked to oxidative stress in autism. Using whole blood or polymorphonuclear leukocytes, we showed that autistic children produced significantly higher levels of oxygen free radicals (OFR).*

Furthermore, we found significantly higher levels of SHH protein in serum in children with mild and severe forms of autism. We also found that the serum BDNF level was significantly reduced in autistic children with mild forms of the disorder but not with severe forms of the disorder. Our results are the first to report a correlation between SHH, BDNF and OFR in autistic children, suggesting a pathological role of oxidative stress and SHH in autism spectrum disorders".

Das's (2013) study focused on oxidative and nutritional phenomena: "*Autism has a strong genetic and environmental basis in which inflammatory markers and factors related to synapse formation, nerve transmission and information processing such as brain-derived neurotrophic factor (BDNF), polyunsaturated fatty acids (PUFAs) arachidonic (AA), eicosapentaenoic (EPA) and docosahexaenoic (DHA) acids and their products and neurotransmitters: dopamine, serotonin, acetylcholine, γ-aminobutyric acid, catecholamines and cytokines are impaired. Antioxidants, vitamins, minerals and trace elements are required for the normal metabolism of neurotrophic factors, eicosanoids and neurotransmitters, supporting reports of their alterations in autism. But the exact relationship between these factors and their interaction with genes and proteins involved in brain development and growth is unclear. It is suggested that maternal infections, inflammation, and adverse events during intrauterine foetal growth could lead to alterations in gene expression profiling and proteomics that result in dysfunction of neuronal function and neurotransmitters, alteration in the metabolism of PUFAs and their metabolites resulting in overproduction of eicosanoids and proinflammatory cytokines, and a deficiency of anti-inflammatory cytokines and bioactive lipids that ultimately results in the development of autism*".

Halepoto and colleagues (2014) stated that "*over the years, research findings have suggested a critical role played by BDNF in the development of autism, including increased serum BDNF concentrations in children with autism and the identification of different forms of BDNF in families of autistic individuals*".

Almeida and colleagues (2014) state that "*in light of BDNF's well-established role in the regulation of neurogenesis and the laminar fate of postmitotic neurons in the developing cortex, an aberrant increase in BDNF expression in the fetal brain may contribute to VPA-induced cognitive impairments by altering brain development*".

According to Zheng and colleagues (2016), "*the meta-analysis provided evidence of higher peripheral BDNF levels in autism compared to controls* *[standardised mean difference (SMD) = 0.63, 95% confidence interval (95% CI) = 0.18–1.08; P = 0.006]. Subgroup analyses revealed higher BDNF levels in autism than in controls for both sera [SMD = 0.58, 95% CI = 0.11–1.04; P = 0.02] and plasma [SMD = 1.27, 95% CI = 0.92–1.61; P < 0.001]. The childhood studies produced a similar cumulative effect size [SMD = 0.78, 95% CI = 0.31–1.26; P = 0.001], whereas this was not true for the adult studies [SMD = 0.04, 95% CI = -1.72–1.80; P = 0.97]. This meta-analysis suggests that peripheral BDNF levels are a potential biomarker of autism*".

Also for Armeanu and colleagues (2017), "*the meta-analysis estimated a significant summary effect size of 0.33 (CI 95% 0.21–0.45, P < 0.001), suggesting higher BDNF in autism compared to controls. The studies showed considerable heterogeneity, but no evidence of publication bias. A higher peripheral BDNF in autism agrees with several neurological and psychological theories about the causes and symptoms of this condition, and contrasts in particular with the lower BDNF levels found in schizophrenia, bipolar disorder and depression*".

Saghazadeh and colleagues in 2017 confirm that "*a meta-analysis of blood BDNF in 887 patients with autism and 901 control subjects demonstrated significantly higher BDNF levels in autism compared to controls with MDS by 0.47 (95% CI 0.07–0.86, P = 0.02). Furthermore, subgroup meta-analyses were performed on the basis of the BDNF sample. The present meta-analysis study led to the conclusion that BDNF might play a role in the initiation/propagation of autism and thus can be considered a possible biomarker of autism*".

Meng and colleagues (2017) state, "*The aim of our study was to illuminate the potential role of brain-derived neurotrophic factor (BDNF) in autism spectrum disorder. We measured circulating levels of BDNF in serum and BDNF gene polymorphisms (Val66Met), where two markers were then compared between autism and normal controls. (...) These results suggested that, rather than the Val66Met polymorphism, it was more likely that it was BDNF that had an impact on the pathogenesis of autism*".

The study by Ohja and colleagues (2018) focused on the immune system. "*Dysregulation of neurotrophins such as BDNF and their signalling pathways have been implicated in autism. Although abnormal cortical formation and autistic behaviours have been described in mouse models of immune activation, no theory linking immune activation to specific aberrant brain signalling pathways in autism has been described. In this article we explore the relationship between neurotrophin signalling, the immune system and autism. To this end, we hypothesise that an interaction between the dysregulated immune system, synaptogenic growth factors and their signalling pathways contribute to the development of autism phenotypes*".

Skogstrand and colleagues (2019) conducted a specific analysis of infants: "*in infants, a lower blood level of BDNF was significantly associated with an increased probability (OR 1.15) of developing ASD (P = 0.001). This difference cannot be explained by genetic variation in the BDNF-coding gene region. A trend towards decreasing levels of all neurotrophic markers and increasing levels of all inflammatory markers was noted. The low blood levels of BDNF in children who develop autism are an important finding, suggesting that lower levels of BDNF in infants contribute to the aetiology of autism and point to new directions for further research. It could also help identify a long-sought marker for a high risk of autism, for example, in younger siblings of children with autism*".

These are the conclusions of the Brazilian researchers led by Barbosa (2020): "*Patients diagnosed with autism often show dysregulation of proteins associated with synaptic plasticity, particularly brain-derived neurotrophic factor (BDNF). The aim of the present study was to analyse BDNF serum concentration levels in children with classical forms of autism and in a healthy control group to determine whether there is a correlation between BDNF serum levels and autism. The mean serum BDNF concentration level was statistically higher for children with autism (P < 0.000) compared to the control group. There is no doubt that BDNF plays a role in the developmental pathophysiology and evolution of autism, but its brain levels may fluctuate depending on several known and unknown factors. The crucial question is not whether BDNF levels can be considered a prognostic or diagnostic indicator of autism, but to determine its role in the onset and progression of this disorder*".

The study by Ahmed and colleagues (2021) "*investigated the role of epigenetic dysregulation of brain-derived neurotrophic factor (BDNF) and glial fibrillary acidic protein (GFAP) genes and oxidative stress as possible mechanisms of autism-like behaviour in the neonatal isolation model in rats and the impact of folic acid administration on these parameters. […] Neonatal isolation induces 'autistic-like' behaviours and these behaviours are reversed by folic acid supplementation. Neonatal isolation induces DNA hypomethylation of BDNF and GFAP, increased markers of oxidative stress and neuroinflammation. All these changes were reversed by daily folic acid supplementation*".

On external factors such as lead, Shin and colleagues (2021) state that "*taken together, the data demonstrate that Pb exposure may contribute to ASD-associated brain developmental abnormalities, particularly in individuals with a genetic predisposition to ASD*".

Robinson-Agramonte and colleagues (2022) confirm that "*serum levels of proBDNF were decreased in subjects with autism compared to non-autistic controls. Medicated but non-medicated subjects with autism showed lower serum levels of proBDNF than controls, while neither IGF-1 nor BDNF levels differed between treatment groups. These data support the involvement of proBDNF and IGF-1 in the pathogenesis and treatment of autism*".

According to Qi, Chen and colleagues (2022) "*the expression of brain-derived neurotrophic factor (BDNF), a neurotrophin that promotes the formation and maturation of glutamatergic and GABAergic synapses during postnatal development, was severely reduced in the group exposed to VPA. Treatment with exogenous BDNF during the critical period of E/I (excitatory/inhibitory) imbalance rescued synaptic functions and autism-like behaviour, such as social defects. With these results, we demonstrated experimentally that the social dysfunction in the VPA autism mouse-model could be caused by E/I imbalance resulting from BDNF deficits during the developmental phase*".

The research group of Camuso and colleagues (2022) in Rome established that "*brain-derived neurotrophic factor (BDNF) is one of the most studied neurotrophins in the mammalian brain, essential not only for central nervous system development but also for synaptic plasticity. BDNF is present in various areas of the brain, but the highest levels of expression are found in the cerebellum and hippocampus. After birth, BDNF acts in the cerebellum as a mitogenic and chemotactic factor, stimulating cerebellar granule cell precursors to proliferate, migrate and mature, while in the hippocampus BDNF plays a key role in synaptic transmission and plasticity, representing a key regulator for long-term potentiation, learning and memory. Furthermore, BDNF expression is highly regulated and changes in its expression are associated with both physiological and pathological conditions. The aim of this review is to provide an overview of the current state of knowledge on the biology of BDNF and its neurotrophic role in the proper development and functioning of neurons and synapses in two important brain areas of postnatal neurogenesis, the cerebellum and hippocampus. Dysregulation of BDNF expression and signalling, resulting in alterations in neuronal maturation and plasticity in both systems, is a common hallmark of several neurodevelopmental disorders, such as autism spectrum disorder, suggesting that the neuronal malfunctioning present in these disorders is the result of excessive or reduced BDNF-supported activity. We believe that the more we highlight the relevance of the pathophysiological actions of BDNF, and its downstream signals, in early postnatal development, the more likely it is that new neuroprotective therapeutic strategies will be identified in the treatment of various neurodevelopmental disorders*".

But, Taylor and colleagues (2023) state "w*e hypothesised that activity-dependent neural signalling is a dysregulated convergent molecular pathway in autism. The causal link between decreased neural activity-dependent signalling and autism remains unclear. Brain-derived neurotrophic factor (BDNF) is a key molecule that mediates activity-dependent neural signalling. We therefore hypothesise that reduced activity-dependent BDNF signalling may confer autism-like behavioural deficits. We investigated the effect of reduced activity-dependent BDNF signalling on autism-like behavioural deficits using mice with genetic knock-in of a human methionine (Met) allele of BDNF, which reduced activity-dependent BDNF release without altering the basal BDNF level. Compared to wild-type (WT) controls, the decrease in activity-dependent BDNF signalling similarly induced anxiety-like behaviour in male and female mice. In particular, decreased BDNF activity-dependent signalling differentially induced autism-like social deficits and increased self-care in male and female mice, and male mice were more severe than female mice. Again, sexually dimorphic spatial memory deficits were observed in BDNF + /Met female mice, but not in BDNF + /Met male mice. Our study not only reveals a causal link between reduced BDNF activity-dependent signalling and autism-like behavioural deficits, but also identifies a previously underestimated sex-specific effect of reduced BDNF activity-dependent signalling in autism. These genetic knock-in mice of the human BDNF Met variant provide a distinct mouse model to study the cellular and molecular mechanisms underlying reduced activity-dependent neural signalling, the common dysregulated molecular pathway in autism*".

The Pisa research group led by Carpita summarised his 2024 research as follows: "*Several studies have investigated circulating levels of serotonin (5-HT) and brain-derived neurotrophic factor (BDNF) in children with autism spectrum disorder. A more limited literature has focused on adults with autism or populations with subthreshold autism spectrum manifestations, such as relatives of probands with autism. This study aimed to investigate 5-HT and BDNF levels in adults with autism spectrum conditions. Correlations between levels of biochemical variables and autism symptoms were also assessed. Adults with autism showed significantly lower levels of 5-HT in platelet-poor plasma (PPP) compared to the BAP and control groups. No significant difference was found between the groups for BDNF PPP and intraplatelet 5-HT levels. It has been reported that 5-HT levels are specifically correlated with some symptoms of autism. Our work showed the presence in adults with autism of reduced 5-HT PPP levels compared to other groups, without significant differences with BDNF levels, supporting the hypothesis that the biochemical correlates of autism in adults might be different from those typically reported in children*".

Stage Seven: Out of the Lab

Elsheikh and colleagues (2023) went further.

According to them, BDNF is not only a central factor that can be adopted as a predictor in conjunction with genetic investigations, but also offers a useful indication in therapy.

According to their study "*biostimulation with low-level laser acupuncture (LLLA) could help improve symptoms and communication in children with autism spectrum disorder. Photobiomodulation could influence the level of brain-derived neurotrophic factor (BDNF) and the expression of miR320*".

The aim of their very recent study was to understand the influence of LLLA biostimulation on severity, language abilities, BDNF levels and miR-320 in a sample of children diagnosed with autism.

Autistic participants ($N = 30$) were randomly divided equally into groups: group I received LLLA therapy twice a week for 12 sessions and group II did not. Before and after the intervention, assessments of severity, language abilities, BDNF level by enzyme-linked immunosorbent assay and miR-320 expression by quantitative reverse transcriptase polymerase chain reaction were performed. A comparison was made between autistic cases ($N = 30$) before starting therapy and neurotypical children ($N = 15$) with regard to miR-320 expression.

These results: "*Following the intervention, the severity of expression of autistic features was reduced and language performance was elevated in both groups. The improvement in Group I was greatest with ($P = 0.002$; 0.03). The plasma level of BDNF was reduced only in Group I ($P < 0.001$). The expression level of miR-320 in Group I showed no change ($P = 0.641$).*

A significant difference in miR-320 expression was observed between the children diagnosed with autism and the neurotypical group (P = 0.000)".

This study would—according to the authors—introduce *"LLLA therapy as a safe and promising therapeutic procedure to improve core manifestations and communication skills and to modulate BDNF levels in children with ASD. Reduced expression of miR-320 has shown good diagnostic value in children with autism"*.

In a different direction is the use of the PONS (*Portable Neuromodulation Stimulator*) (Tyler et al. 2014), devised at the University of Wisconsin, Madison with the collaboration of Russian technical specialists who had already developed the same concepts before the fall of the Soviet Union.

The basic idea is the reactivation of the basal ganglia and the 'ancient brain', as well as the brainstem via the 'shortest and most direct route', i.e. the tongue, in a non-invasive manner.[1]

The technique has proved very effective for multiple sclerosis and walking problems, even after very few operations, but stimulation of the basal ganglia can have more amplified effects and in different directions.

Michael Merzenich is a neuroscientist at the University of California San Francisco who won the Kavli Prize in 2016 for his numerous studies on brain plasticity. Merzenich has helped identify two distinct periods of brain plasticity: The critical infant period is when the child's brain establishes neural processes for the stimuli it is subjected to. The adult plasticity period is when the brain refines its neural processes as it masters a range of tasks (Merzenich 1995).

Understanding how the brain can rewire itself has allowed Merzenich, Tallal and other colleagues to develop strategies to correct individuals with deficits in language, reading and speech (Merzenich et al. 1999).

Through research on experience-dependent learning in non-human primates, neurophysiologists, including Merzenich, have shown that neuroplasticity persists into adulthood. Further studies in monkeys have suggested that the principles of Hebbian learning that drive neuroplasticity can be used to treat children with learning and language problems (Merzenich et al. 1996).

Merzenich and his team used micro-mapping to show how the maps of some newborn mice were formed in the critical period. Immediately after birth, at the beginning of the critical period, the auditory maps are undifferentiated and consist of only two large regions in the cortex. One half of the map responds to all higher sound frequencies, the other to all lower ones.

When the animal was exposed to a specific frequency during the critical period, that simple organisation changed. If the animal was repeatedly exposed to a sharp 'do', after a very short time only a few neurons were activated, which became selective for that sharp 'do'. The same happened when the animal was exposed to a 're', 'mi', 'fa' and so on. Now the map, instead of being structured into two larger areas, had many smaller areas, each of which responded to the various notes. In other words, it had differentiated.

What is remarkable about the cortex during the critical period is that its plasticity is such that its structure can be changed simply by exposure to novel stimuli. This sensitivity allows infants and very young children in the critical period of language development to pick up new words and sounds effortlessly, simply by listening to their parents talk; simple exposure to stimuli causes the brain maps to 'hardwire' the changes. After the critical period, older children and adults can, of course, learn languages, but they really have to make an effort to maintain the necessary concentration.

According to Merzenich, the difference between the plasticity typical of the critical period and that of the adult is that in the former case the brain maps can be changed simply by exposure to the outside world, since 'the learning machine is always switched on'.

There is a good biological reason why this 'machine' is always switched on: in all probability, infants do not know what will be important in life, so they pay attention to everything. Only a brain that is already somewhat organised could select what is worth focusing on.

So Merzenich began to wonder about the role of a new environmental risk factor that could affect anyone, but with a more damaging effect on genetically predisposed children: the constant background noise caused by cars, sometimes called 'white noise'. White noise is the sum of many frequencies and is a very powerful stimulus for the auditory cortex.

"Babies are raised in environments that are increasingly immersed in constant noise" says Merzenich. Today white noise is everywhere, coming from the fans of electronic devices, air conditioners, heating, car engines. How does this noise affect brain development?

To test his hypothesis, Merzenich's team exposed rat pups to pulses of white noise for the duration of the critical period, and found that the pups' cortex was devastated.

With each impulse, the entire auditory cortex, every single neuron, is excited. Such neuronal activation leads to an abnormal release of BDNF. And as Merzenich's model predicted, this exposure leads to a premature end of the critical period. The animals were left with undifferentiated brain maps and neurons that activated indiscriminately in response to any frequency.

Merzenich discovered that rat pups were predisposed to epileptic seizures that were induced by exposing them to simple human voices, like autistic children (in humans, epileptic seizures can be triggered by strobe lights that consist of pulsed white light made up of numerous bright frequencies). Merzenich had an animal model of autism.

Now the question was whether anything could be done to bring the undifferentiated brain maps back to normal after the critical period.

And here we return to what we said in the previous chapter. In fact, it is no coincidence that Merzenich too has always dealt with acoustics and learning problems, and arrived by this route first at neuronal plasticity and then at autism.

First, Merzenich and colleagues used white noise to 'dedifferentiate' the auditory map of rats. Then, once the lesion had been produced, they normalised and 'redifferentiated' the maps using very simple sounds, one at a time. With the exercise, the maps were brought back above normal.

What if it was possible to reopen the critical period for plasticity, so that adults could learn languages like children, simply by coming into contact with them? Merzenich had already shown that plasticity extends into adulthood, but now he wondered whether it was possible to extend the critical period for spontaneous learning.

Merzenich and Kilgard implanted microelectrodes in the nucleus basalis of rats and applied an electric current to activate it. They then exposed the rats to a 9 Hz sound frequency to test whether they were able to spontaneously develop a specific brain map, just like puppies during the critical period. After a week, the two researchers discovered that they could greatly extend the brain map for that particular sound frequency. They had found a way to artificially reopen the critical period in the adult brain.

They then used the same technique to speed up the brain's processing time. Normally, neurons in the auditory cortex of an adult rat can fire in response to up to twelve sound impulses per second. By stimulating the nucleus basalis, it was possible to 'train' the neurons to respond to even faster inputs (Doidge, 2015).

This research opens up the possibility of very fast learning even in adulthood. The nucleus basalis can be activated by an electrode, by microinjections of specific substances, or by drugs. Of course, such aggressive interventions could have unforeseen adverse effects on the brain—not to mention our ability to regulate its use—but they would probably be pioneering in cases of serious medical problems. Reactivation of the nucleus basalis could help brain-injured patients, many of whom can no longer relearn impaired functions, such as reading, writing, speaking or walking, because they cannot concentrate sufficiently.

A Unified Proposal

In Chapter 3 and Chapter 11 I described the various aetiological theories of autism, at least those that are based on cross-checked scientific evidence, and in Chapters 9 and 10 we went into the theoretical foundations based on recent evidence from neuroscience and genetics.

These insights allowed us to have a unified picture of the characteristic and symptomatological complex summarised in Chapter 4.

In the conclusions of Chapter 11 I mentioned the martial arts story.

In the metaphor, a fist may ultimately be just a fist, but we must also clarify what it is. There is no fist if all the fingers are not clenched together, if the palm in which they close does not have a certain solidity, if there is no unity of gesture, and if this gesture is not strong enough to be also effective.

Out of metaphor, the unitary thesis I propose is, in short, that there is, as a prerequisite, a genetic predisposition in terms of susceptibility.

The whole panel of genes involved is not essential, but there certainly needs to be a critical number of them, which may also vary from person to person (Chapter 9).

At a critical stage of NGF and BDNF expression in neurogenesis, and in particular at the moment of specialisation, an environmental factor, in the intrauterine developmental phase, intervenes as a trigger in a chain that leads to atypical brain development: a total of 974 abnormalities in neural connections associated with autistic behaviour have been identified (Chapter 6); fewer may occur as well as we may discover more or different ones.

It is necessary to be clear on this point.

One does not expect a specific location (locus) or a specific damaged brain organ. It is not a stroke-like damage to the left frontal or frontoparietal dominant area, which includes Broca's area, the effect of which is the so-called expressive aphasia (Broca's aphasia) with the consequence that the person cannot speak or write well (but understands what is said to him and is able to process concepts), nor conversely a lesion of Wernicke's area, whereby the resulting aphasia will impair the interpretation of the meaning of words. In these cases there is a precise damage, in a precise place (visible by NMR).

The abnormalities we are talking about are mainly connectional: something in the circuit and in the process of connection between neurons has gone wrong.

This is confirmed by the cranio-cerebellar anomalies that we spoke about and which we described in Chapter 6.

This is confirmed by the quantitative differences in the brain's white and grey masses and their size.

This is confirmed by the consequences of damage to the neurobiological circuit of attention.

This is confirmed by the effects of damage to the neurophysiological circuit related to the theory of mind.

This is confirmed by the dysfunctions implicit in the malfunctioning of the mirror neuron complex (Chapters 8, 9 and 11.

The entire characteristic and symptomatic picture and comorbidity (Chapter 4) is consistent with this general explanation.

And the intensity and characteristics depend on where, how, when and how much environmental interference has intervened in the neurogenesis process.

Let us reason in reverse.

What happens if the genetic susceptibility base is missing?

What already happens. Malformations, deletions, functional abnormalities, motor dysfunctions, delays or problems in psychic development.

My friend Bruno, with whom we started our journey, suffered a powerful anoxia at birth, suffered all the physical-motor consequences, but probably not having an atypical genetic susceptibility base, he did not have autistic traits.

Magdalena is a young girl of 16, at birth her placenta was dark green and her mother had been exposed to plastic fumes (dioxin) for 20 years: today she has severe learning and concentration problems, but has no obvious symptoms that would qualify for an autism diagnosis. In the same hospital Gabriele was born, two years earlier, same problems as Magdalena, but with a diagnosis of severe autism.

It is likely that the difference between Magdalena and Gabriele is the initial genetic picture and the 'protection' that reduces the incidence in women. But it is common experience that this is not always the case, indeed, when this 'protection' is not effective for girls the autistic condition is often severe.[2]

In addition to opening up and consolidating a research front, this thesis offers us a powerful diagnostic tool related to genetic testing; following the cited reasoning of Skogstrand's research group, "*it could help identify a long-sought marker for a high risk of autism*", as Zheng also confirms, according to whom "*peripheral BDNF levels are a potential biomarker of autism*".

Regarding therapies, the cited study by Camuso and colleagues is clear: "*We believe that the more the relevance of the pathophysiological actions of BDNF, and its downstream signals, in early postnatal development is highlighted, the more likely it is that new neuroprotective therapeutic strategies will be identified in the treatment of various neurodevelopmental disorders*".

The position of savants diagnosed with autism remains to be clarified.

People on the autism spectrum approach creative problem-solving differently from neurotypical people. In a study conducted on both neurotypical and autistic spectrum people, Martin Doherty (Best et al. 2015) of the University of East Anglia, England, and his colleagues found that people with a high number of autistic traits produce fewer but more original ideas. Their hypothesis is that autistic people are more likely to go straight to less common ideas because they rely less on associations or memory, which would limit their creative thinking. A critical analysis of numerous studies conducted by Francesca Happé and Uta Frith (2009) suggests that greater sensory acuity, concentration on detail, visual memory and pattern recognition may be involved, along with an obsessive need for training. Almost 30 per cent of people on the autistic spectrum show special skills in music, memory, calculating numbers and dates, drawing or language. These results suggest that the biological basis of talent, for example, in numerical calculations or calculating dates, does not differ significantly from what talent in art or music is based on, a conclusion that may also extend to neurotypical persons. Allan Snyder (2009), director of the Centre for the Mind at the University of Sydney in Australia, has supported the idea that the left hemisphere's control over the right hemisphere's creative potential is reduced in people with autism.

How to explain these extraordinary skills in some autistic people?

In the words of Oliver Sacks (2007) '*The rapidity with which savant talents can emerge in such circumstances indicates the disinhibition or liberation of right-hemisphere functions, which are released from the inhibition or suppression normally exerted by the left temporal lobe. (…) In 1999, Allan Snyder and D.J. Mitchell reversed the terms of the usual question—why are savants so rare?—and asked instead: why don't we all have those talents? Snyder and Mitchell hypothesised that at an early age, the mechanism underlying such abilities is most likely present in all of us, but that as the brain matures, they are inhibited, or at least removed from conscious awareness. These authors theorised that savants might have "privileged access to lower levels of information not available through introspection. (…) The existence of such potentialities can only be understood in terms of phylogeny and ontogeny, i.e. as ancestral forms of perception and cognition once endowed with adaptive value but now suppressed and replaced by others*"'.

I believe this phenomenon fits into the proposed theory.

Certain abilities ineluctably rest on the functioning of a precise neuronal circuit that is in charge of a function (in integrated coordination with other circuits).

If one accepts that an 'excess' of BDNF undifferentiatedly consolidates an excessive number of connections, it is probable that in some people, despite their overall autistic condition, some of these circuits are more functionally amplified, even greatly increasing their relative performance.

A New Definition of Autism

In this context, the definition of autism as a spectrum makes further sense, and perhaps even more sense to speak of the non-existence of one autistic person identical to another.

There is a subjective-individual difference, and there is an individual differentiation of the 'damage-end' as the physiological basis of the condition.

It is likely that we need another definition of autism at this point.

While neurodiversity may well represent the macro framework, for autism, rather than "spectrum disorder" it is perhaps more appropriate—accepting the thesis proposed here—to speak of "autistic complex condition".

If the "autistic state" represents the person's "here and now" condition, the condition of the complex allows us to glimpse a "dynamic".

Depending on the type of impairment, the "autistic state" and the "condition of the complex", personalised and individualised interventions can in fact be imagined that can work and affect the person in an individualising way to alleviate, reduce, contain the effects present in the state.

The 'complex condition' in my opinion renders both the idea of a heterogeneity of characteristics and symptomatologies and of the possible different severities, but it leaves the univocal reference to the clinic to return to the person with a greater individuality.

As Oliver Sacks (1973) wrote, "Dialogue about how one is can only be expressed in human terms, in familiar terms, which come easily and naturally to all of us; and it can only take place if there is a direct human confrontation, an 'I-you' relationship, between the world of discourse of the doctors and that of the sick".

I therefore propose, briefly, the definition of 'autistic state' (if anything with a descriptive characteristic-symptomatic degree) for the moment of diagnosis.

I propose 'autistic complex condition' for the individual description that can provide an indication of possible intervention.

Obviously it is not as important whether or not these definitions are accepted as what they are meant to represent.

A Jump to 1925

Only after 'the fall of the wall' were Vygotskij's essays on 'defectology' published in English, including the 1929 study on the special education of deaf children (Vygotskij, 1929).

It was in general a very serious detriment to scientific research to restrict the circulation of studies and ideas from that extraordinary forge that was the Soviet Union and 'the other half of the world', the communist one (even today, pioneering studies on vision and opiate addiction therapies carried out in Cuba are little known in the West).

Vygotskij passionately fought the evaluation of handicapped children in terms of defects or deficits; he evaluated them, instead, according to their integrity, their being intact, their 'pluses'. They had to be seen not as deficient, but as different: "*A child represents a qualitatively different, unique type of development'. And it was precisely this qualitative difference, this uniqueness, that the work of education or rehabilitation had to address, according to Vygotskij: 'If a blind or deaf child reaches the same level of development as a normal child, he does so in another way, following another path, by other means; and for the pedagogue it is essential to know the uniqueness of the path on which he has to guide the child. This uniqueness transforms the less of handicap into the more of compensation*".

The development of higher psychological functions is not, for Vygotskij, something that happens 'naturally' automatically: it requires mediation, culture and a cultural instrument. And the most important of these cultural tools is language. But cultural tools and language, Vygotskij argues, were developed for 'normal' people, who have their sense organs intact, their biological functions intact. The key to its development will be compensation, the use of an alternative cultural tool.

Thus Vygotskij arrives at special education for the deaf: for them the alternative cultural tool is sign language, which was created for them and by them.

It addresses the functions that remain intact, the visual ones; and it is the most direct way to reach deaf children, the simplest means to enable their complete development, the only one that respects their difference, their uniqueness (it is interesting to note that in some cognitive-behavioural therapies one of the facilitators of relational communication of autistic children and adolescents is precisely the teaching of sign language elements).

Lev Vygotskij died at only 38 years of age, but he left us powerful publications and research and teachings, today, almost one hundred years later, of a shocking modernity and topicality.

Notes

1. Electrical stimulation of the tongue activates two cranial nerves: the facial nerve (mixed nerve responsible for the contraction of the muscles that regulate facial expressions) and the trigeminal nerve (nerve responsible for the sensory function).

   Once the device has been put into operation by the therapist, the control unit emits signals that reach the mouthpiece. Here, dense nerve endings in the tongue send these impulses to the brainstem, the component that controls sensory perception and movement.

   The mouthpiece consists of 144 circular, gold-coated electrodes, arranged on a 3 cm × 3 cm square array placed in turn on a circuit covered with an epoxy resin. The grid of electrodes (12 × 12) that is formed is in turn divided into 9 4 × 4 portions.

   As the impulse is delivered, one electrode at a time is stimulated in each portion, while the remaining electrodes work together to create a path for the return current.

   This generates a triplet of positive pulses of 50 μs amplitude, delivered at 200 pulses/s every 20 ms.
2. I use pseudonyms here, for reasons of people's privacy, but the cases and people are real. Bruno's name is real.

References

Ahmed OG, Shehata GA, Ali RM, Makboul R, Abd Allah ESH, Abd El-Rady NM. (2021) Folic acid ameliorates neonatal isolation-induced autistic like behaviors in rats: epigenetic modifications of *BDNF* and *GFAP* promotors. Appl Physiol Nutr Metab.

Al-Ayadhi LY. (2012) Relationship between Sonic hedgehog protein, brain-derived neurotrophic factor and oxidative stress in autism spectrum disorders. Neurochem Res.

Almeida LE, Roby CD, Krueger BK. (2014) Increased BDNF expression in fetal brain in the valproic acid model of autism. Mol Cell Neurosci.

Armeanu R, Mokkonen M, Crespi B. (2017) Meta-Analysis of BDNF Levels in Autism. Cell Mol Neurobiol

Barbosa AG, Pratesi R, Paz GSC, Dos Santos MAAL, Uenishi RH, Nakano EY, Gandolfi L, Pratesi CB. (2020) Assessment of BDNF serum levels as a diagnostic marker in children with autism spectrum disorder. Sci Rep.

Best, C., Arora, S., Porter, F., Doherty, M. (2015), "The relationship between subthreshold autistic traits, ambiguous figure perception and divergent thinking". In Journal of Autism and Developmental Disorders, 45, 12, pp. 4064–4073.

Camuso S, La Rosa P, Fiorenza MT, Canterini S. (2022) Pleiotropic effects of BDNF on the cerebellum and hippocampus: Implications for neurodevelopmental disorders. Neurobiol Dis.

Carpita B, Stagnari R, Palego L, Baroni D, Massimetti G, Nardi B, Cremone IM, Betti L, Giannaccini G, Dell'Osso L. (2024) Circulating Levels of 5-HT and BDNF in Adults with Autism Spectrum Conditions: An Investigation in a Sample of Subjects with Autism Spectrum Disorder, their First-degree Relatives and Controls. Curr Med Chem.

Correia CT, Coutinho AM, Sequeira AF, Sousa IG, Lourenço Venda L, Almeida JP, Abreu RL, Lobo C, Miguel TS, Conroy J, Cochrane L, Gallagher L, Gill M, Ennis S, Oliveira GG, Vicente AM. (2010) Increased BDNF levels and NTRK2 gene association suggest a disruption of BDNF/TrkB signaling in autism. Genes Brain Behav.

Das UN. (2013) Autism as a disorder of deficiency of brain-derived neurotrophic factor and altered metabolism of polyunsaturated fatty acids. Nutrition

Doidge, N. (2015). The Brain's Way of Healing: Remarkable Discoveries and Recoveries from the Frontiers of Neuroplasticity - Penguin

Elsheikh MS, Ashaat EA, Ramadan A, Mohamed NH, Elaraby NM, El-Hariri HM, Hashish AF, Nashaat NH. (2023) Efficacy of Laser Acupuncture for Children With Autism Spectrum Disorder: Clinical, Molecular, and Biochemical Study. Pediatr Neurol.

Frith, U. (1989). Autism: Explaining the enigma. Basil Blackwell.

Halepoto DM, Bashir S, A L-Ayadhi L. (2014) Possible role of brain-derived neurotrophic factor (BDNF) in autism spectrum disorder: current status. J Coll Physicians Surg Pak

Happe, F., Frith, U. (2009), "The beautiful otherness of the autistic mind" In Philosophical Transactions of the Royal Society B: Biological Sciences, 364, 1522, pp. 1346–1350.

Jung, C.G. (1929) Wirklichkeit der Seele Cap. 5

Konopko MA, Densmore AL, Krueger BK. (2017) Sexually Dimorphic Epigenetic Regulation of Brain-Derived Neurotrophic Factor in Fetal Brain in the Valproic Acid Model of Autism Spectrum Disorder. Dev Neurosci.

Lai KO, Ip NY. (2013) Structural plasticity of dendritic spines: the underlying mechanisms and its dysregulation in brain disorders. Biochim Biophys Acta

Ma K, Taylor C, Williamson M, Newton SS, Qin L. (2023) Diminished activity-dependent BDNF signaling differentially causes autism-like behavioral deficits in male and female mice. Front Psychiatry.

Meng WD, Sun SJ, Yang J, Chu RX, Tu W, Liu Q. (2017) Elevated Serum Brain-Derived Neurotrophic Factor (BDNF) but not BDNF Gene Val66Met Polymorphism Is Associated with Autism Spectrum Disorders. Mol Neurobiol

Merzenich, M.; Jenkins, W.M. (1995). "Cortical plasticity, learning, and learning dysfunction". Maturational Windows and Adult Cortical Plasticity.

Merzenich, M.M.; Jenkins, W.M.; Johnston, P.; Schreiner, C.; Miller, S.L.; Tallal, P. (1996). "Temporal processing deficits of language-learning impaired children ameliorated by training". Science.

Merzenich, M.; Tallal, P.; Peterson, B.; Miller, S.L.; Jenkins, W.M. (1999). "Some neurological principles relevant to the origins of - and the cortical plasticity based remediation of - language learning impairments". Neuroplasticity: Building a Bridge from the Laboratory to the Clinic.

Nishimura K, Nakamura K, Anitha A, Yamada K, Tsujii M, Iwayama Y, Hattori E, Toyota T, Takei N, Miyachi T, Iwata Y, Suzuki K, Matsuzaki H, Kawai M, Sekine Y, Tsuchiya K, Sugihara G, Suda S, Ouchi Y, Sugiyama T, Yoshikawa T, Mori N. (2007) Genetic analyses of the brain-derived neurotrophic factor (BDNF) gene in autism. Biochem Biophys Res Commun.

Ohja K, Gozal E, Fahnestock M, Cai L, Cai J, Freedman JH, Switala A, El-Baz A, Barnes GN. (2018) Neuroimmunologic and Neurotrophic Interactions in Autism Spectrum Disorders: Relationship to Neuroinflammation. Neuromolecular Med.

Qi C, Chen A, Mao H, Hu E, Ge J, Ma G, Ren K, Xue Q, Wang W, Wu S. (2022) Excitatory and Inhibitory Synaptic Imbalance Caused by Brain-Derived Neurotrophic Factor Deficits During Development in a Valproic Acid Mouse Model of Autism. Front Mol Neurosci.

Raznahan A, Toro R, Proitsi P, Powell J, Paus T, F Bolton P, Murphy DG. (2009) A functional polymorphism of the brain derived neurotrophic factor gene and cortical anatomy in autism spectrum disorder. J Neurodev Disord.

Robinson-Agramonte MLA, Michalski B, Vidal-Martinez B, Hernández LR, Santiesteban MW, Fahnestock M. (2022) BDNF, proBDNF and IGF-1 serum levels in naïve and medicated subjects with autism. Sci Rep.

Sacks, O. (1973) Awakenings - Duckworth & Co NY.

Sacks, O. (2007) - Musicophilia - Tales of Music and the Brain.

Saghazadeh A, Rezaei N. (2017) Brain-Derived Neurotrophic Factor Levels in Autism: A Systematic Review and Meta-Analysis. J Autism Dev Disord.

Sanders, l. (2023) What a look at more than 3,000 kinds of cells in the human brain tells us – Science News - https://www.sciencenews.org/article/human-brain-cells-new-census

Sheikh AM, Malik M, Wen G, Chauhan A, Chauhan V, Gong CX, Liu F, Brown WT, Li X. (2010) BDNF-Akt-Bcl2 antiapoptotic signaling pathway is compromised in the brain of autistic subjects. J Neurosci Res.

Shin K, Lim G, Hong YS, Kim S, Hwang S, Lee J, Sin S, Cho A, Kim Y, Gautam R, Jo J, Acharya M, Maharjan A, Lee D, K C PB, Kim C, Heo Y, Kim HA. (2021) Exposure to lead on expression levels of brain immunoglobulins, inflammatory cytokines, and brain-derived neurotropic factor in fetal and postnatal mice with autism-like characteristics. J Toxicol Environ Health A.

Skogstrand K, Hagen CM, Borbye-Lorenzen N, Christiansen M, Bybjerg-Grauholm J, Bækvad-Hansen M, Werge T, Børglum A, Mors O, Nordentoft M, Mortensen PB, Hougaard DM. (2019) Reduced neonatal brain-derived neurotrophic factor is associated with autism spectrum disorders. Transl Psychiatry.

Snyder, A. (2009), "Explaining and inducing savant skills: Privileged access to lower level, less-processed information". In Philosophical Transactions of the Royal Society B, 364, 1522, pp. 1399–1405.

Tyler, M.E., Kaczmarek, K.A., Rust, K.L. et al. (2014) Non-invasive neuromodulation to improve gait in chronic multiple sclerosis: a randomized double blind controlled pilot trial. J NeuroEngineering Rehabil

Vygotskij, L. (1929) The Fundamental Problems of Defectology. In it. Fondamenti di difettologia, cura e prefazione di Guido Pesci, Roma, Bulzoni, 1986.

Zheng Z, Zhang L, Zhu T, Huang J, Qu Y, Mu D. (2016) Peripheral brain-derived neurotrophic factor in autism spectrum disorder: a systematic review and meta-analysis. Sci Rep

CHAPTER 16

Bringing Everything Home; Autism and Neuroscience: A Possible Future

Our journey ends here.

With the new frontiers of neuroscience and new perspectives still open. All to be verified and all to be refuted. That is the power of the scientific method.

As I said at the beginning, this is not a book that gives answers.

It is the story of a quest, and the quest is to ask questions first.

As Groddeck (1920) said, "*it is worth considering, for once, universally known things through the eyes of an outsider, and even if the result were only to provoke a more lively clash of opinions, I would already be satisfied*".

I would like to make my own the words of a great physicist whom I have often quoted,

> "In the honest search for truth, one often has to resign oneself to ignorance for an indefinite period of time. Instead of filling a gap with conjecture, true science prefers to suffer it; and this is not so much out of the scruple of not saying false things, as out of the consideration that, even if this gap is unpleasant, its suppression by means of a fallback may nullify the impulse to seek a convenient solution. [...] Gaps in science can easily suggest weakness and indefensibility. They are sometimes exploited by people who welcome them not as a stimulus for further research, but as an antidote against the fear that science, by 'explaining everything', will deprive the world of its metaphysical interest. [...] Science finds an obstacle in certain deeply ingrained habits of thought, some of which are very

M. Di Salvo, *Autism Research between Psychology and Neuroscience*,
https://doi.org/10.1007/978-3-031-68338-1_16

> difficult to make manifest, while others have already been made evident" (Schrödinger, 1948).

And perhaps it is no coincidence that these statements agree perfectly with those of another physicist, whom I have quoted elsewhere, also one of the fathers of the quantum revolution:

> "Throughout the ages mankind has sought to unravel the meaning of life, in the belief that if we could give direction and meaning to our actions, and to everything else, great human forces would be unleashed. The most diverse answers have been given to this great question over the millennia. Proponents of one theory looked on in horror at the proponents of another, convinced that a different point of view would channel all the energies of mankind in the wrong direction, leading us down a blind alley. Precisely by studying the history of the enormous monstrosities created by false beliefs, philosophers realized the fantastic potential of human beings. Humanity's dream is to find the right channel. Looking back, one gets the impression that the worst periods in our history are those in which there was a stronger presence of people who believed in something with blind faith and absolute dogmatism, taking themselves so seriously that they expected the whole world to think as they did. And then they did things expressly contrary to their own principles in order to prove the truth of their doctrine. As I have said before, and I will say it again here, the only hope for the progress of mankind in a direction that does not lead us down a blind alley (as has happened so many times before) lies in the admission of ignorance and uncertainty. I say that we do not know what the meaning of life is and what the right moral values are, and we have no way of choosing them" (Feynman, 1963).

As always at the end of a trip, one returns home and makes a selection of things to keep.

I set out at the outset to save the good of some 100 years of autism research, which as I suggested is far more than the 10% suggested by Sturgeon's law.

We owe the first definition of autism in the medical sense to Kanner.

We owe to him the descriptive rigour of those first eleven cases that set the course. Kanner was a man and a doctor and a scientist of his time, and to impute his limitations to him, without considering those of his time, is ungenerous and unfair.

You may ask me what I save of Bettelheim's theory that assumed that autism was caused by an inadequate relationship of the child with the 'refrigerator mother'. Well, I believe that the work to disprove that thesis, which we now know to be absurd and unfounded, has instead shown how love, affection, is a very powerful weapon (much more effective than many antidepressants and psychotropic drugs) for the cure of all of us: neuronormal and neurodiverse.

Those were the years when 'uranium toothpaste' was being sold in the United States as a miracle antibacterial and whitening agent (FDA approved).

Among Bernard Rimland's many mistakes, we save an enormous heritage: the testimony of the value of parent-researchers' commitment, activism, never giving up, continuous research on all fronts, and many he opened; among the many doors that were later closed, many have come back to life, in a different form and with unexpected results.

Science and research are sometimes slow, but in the long run they also know how to retrace their steps when technology and new studies open up new scenarios.

Think of how many mental experiments Einstein had to perform in order not to wait a hundred years for technology to show what he had intuited.

Even with the theory of correlation with vaccines we save much, however scientifically unfounded it may be; it has offered us extreme proof of what can happen if we confuse correlation and causation, causality and randomness, and when science is not entirely independent. This is a huge and very topical issue to be vigilant about, especially in a world where information is likely to circulate and amplify without control and with extreme speed.

From the 'English school' we save unparalleled work. So many theories have come out of that school, and they all constitute a fragment of the construction of the greatest jigsaw puzzle that lies before us.

The theory of mind, the theory of levels of empathy, the theory of central coherence, the theory of systemisation, and the theory of heightened perceptual processing, all of which were created to explain autism, allowed us to discover transverse circuits that we could at best intuit but of which we had no knowledge.

Recent discoveries in the neuroscientific field, such as Gazzaniga's interpreter theory and Ramachandran's mirror neuron theory, based on the all-Italian discovery by the Parma research group led by Rozzolatti in the 1990s, have gone in this same direction: understanding the systemic aspect of dysfunction in the brain.

Admirable discoveries in neuroscience owe a great deal to extraordinary people like Edelman and great popularisers like Sacks.

A true epic that is intertwined with that of genetic research, which has revealed so much, and so much has yet to show us.

While it is true that he has not fulfilled his mystical promise to find 'the' autism gene, he has so far given us a picture, at least a preliminary one, of the genetic susceptibility complex, and in the course of a few decades he has unravelled the mysteries of many rare diseases, these being entirely genetic in origin.

In giving us answers about those diseases, science has also given us answers about all of us as a species homo sapiens. In its mission, it has opened doors to therapies that were until a few years ago unhoped for.

While it is true that theories such as 'male extreme' are weak, the correlation with testosterone in the gestation period has its wider significance.

In the end, if we think about it, we throw away very little of the autism research of this century. And much of what has been discovered goes far beyond autism.

The parental advocacy movement has shown new avenues for participation in both research and government policy-making, stimulating public assistance and enhancing the rights of children, all of them.

Without that movement, self-advocacy would probably not have been stimulated, which made neurodiverse people as actors of their own social existence, in the first person.

And even this achievement is the heritage of everyone, not just of one category of people.

Because by intervening in society they have transformed it in its complexity.

It is true that some claims seem excessive, such as the assumption that autistic people are almost a sub-species of independent homo sapiens who would constitute the hard core of 'inventor-systematisers' to whom we owe our technological progress. This thesis triggered the 'genius hunt': from Newton to Bill Gates.

The vast majority of comments on Silverman's (2016) book are positive and enthusiastic. But there are 1% of comments that deserve some reflection. From these I chose one: '*As someone on the mild end of the autism spectrum who despises neurodiversity and desires a cure, I read this book with enormous interest. The book [...] claims or at least implies that historical figures such as Cavendish and others were on the autistic spectrum. The writer cannot have met people who lived in the eighteenth century who he posthumously diagnosed....pursuing on IT people who, according to him, were autistic, without evidence. He trivialises autism by claiming it is similar to a different operating system like Linux than Windows.*"

These are opinions, of course, and as we saw in Chapter 12 the internal debate is very heated and animated.

From my point of view, the strengths of the neurodiversity movement are significant, first and foremost to move autism out of the purely pathological dimension and into that of 'one of the states of man' and to understand a broader idea of human heterogeneity.

However, extolling the savant and genius aspect of some people can generate the misunderstanding of 'pretending' that one's child is a genius, and this is not fair to the child; or recriminating as to why that very child is not the new Elon Mask (*ubiquitous in lists of alleged Asperger-Boomers*), distracting attention from the unquestionable painful aspects that can often accompany the autistic condition, such as the one expressed in the comment I reported and which is rightly accounted for.

Unfortunately, in today's system of communication, where a shout is needed to make a voice heard in the confused and deafening noise that surrounds us all, it is necessary to shout, and Silberman's book (like others of course) I read it like this and save this: let us also think of the incredible potential that an autistic person can express with profit for himself and for his own realisation and for society, in cases where the impairment does not overwhelm the person, and in cases where he is adequately supported and placed in the right context.[1]

Karl Popper said that 'science does not aim at simplicity: it aims at parsimony' (Changeux, 2008).

Thrift is apparently a simple concept that etymologically means 'to save'. In reality it implies something else: prudence, thoughtfulness, moderation and sobriety. And in fact, the first meaning befits little to scientific research, which spares no means, commitment, human and personal sacrifice to achieve the answer to even one apparently small 'why'.

But the other characteristics, prudence, thoughtfulness, moderation and sobriety, are all characteristics that a researcher should never lack.

In Popper's intentions, his statement was meant to indicate that the explanations and results that science arrives at are not necessarily simple answers (*indeed, we know that things are sometimes very complex*), but these answers are not lost in a thousand rivulets, but have a well-defined core.

Indeed, the greatest discoveries, even the most complex ones, have their own implicit 'parsimonious elegance'.

Think of the formula $E = mc^2$.

Or to Dirac's formula $(\partial + m)\,\psi = 0$.

This formula basically states that "if two systems interact with each other for a certain period of time and are then separated, they can no longer be described as two separate systems, but somehow become one system".

Dirac's formula (known as 'the love formula') is designed for the world of quantum physics, but I like to think that it fits (not mathematically of course, but as a concept) perfectly with the interactions between the systems that make up the human brain.

Once neurogenesis has begun, systems interact with each other for (more than) a certain period of time, and if they are separated (in our case if they suffer damage) they can no longer, however, be described and considered as separate systems, but in some way, they become a single system: the damage affects the whole system as a whole. And what applies to the brain, in its various forms and declinations, concerns the person's entire body and mind.

In Schrödinger's (1935) definition, 'the natural sciences are not only based on experience, but also on a certain fundamental hypothesis, on a hypothesis that is extremely obvious and accepted by everyone, by every sensible person, but which cannot be verified with the help of exact scientific methods.

Science is therefore not sufficient on its own, but needs a fundamental axiom that comes from outside. [...] any 'natural law', whether physical or chemical, duly established and which we consider to be universal in scope, can be subjected to a new test at any time, as soon as someone expresses a doubt as to its validity. This is how a scientific truth is distinguished from a historical truth, for example."

Everything I have set out in this book, far beyond the necessarily contained bibliographical limits, is what I have found to be scientifically well-founded, established and reliable.

As I said in the introduction, I have included in the bibliographical references of each chapter only those directly cited.

It is common experience, however, that when writing many are the stimuli, ideas, suggestions that come from different and transversal readings, which are somehow present in the book, even without direct quotation.

Dealing with autism has led us to address many issues of extreme relevance even if apparently not directly related to the subject; among these, we have extensively addressed the issue of the definition (and measurement) of human intelligence (for example, but not only, in note 6 of Chapter and we have critically examined and re-evaluated the methodologies of study and data collection, statistics and judgement on analyses and diagnoses, as we saw in Chapter 3 affected by errors, noise and bias, borrowing notions and studies and research typical of neureconomics and decision psychology.

A demonstration that in this total dimension, no field of study is excluded: and this is certainly another great advance and lesson we owe to autism research: opening minds and (hopefully) breaking down the frontiers of knowledge, knowledge and comparison.

Again quoting Schrödinger (1958): "I recall a brief paradoxical remark by Einstein: 'Certainly every theory is true, provided its symbols are appropriately associated with the observed quantities' ... What is at stake today—far more important than any decision in favour of this or that view—is the danger of a progressive narrowing of the visual field, a kind of mental glaucoma".

Kurt Goldstein (1934), studying soldiers who had suffered brain injuries during the First World War, moved away from his original deficit-based viewpoint and towards a more holistic view. There was never just a deficit or a loosening of the brakes; there were always processes of reorganisation, which he saw as strategies (albeit unconscious and almost automatic) by which the organism with the damaged brain tried to survive, albeit perhaps in a more rigid and impoverished way.

For Hughlings Jackson (1894), in the case of neurological damage there were never only deficits, but also 'hyperphysiological' or 'positive' symptoms, as he called them, that 'unlocked' or enhanced normally restrained or inhibited neural functions. He spoke of 'dissolution', which for him was characterised by regression or reversion to more archaic levels of neural function—the reverse of evolution.

Ivy Mackenzie (Sacks, 2019), a Scottish physician who worked with post-encephalitic patients, described the remote effects that follow the primary insult. In studying such phenomena, he wrote, we see 'organised chaos': the various ways in which the organism, the brain, comes to terms with itself, re-establishing itself at other levels. '*The physician, unlike the naturalist, has to deal [...] with a single organism, the human subject struggling to preserve its identity in adverse circumstances*'.

Sacks (2019) wrote '*Brain function is not like heart or kidney function, which proceeds throughout life in an autonomous, almost mechanical and relatively uniform manner. Instead, the brain/mind is anything but automatic, because it is always engaged—at every level, from the perceptual to the philosophical—in an attempt to categorise and recategorise reality, to understand its own experience and give it meaning. The fact that it is not uniform, but constantly changing, stimulating and in need of an ever more comprehensive integration process, is in the very nature of living an authentic life.*'

In the last chapter, I put forward a unifying proposal, a recent research development that in my opinion underpins and integrates and connects together virtually all the evidence to date Oliver Sacks.

It is not a theory of the whole, let alone closing the game.

If anything, it opens it up with new horizons and towards new perspectives.

This opens up important early prognostic possibilities; it opens up scenarios of the relationship between genetics and epigenetics, between genetic susceptibility and environment, on a solid scientific basis. Especially on bases that bring together genetic, medical, biological, physiological and neuroscientific evidence.

Finally unifying bases for transversal knowledge, and no longer closed in narrow disciplinary universes.

This opens up scenarios for therapy and possible interventions that can really improve people's quality of life.

I think these are roads worth pursuing.

In his book, Robert Wolke recalls an anecdote. When Albert Einstein was residing at the Institut for Advanced Study at Princeton University, an enthusiastic young journalist approached him one day, notebook in hand.

"Well, Professor Einstein", he asked, "what's new in science?".

Einstein looked at him and replied: "Ah, have you already written about all the old science?".

He meant that science should not only be characterised by the latest discoveries that make the headlines. Scientific observation has been going on for centuries, and in all that time we have learnt an extraordinary amount about the world around us. There is a vast legacy of knowledge that explains common and familiar events.

He comments, 'The physical world is a complex web of events, and nothing happens for a single, simple reason. In science, every answer gives rise to new questions, and no explanation can ever be considered complete." (Wolke, 2000).

Daniel Dennett (2013) wrote "we honour scientists who err in a useful way".

First of all, we honour all those who for nearly a century have committed themselves by making essential contributions to research, even where they have made useful mistakes, because from every mistake we have had something to learn, and every road that has closed is one less road to travel for those who continue the journey.

As for me, I make my own the words of two fathers of psychoanalysis.

As Sigmund Freud (1912) wrote: "*If one day it is proved that I have been mistaken about the theoretical problem of the 'present neurosis', I will be able to console myself with the progress of our knowledge, which does not have to take into account the opinions of individuals*".

As Carl Jung wrote (1917) '*Work in this field is pioneering work. Often I have been wrong and often I have had to change my mind. But I know, and therefore I have settled for this rule: just as day can only be born out of night, so too truth is born out of error. I have taken Guglielmo Ferrero's words on the *miserable vanity of the scientist* as a warning, and therefore I have neither feared error nor seriously regretted it. For me, scientific research has never been a cow to be milked or an instrument of prestige, but rather a confrontation imposed by daily psychological experience in contact with the sick, a confrontation that is often bitter. Not everything I write therefore comes from the brain, something also comes from the heart; the benevolent reader should not forget this, if in following the intellectual thread he or she finds here and there a few fractures that are sometimes not perfectly rhymed. Harmonious, flowing descriptions can only be had when writing about things one already knows. But when, driven by the need to help and heal, one seeks the way, one is forced to speak even of things one does not yet know. [...]* G. Ferrero says in the preface to the French edition (1895) of his book I simboli in rapporto alla storia e filosofia del diritto, alla psicologia e alla sociviogia (Torino 1893): 'It is therefore the moral duty of the man of science to expose himself to making mistakes and suffering criticism as long as science continues to progress (...)....) Those who are endowed with a sufficiently serious and detached spirit not to believe that everything they write is an expression of absolute and eternal truth, approve of this theory that places the reasons of science far above the miserable vanity and petty self-love of the scientist*".

There is a popular saying that by making mistakes we learn.
Thanks to neuroscience, we now know that this saying has an extraordinarily deep scientific basis in our brains.

It is necessary to try, even at the cost of failure, because the size and direction of the error indicate correction. Data has accumulated in favour of this hypothesis. Two American researchers, Robert Rescorla and Allan Wagner (1972), put forward this hypothesis: the brain only learns when it detects a gap between what it predicts and what it receives. No learning is possible in the absence of an error signal: 'Organisms learn only when events violate their expectations'. Surprise is one of the fundamental drivers of learning.
The theory manages to explain many aspects of learning by conditioning. You are probably familiar with Pavlov's dog experiment: a bell, initially neutral and without effect, ends up triggering a conditioned salivation reflex when the dog realises that the sound systematically precedes food. How do the two researchers explain this type of learning? The Rescorla–Wagner rule assumes that the brain uses sensory input (the sensations generated by the bell) to predict the probability of another stimulus (food) occurring. It stipulates that: the brain creates a prediction by weighing the sensory input; it then calculates the difference between the prediction and the stimulus actually received: it is the prediction error, a fundamental concept of the theory, that determines the degree of surprise associated with the stimulus; the brain corrects its internal representation, in direct proportion to the strength of the stimulus and the value of the prediction error, so that the subsequent prediction is closer to reality.
Thus this theory explicitly includes the three pillars of learning that we have postulated: learning only occurs if the brain amplifies the appropriate sensory input (attention); if it uses it to produce a prediction (active engagement) and if it can determine its correctness (error detection). The equation proposed by Rescorla and Wagner is practically identical to the one that will later be used extensively in artificial neural networks under the name 'delta rule'. This is a special case of the error back-propagation rule that practically all current supervised learning systems employ (in which the network is explicitly taught an exact response). Moreover, even in the case of reinforcement learning (in which the network is only told when it makes a mistake), a similar equation is used: the network predicts the reward, and it is the difference between the prediction and the actual reward that is used to update the internal representation.

The human brain goes even further: to extract as much information as possible from each learning episode, it probably uses a much more sophisticated language of thought and statistical models than today's neural networks. Nevertheless, Rescorla and Wagner's basic idea remains correct: the brain tries to anticipate the input it receives and adjusts its predictions according to the surprise, improbability and error associated with the information it receives. Learning means reducing the unpredictability. Rescorla and Wagner's theory was highly influential, as it represented a major step forward from earlier theories based on the concept of associative learning. Previously, the accepted idea was that the brain learned to 'associate' the bell with food: it was content to passively register all coincidences between stimuli and responses. But even in the case of Pavlovian conditioning, this view is completely wrong (Balsam, Gallistel, 2009; Gallistel 1990).
The brain is not a passive organ that catalogues associations. Learning is active and depends on the degree of surprise associated with the violation of our expectations. One of the most remarkable experiments, which demolishes the notion of purely associative learning, is that of the block.[2]

The animal is given not one but two sensory stimuli that both predict food. The trick is to present the stimuli one after the other. One starts with light, and the animal learns that it predicts food. Next, the animal is subjected to dual stimuli, so that both the bell and the light predict food. In the end, the effect of the bell alone is tested: it has no effect! The light-based prediction already explains everything. The animal already knows that light predicts food, its brain does not generate any prediction errors during the bell + light test. The block experiment shows that learning does not work by association: in the absence of surprise, no learning occurs: the prediction error is essential for learning. We are talking about an internal error signal, not necessarily an actual error. What is needed to learn is not to be wrong every time. What is important is the surprise, i.e. the gap between prediction and reality. And it is this gap that we call the error signal.
Learning from error correction is universally present in the animal world and there is every reason to think that surprise governs learning from the very beginning of life. Error plays such a fundamental role in learning that virtually all areas of the brain emit and exchange error messages.

The brain circuit in which prediction and error have been best demonstrated is the reward circuit (Pessiglione et al. 2006; Schultz et al. 1997; Waelti et al. 2001).

The dopamine circuit is not content to respond to rewards: it constantly anticipates them. Dopaminergic neurons, located in a small group of cells called the ventral tegmental area, do not respond directly to the pleasures of sex, food or drink; in fact, they are activated when there is a difference between the expected reward and the reward obtained, i.e. when there is a prediction error. So if we give the animal a reward out of the blue, say a drop of sugar water, the nice surprise will lead to a neural discharge. But if this reward is preceded by a signal that makes it predictable, the same sugary syrup will no longer provoke the slightest reaction; on the contrary, it is the signal itself that triggers a dopamine release: learning shifts the response as close as possible to the signal that predicts the reward.

Because the dopamine circuitry begins to respond in advance, arbitrary signals can become carriers of their own reward. This has been demonstrated in humans with regard to coins or banknotes, or in the case of the simple sight of a syringe in a drug addict. This predictive signal is extremely useful for learning because it allows the system to learn to criticise itself and anticipate the error or success of an action without having to wait for external confirmation.

The time has indeed come to say goodbye.

I hope that this journey has passionate you, as it has passionate me, that you take home, inside your true home, positive feelings and that your point of view has at least changed a little.

I hope you do not have certainties but many questions that impetuously ask and seek answers.

My wish, for all of us, is in the verses of Eliot that I have already quoted.

We shall not cease from exploration
And the end of all our exploring
Will be to arrive where we started
And know the place for the first time.

Notes

1. On the subject see Victor Klemperer (1947) "LTI. Notizbuch eines Philologen" Aufbau Verlag, Berlin 1947 - in particular Chapter 30 "The Curse of the Superlative".
2. On blocking associative learning, see Beckers et al. (2006); Fanselow (1998); Waelti, Dickinson, Schultz (2001).

References

Balsam P. D. , Gallistel C. R. (2009) Temporal maps and informativeness in associative learning. Trends Neurosci. 2009 Feb; 32 (2): 73–8. https://doi: 10.1016/j.tins.2008.10.004.

Changeux, JP. (2008) Du vrai, du beau, du bien : Une nouvelle approche neuronale—citation from English edition p.46.

Dennett, D. (2013) Intuition Pumps And Other Tools for Thinking.W. W. Norton

Eliot, T.E. (1943)—Four Quartets

Feynman, R. (1963, 1998) The Meaning of it All—a collection of three lectures in April 1963 at the University of Washington as part of the John Danz Lectures Series.

Freud S.(1912) 'Zur Einleitung der Onanie. Diskussion'. in Die Onanie dossier di: Diskussionen der Wiener Psychoanalytischen Vereinigung, Bergmann, Wiesbaden.

Gallistel, C. R. (1990). The organization of learning. The MIT Press. Pessiglione M, Seymour B, Flandin G, Dolan RJ, Frith CD. Dopamine-dependent prediction errors underpin reward-seeking behaviour in humans. Nature. 2006 Aug 31; 442 (7106): 1042–5. https://doi:10.1038/nature05051

Goldstein, K. (1934) Der Aufbau des Organismus. Einführung in die Biologie unter besonderer Berücksichtigung der Erfahrungen am kranken Menschen. Den Haag, Nijhoff.

Groddeck, G. (1920) Psychoanalytische Schriften zur Literatur und Kunst

Hughlings J. J. The Factors of Insanities. In History of Psychiatry 12–2001.

Jung, C. G. (1917) Über die Psychologie des Unbewussten

Recorla, R. A., & Wagner, A. R. (1972). A Theory of Pavlovian Conditioning: Variations in the Effectiveness of Reinforcement and Nonreinforcement. In A. H. Black, & W. F. Prokasy (Eds.), Classical Conditioning II: Current Research and Theory (pp. 64- 99). New York: Appleton-Century-Crofts.

Sacks, O. (2019) Everything in Its Place: First Loves and Last Tales—published posthumously

Schrödinger, E. (1935) Quelques remarques au sujet des bases de la connaissance scientifique, Scientia, vol. 57, 181
Schrödinger, E. (1948) Nature and the Greeks, 'Shearman Lectures' University College, London, 1948; Cambridge University Press 1954.
Schrödinger, E. (1958) Is energy a purely statistical concept?, Conference—Vienna
Schultz W, Dayan P, Montague PR. (1997) A neural substrate of prediction and reward. Science. 1997 Mar 14; 275 (5306): 1593–9. https://doi:10.1126/science.275.5306.1593
Silverman, S. (2016) Neurotribes: The Legacy of Autism and the Future of Neurodiversity—Avery
Waelti P, Dickinson A, Schultz W. (2001) Dopamine responses comply with basic assumptions of formal learning theory. Nature. 2001 Jul 5; 412 (6842): 43–8. https://doi:10.1126/science.275.5306.1593
Wolke R. L. (2000) What Einstein told his barber. More Scientific Answers to Everyday

Correction to: Autism Global Emergency: The New Epidemiology, Parenting, Self-Advocacy and Neurodiversity Movement

Correction to:
Chapter 12 in: M. Di Salvo, ***Autism Research between Psychology and Neuroscience*****,**
https://doi.org/10.1007/978-3-031-68338-1_12

The original version of Chapter 12 was inadvertently published with the incorrect author names in the references, which have now been corrected. The corrections to the book have been updated with the changes.

The updated version of this chapter can be found at
https://doi.org/10.1007/978-3-031-68338-1_12

M. Di Salvo, *Autism Research between Psychology and Neuroscience*,
https://doi.org/10.1007/978-3-031-68338-1_17

Name Index

M. Di Salvo, *Autism Research between Psychology and Neuroscience*, https://doi.org/10.1007/978-3-031-68338-1

M

N

O

S

T

Y

Z

Subject Index

M. Di Salvo, *Autism Research between Psychology and Neuroscience*,
https://doi.org/10.1007/978-3-031-68338-1

B

F

G

H

I

L

M

N

The manufacturer's authorised representative in the EU is Springer Nature Customer Service Centre GmbH, Europaplatz 3, 69115 Heidelberg, Germany. If you have any concerns regarding our products, please contact ProductSafety@springernature.com

Printed and bound by CPI Group (UK) Ltd, Croydon, CR0 4YY

01/10/2026

02225852-0008